IFIP Advances in Information and Communication Technology 365

T0189958

IFIP – The International Federation for Information Processing

IFIP was founded in 1960 under the auspices of UNESCO, following the First World Computer Congress held in Paris the previous year. An umbrella organization for societies working in information processing, IFIP's aim is two-fold: to support information processing within ist member countries and to encourage technology transfer to developing nations. As ist mission statement clearly states,

> *IFIP's mission is to be the leading, truly international, apolitical organization which encourages and assists in the development, exploitation and application of information technology for the bene t of all people.*

IFIP is a non-profitmaking organization, run almost solely by 2500 volunteers. It operates through a number of technical committees, which organize events and publications. IFIP's events range from an international congress to local seminars, but the most important are:

- The IFIP World Computer Congress, held every second year;
- Open conferences;
- Working conferences.

The flagship event is the IFIP World Computer Congress, at which both invited and contributed papers are presented. Contributed papers are rigorously refereed and the rejection rate is high.

As with the Congress, participation in the open conferences is open to all and papers may be invited or submitted. Again, submitted papers are stringently refereed.

The working conferences are structured differently. They are usually run by a working group and attendance is small and by invitation only. Their purpose is to create an atmosphere conducive to innovation and development. Refereeing is less rigorous and papers are subjected to extensive group discussion.

Publications arising from IFIP events vary. The papers presented at the IFIP World Computer Congress and at open conferences are published as conference proceedings, while the results of the working conferences are often published as collections of selected and edited papers.

Any national society whose primary activity is in information may apply to become a full member of IFIP, although full membership is restricted to one society per country. Full members are entitled to vote at the annual General Assembly, National societies preferring a less committed involvement may apply for associate or corresponding membership. Associate members enjoy the same benefits as full members, but without voting rights. Corresponding members are not represented in IFIP bodies. Affiliated membership is open to non-national societies, and individual and honorary membership schemes are also offered.

Scott A. Hissam   Barbara Russo
Manoel G. de Mendonça Neto   Fabio Kon (Eds.)

Open Source Systems: Grounding Research

7th IFIP WG 2.13 International Conference, OSS 2011
Salvador, Brazil, October 6-7, 2011
Proceedings

 Springer

Volume Editors

Scott A. Hissam
Carnegie Mellon University, Software Engineering Institute
Pittsburgh, PA 15213, USA
E-mail: shissam@sei.cmu.edu

Barbara Russo
Free University of Bolzano-Bozen, Center for Applied Software Engineering
Piazza Domenicani 3, 39100 Bolzano-Bozen, Italy
E-mail: barbara.russo@unibz.it

Manoel G. de Mendonça Neto
Universidade Federal da Bahia, Laboratório de Engenharia de Software
Av. Adhemar de Barros, s/n, Campus de Ondina
40170-110 Salvador, Bahia, Brazil
E-mail: manoel.g.mendonca@gmail.com

Fabio Kon
University of São Paulo, Department of Computer Science
Rua do Matão, 1010, 05508-090 São Paulo, SP, Brazil
E-mail: fabio.kon@ime.usp.br

ISSN 1868-4238   e-ISSN 1868-422X
ISBN 978-3-642-26956-1   ISBN 978-3-642-24418-6 (eBook)
DOI 10.1007/978-3-642-24418-6
Springer Heidelberg Dordrecht London New York

CR Subject Classification (1998): D.2, D.3, C.2.4, D.1, K.6.3, D.2.4, H.5

Typesetting: Camera-ready by author, data conversion by Scientific Publishing Services, Chennai, India

Printed on acid-free paper

Springer is part of Springer Science+Business Media (www.springer.com)

General Chair's Foreword

Welcome to the 7th International Conference on Open Source Systems of the International Federation for Information Processing (IFIP) Working Group 2.13. The people and city of Salvador, Brazil, could not have been a more generous and exciting host for our inaugural South American venue. Within these proceedings you will find papers, panels, and workshops of practitioners and researchers sharing their experiences, lessons, and discoveries as they pertain to creating, distributing, acquiring, and using software and software-based services based on free, libre, and open source software (FLOSS).

Selecting reliability as the theme for this year's conference was timely and appropriate given the pervasiveness of FLOSS in seemingly everything from academia to industry, from scientific computing to entertainment, and from consumer electronics to safety-critical systems—the reach of FLOSS seems unbounded. I continue to be amazed as to where I find FLOSS, such as my television, navigation unit, picture frame, WiFi access point, multimedia set-top box, and smartphone. However, thinking more broadly, perhaps I should not be so surprised. As a community that stands on the shoulders of giants, the limits of innovations are bounded only by our imagination, evidenced by the permeation of FLOSS today. But, for FLOSS to maintain and expand its status in these industries, as well as those not explicitly mentioned here, we as a community must continue to reflect appropriately in order to repeat our successes, and to reach beyond that which is comfortable and ask the hard questions so that we may address any remaining barriers impeding trust, adoption, and use of FLOSS.

Putting together this international conference from geographically disparate locales was very challenging. Sincere thanks go to the Organizing Chairs Fabio Kon and Manoel G. de Mendonça Neto for their relentless drive to get every detail just right for this conference. For filling this wonderful venue with a top-notch technical program, I thank Barbara Russo for selecting the theme of the conference, publishing the call for papers, and marshaling all the submissions through the review process and selecting the best of the best for the papers accepted this year. And without skipping a beat, I thank Imed Hammouda and Bruno Rossi for their tireless job of assembling and preparing the conference proceedings.

It was only possible to accomplish all this work because others took on important duties to help make this conference a success. From tutorials chaired by Stefan Koch and Hyrum Wright, workshops chaired by Roberto Di Cosmo, to panels chaired by Jay Kesan, John Noll, and Marcos Sfair Sunyé—thank you. The Publicity Chair, Greg Madey, and Co-chairs who worked to spread word of the conference, which included Africa (Imed Hammouda), Asia (Antonio Cerone), Central and South America (Carlos Denner Santos Jr.), Eastern Europe (Sulayman K. Sowe), Middle East (Faheen Ahmed), and Western

Europe and the Nordic regions (Björn Lundell), are to thank for the international diversity of the conference. For generating industry participation, I thank Stefano De Panfilis and Rafael Prikladnicki for their efforts. I especially would also like to thank the Web Portal team, Antonio Terceiro, Beraldo Leal, and Paulo Meirelles for the complete revamp of the conference website.

A very special thanks goes to the Chairs and organizers of the Doctoral Consortium, Paula Bach, Charles M. Schweik, and Alberto Sillitti, for their efforts to continue this very special and important service to newly emerging researchers embarking on the next discoveries related to FLOSS. I specifically want to thank Charles Schweik for obtaining National Science Foundation support for US-based doctoral student participation in the consortium. I also thank the IFIP Working Group 2.13 and past conference organizers for their help and advice provided to me, Pär J. Ågerfalk, Ernesto Damiani, Björn Lundell, Greg Madey, Walt Scacchi, Giancarlo Succi, and Tony Wasserman.

Lastly, I humbly thank the authors and members of the Program Committee without whom there would be no technical program.

August 2011   Scott Hissam

Program Chair's Foreword

Welcome to the proceedings of the 7th International Conference on Open Source Systems—Open Source Systems: Grounding Research (OSS 2011)—which was held in Salvador, Brazil!

This year we had a very polyhedral technical program originated by open source software (OSS) reliability the major theme of the conference and intended to ground theory on OSS. The program had a special focus on OSS adoption in industry, as quality and reliable software has better chances of being selected and adopted in enterprises. Quality software also provides value to the final user, and one of the conference sessions was dedicated to OSS value and economics. Adopting OSS needs the support of an extensive OSS technology review and building that follow the principles of open innovation, reuse, integration, and compliance as proposed in the first-day session of the conference. Availability of data is the wealth of the OSS community and OSS 2011 papers on mining software repositories illustrate this through methods, practices, and results. Finally, OSS research needs to reflect on its own evolution. A session on knowledge and research building concluded the first day. The majority of sessions included lightning talks. These short and focused presentations were intended to increase the debate and research discussion at the conference.

There were three great keynotes: Linda Northrop of the Software Engineering Institute opened the conference at the welcome reception, James Herbsleb of Carnegie Mellon University on the first day, and Christiana Soares de Freitas of the University of Brasilia and Corinto Meffe of the Brazilian Ministry of Planning on the second conference day.

There were workshops, tutorials, and a doctoral consortium event collocated with the main conference. Panels helped increase the research discussion, for example, on education in OSS and eGovernement and OSS.

The call for research papers attracted 56 paper submissions. The review was intensive and several online discussions increased the quality of the review process. Four full papers underwent a major revision process. We accepted 20 papers as full research papers, 4 as industrial full papers, and 8 lightning talks.

I would like to join Scott in thanking the authors who provided the content for the program, expressing my sincere gratitude to the Program Committee members and the external reviewers who volunteered time and resources to make this program possible. Finally, I thank all the people that made this conference a success. In particular, I sincerely thank Scott, who was great in making this conference meet the highest standards.

We hope that you will enjoy the proceedings.

Barbara Russo

Organization

Conference Officials

General Chair
Scott Hissam   Software Engineering Institute,
                                Carnegie Mellon, USA

Program Chair
Barbara Russo   Free University of Bolzano-Bozen, Italy

Organizing Chairs
Fabio Kon   University of São Paulo, Brazil
Manoel G. de Mendonça Neto   Federal University of Bahia, Brazil

Proceedings Chairs
Bruno Rossi   Free University of Bolzano-Bozen, Italy
Imed Hammouda   Tampere University of Technology, Finland

Industry Chairs
Rafael Prikladnicki   PUC-RS, Brazil
Stefano De Panfilis   Engineering Ingegneria Informatica, Italy

Doctoral Consortium Chairs
Alberto Sillitti   Free University of Bolzano-Bozen, Italy
Charles M. Schweik   University of Massachusetts, USA
Paula Bach   Microsoft, USA

Workshops Chair
Roberto Di Cosmo   University of Paris, France

Tutorial Chairs
Hyrum Wright   University of Texas at Austin, USA
Stefan Koch   Bogazici University, Turkey

Panel Chairs
Jay Kesan   University of Illinois, USA
John Noll   University of Limerick, Ireland
Marcos Sfair Sunyé   Federal University of Paraná, Brazil

Publicity and Social Media Chair

Greg Madey   University of Notre Dame, USA

Publicity Chairs

(Africa) Imed Hammouda   Tampere University of Technology, Finland
(Asia) Antonio Cerone   United Nations University Macau, SAR China
(Central and South America)
   Carlos Denner Santos Jr.   University of São Paulo, Brazil
(E. Europe) Sulayman K. Sowe   United Nations University Institute of
                                       Advanced Studies, Japan
(Middle East) Faheen Ahmed   United Arab Emirates University,
                                       United Arab Emirates
(W. Europe, Nordic)
   Björn Lundell   University of Skövde, Sweden

Web Portal

Antonio Terceiro   Federal University of Bahia, Brazil
Beraldo Leal   University of São Paulo, Brazil
Paulo Meirelles   University of São Paulo, Brazil

Advisory Committee

Giancarlo Succi   Free University of Bolzano-Bozen, Italy
Walt Scacchi   University of California, Irvine, USA
Ernesto Damiani   University of Milan, Italy
Scott Hissam   Software Engineering Institute, USA
Pär J. Ågerfalk   Uppsala University, Sweden

Program Committee

Alberto Sillitti   Free University of Bolzano-Bozen, Italy
Antonio Cerone   United Nations University Macau SAR, China
Barbara Russo   Free University of Bolzano-Bozen, Italy
Björn Lundell   University of Skövde, Sweden
Bruno Rossi   Free University of Bolzano-Bozen, Italy
Charles Knutson   Brigham Young University, USA
Chintan Amrit   Twente University, The Netherlands
Cleidson de Souza   Universidade Federal do Pará, Brazil
Cornelia Boldyreff   University of East London, UK
Daniela Cruzes   Norwegian University of Science and
                                       Technology, Norway
Davide Tosi   University of Milano Bicocca, Italy
Diomidis Spinellis   Athens University of Economics and Business,
                                       Greece

Table of Contents

Part I: Papers

OSS Quality and Reliability

OSS Products

Review of Technologies of and for OSS

Knowledge and Research Building in OSS

OSS Reuse, Integration, and Compliance

OSS Value and Economics

OSS Adoption in Industry

Mining OSS Repositories

Part II: Lightning Talks

Part III: Industry Papers

Part IV: Workshops

Impact of Stakeholder Type and Collaboration on Issue Resolution Time in OSS Projects

Anh Nguyen Duc[1], Daniela S. Cruzes[1], Claudia Ayala[2], and Reidar Conradi[1]

[1] Norwegian University of Science and Technology, Department of
Computer and Information Science,
Trondheim, Norway
{anhn,dcruzes,conradi@idi.ntnu.no}
[2] Technical University of Catalunya, Department of Service Engineering
and Information Systems,
Barcelona, Spain
{cayala@essi.upc.edu}

Abstract. Initialized by a collective contribution of volunteer developers, Open source software (OSS) attracts an increasing involvement of commercial firms. Many OSS projects are composed of a mix group of firm-paid and volunteer developers, with different motivations, collaboration practices and working styles. As OSS development consists of collaborative works in nature, it is important to know whether these differences have an impact on collaboration between difference types of stakeholders, which lead to an influence in the project outcomes. In this paper, we empirically investigate the firm-paid participation in resolving OSS evolution issues, the stakeholder collaboration and its impact on OSS issue resolution time. The results suggest that though a firm-paid assigned developer resolves much more issues than a volunteer developer does, there is no difference in issue resolution time between them. Besides, the more important factor that influences the issue resolution time comes from the collaboration among stakeholders rather than from individual characteristics.

1 Introduction

Open source software (OSS) development is a highly distributed and collaborative activity. In OSS projects, stakeholders, who are people involve in software development project such as developers, project leader, tester and end-users, collaborate with each other in various ways to accomplish development tasks. Although OSS was born as a movement mainly based on contributions of volunteer stakeholders, an increasing number of firms are getting involved in OSS projects [21][31]. Lakhani et al. found that around 40% of programmers are paid by companies to contribute to OSS projects [24]. Hars and Ou obtained similar results in a survey on the developers of the Linux kernel [29]. Consequently, many open source projects contain both types of stakeholder (firm-paid and volunteer), which have different motivations, collaboration practices and working styles. For instance, firm-paid developers contribute to the OSS community as part of their jobs, which provide

S.A. Hissam et al. (Eds.): OSS 2011, IFIP AICT 365, pp. 1–16, 2011.
© IFIP International Federation for Information Processing 2011

them a financial motivation. In addition, they often also work on proprietary software since it constitutes a part of the business model of their sponsor firm [2][9][25]. Therefore, they have to learn the community working style and adjust to the rhythms and the demands of OSS development [2]. In contrast, volunteer developers are usually motivated by social or technical reasons to demonstrate or improve their technical skills [9][25].

Several studies have investigated the potential differences among firm-paid and volunteer developers in OSS projects [2][21][24][29][31]. However, these studies did not address whether these differences actually have an impact on the OSS project outcomes such as quality of the source code, productivity of developers, activeness of the community and time to accomplish a software evolution task.

A software evolution task (or software issue) is normally referred as a unit of work to accomplish an improvement in the system. Dealing with a software issue includes fixing defects, implementing new feature requests and enhancing current system features. With a large amount of issues that occur from time to time, resolving them in a cost-effective manner is essential to achieve a high user satisfaction with less working effort.

Besides the impact of some special characteristics of stakeholders (in the issue resolving process, they are usually reporters and assignees), the issue resolution time can be influenced by a collaborative working process between reporters and assignees. Pinzger et al. mention the Coordination theory in OSS, which state that the interaction among stakeholders can impact software quality (such as mean time between failure) and work performance (such as defect removal effectiveness and problem fixing time) [30]. In the issue resolving process, stakeholders often use electronic media such as mailing list, IRC and issue tracking system to discuss, comment and clarify about an assigned task [23][26]. The collaboration among stakeholders, such as discussion, instruction and clarification on an issue, is important to the completion of the issue-resolving task.

This study has three main objectives. First, we characterize the difference in the average amount of resolved issues and issue resolution time between a volunteer assignee and a firm-paid assignee. To best of our knowledge, there is no study that empirically investigates the influence of volunteers versus firm-paid developers on issue resolution time. Second, we investigate collaboration among stakeholders in OSS projects by using Social network metrics and analysis. Last, we explore the impact of the collaboration measures on issue resolution time. While there are several studies using Social network metrics investigating software quality (as described in Section 2.1), this is among the first attempts to apply these metrics on studying issue resolution time.

The rest of the paper is organized as follows. Section 2 presents a construction of stakeholder collaboration measure using Social network analysis (SNA). While Section 3 states our hypotheses, Section 4 describes our case study and data collection procedure. Section 5 provides the hypotheses testing results. Section 6 discusses the findings and Section 7 identifies the threats to validity. The paper ends with a conclusion and future works.

2 Stakeholder Collaboration Measure by Social Network Analysis (SNA)

2.1 Impact of Collaboration on Software Development

Table 1 presents several studies exploring the impact of collaboration on software development outcomes. Bettenburg et al. studied the impact of social structure on software quality and find a statistical relation between a communication flow between

Table 1. Studies about collaboration

Studies	Dependent Variable	Collaboration Variable	Exploring Method	Test Results
Bettenb urg et al. [6]	Number of post-released defects	Participant reputation (number of contributed messages)	Multiple linear regression model	Increase a predictive power of prediction model 11.66%
Abreu et al. [1]	Number of code changes	Number of messages in mailing list Number of messages from high-centrality-degree developers	Spearman's correlation	R = 0.1 to 0.45 p < 0.001 R = 0.06 to 0.16 p < 0.05
Bird et al. [7]	Post-released defect proneness	Developer-component network measures, e.g.: centrality degree	Release-cross Multiple Logistic regression	Recall: 0.705 to 0.859. Precision: 0.747 to 0.827
Wolf et al. [32]	Build failure likelihood	Developer-developer network measures, e.g.: density, centrality, betweenness and structural holes	Bayesian classifier	Recall:0.62, Precision: 0.75
Pinzger et al. [30]	Number of failure	Number of authors, number of commits, networks measures e.g.: Freeman centrality degree and betweenness	Spearman correlation Multiple linear regression model	R= 0.503 to 0.747, p<0.01 R^2= 0.698 to 0.746
Andrew et al. [5]	Vulnerability of software files	Betweenness measures, number of developers and number of commits	Mann-Whitney-Wilcoxon (MWW) test	Higher values for vulnerable file, p<0.0001
Feczak et al. [14]	Bug fixing time	Stakeholder network measures, e.g: Freeman centrality degree	Spearman correlation	R = 0.13 to 0.35 p <0.05
Anbalag an et al. [2]	Defect resolution time	Number of unique participants	Spearman correlation	R = 0.22 p < 0.0001
Guo et al. [16]	Likelihood of fixed defect	Defect opener reputation, number of defect report editors and assignees	Chi square test Correlation test	p < 0.0001 Not reported

developers and users and post-release defects [6]. Abreu et al. investigated Eclipse sub-projects and found a significantly positive correlation between communication frequency between developers and number of injected defects in the software [1]. Bird et al. showed that a socio-technical network of software modules and developers is able to predict software failure proneness with greater accuracy than other prediction methods [7]. Wolf et al. formed a developer-task network to explore the impact of developer communication on software build integration fail [32]. Pinzger et al. constructed a developer-module network to predict the software failures [30].

More relevant to our focus are studies about relationship between developer collaboration and defect fixing time. Feczak et al. empirically validated the Coordination theory in open source projects and found that collaboration among stakeholders, measured by social network metrics, has a positive influence on software defect fixing time [14]. Anbalagan et al. also found a significant correlation between number of participants in editing a defect report and median time taken to correct it [2]. Guo et al. used collaboration measures to predict which defect will get fixed in Windows 7 and concluded that the defects that have more people involved in defect report editing will be more likely to be fixed [16]. While these studies show that developers collaboration, measured by a developer-artifact network metrics is useful for predicting software defects and fixing time, a similar approach can be applied to discover the impact of developers collaboration on issue resolution time.

2.2 Issue-Stakeholder Network Measures

Social network analysis (SNA) considers social relationships in term of network theories, which focus on social nodes, such as people, groups, organizations and measures relationships and information flows among them [15]. In this study, we construct an undirected graph to represent a network of issue-stakeholders. The graph employs two types of nodes: stakeholders and issues. Stakeholders include a *reporter* (who reports the issue), an *assignee* (who is assigned to resolve the issue) or a *commenter* (who comments or discuss about the issue). A link occurs only between a stakeholder and an issue, which represents for a stakeholder's action on the issue, such as an issue report, a report update, a comment on the issue and an issue assignment.

To establish the issue-stakeholder network, we use a social network analysis tool, namely ORA[1]. The most common measure in SNA is centrality, which denotes the structural power position of a node in a given network. There are three centrality measures in SNA, namely Freeman Degree Centrality, Closeness and Betweeness. In the scope of this study, we investigate Freeman Centrality Degree since this metric is successfully applied in relevant studies [14][30][32]. In our network, the Freeman Degree Centrality of an issue represents the number of unique stakeholders that involve in the issue. For each issue, the high value of a centrality degree shows a large number of stakeholders working on it (reporting, commenting or resolving it). The centrality degree of an issue is calculated as in Formula 1:

---

[1] http://www.casos.cs.cmu.edu/projects/ora/

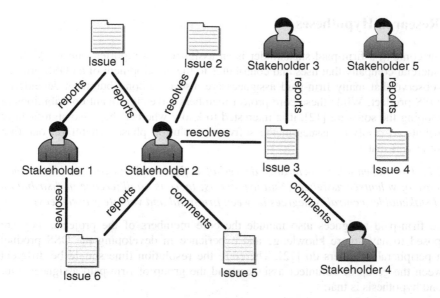

Fig. 1. Issue-stakeholder network in issue resolution

$$Gd(i) = \frac{d(i)}{n-1} \qquad (1)$$

with d(i) is the node degree of a issue,
n is the total number of stakeholders and issues

Similarly, the Freeman Degree Centrality of a stakeholder is the number of issues directly linked to the stakeholder. We also want to explore whether stakeholder centrality has an impact on issue resolution time. For each issue, we calculate the accumulative stakeholder centrality degree (Cs) as a sum of centrality degrees of all involved stakeholders, as in Formula 2:

$$Cs(i) = Gdass(i) + Gdrep(i) + \sum Gdcom(i) \qquad (2)$$

with Gdass(i), Gdrep(i) and Gdcom is the centrality degree of assignee,
reporter and commenter correspondently.

The meaning of Cs(i) is that the issue is important when they are resolved by many stakeholders and by important stakeholders, who involved in many other issues. Illustrated by Figure 1, the Freeman centrality degree of Stakeholder 2 is 5/11 and the degree of Stakeholder 3 is 1/11, which shows that Stakeholder 2 involves in more issues than Stakeholder 3 does. Issue 3's centrality degree is 3/11 and Issue 4's centrality degree is 1/11, which shows that Issue 3 is involved by more stakeholders than Issue 4 is. The accumulative stakeholder centrality degree of Issue 3 is 7/11.

3 Research Hypotheses

In our context, a firm-paid stakeholder is an assignee or a reporter who works for a commercial company that uses and contributes to the development of an OSS project. We observe that many firm-paid assignees are also core contributors in developing the OSS product. While these core project members have significant contributions in developing the software [12], it is interested to know whether they also significantly contribute to resolving issues in the software evolution phase. Therefore, our first hypothesis is that:

H1: The stakeholder's centrality degree of a firm-paid assignee is higher than those of volunteer assignee. (Null hypothesis: there is no difference in distribution of stakeholder centrality degrees between firm-paid and volunteer assignees).

Since firm-paid assignees also include the core members of the projects, they are supposed to have more knowledge and experience in developing the OSS product than peripheral members do [12]. Therefore the resolution time should be different between the group of volunteer assignees and the group of firm-paid assignees. Our second hypothesis is that:

H2: There is a difference in mean issue resolution time between a firm-paid assignee and a volunteer assignee. (Null hypothesis: there is no difference in mean issue resolution time between firm-paid and volunteer assignees).

An issue with many stakeholders involved might relate to many different software modules or different development tasks. Therefore, the complexity of such issues is higher and thus, it takes the assignee longer time to resolve. Our third hypothesis is that:

H3: The larger number of stakeholders involve in an issue is, the longer the issue resolution time is. (Null hypothesis: there is no correlation between the number of stakeholders involved in an issue and the issue resolution time).

A large number of comments and discussions on an issue may be caused by problems on the issue description (which leads to confusion or dissensus among stakeholders) or by the complexity of the resolving task and could lead to longer resolution time. Our last hypothesis is that:

H4: The larger number of exchanged messages on an issue is, the longer the issue resolution time is. (Null hypothesis: there is no correlation between the number of message exchanged in an issue and the issue resolution time).

4 The Case Study

4.1 Projects Context and Selection

Three OSS projects were selected for our study, namely Qt, Qpid and Geronimo. The reasons for selecting these projects were: (1) these projects are active and ongoing for at least 4 years, which ensure the scale of the datasets; (2) there are similar issue tracking system used in these projects, which facilitate the data collection; (3) these projects are similar in business domain and technical level, thus reducing the variability of the results, and, (4) these projects are significantly influenced by firm-paid developers, which enable the investigation of the impact of different stakeholder types.

Qt is an Open Source cross-platform framework developed by Qt Development Framework (Nokia) based on the programming language C++. The framework offers common components such as networking, OpenGL, multimedia and a widget toolkit[2]. **Qpid** is an cross-platform Open Source enterprise messaging system developed around the open standard Advanced Message Queuing Protocol (AMQP). It is implemented in many programming languages, such as: C++, C#, Python, Ruby and Java[3]. The project originated from a joint venture mostly consisting of code by Red Hat, Iona and JP Morgan. **Geronimo** is a server runtime framework that pulls together the Open Source alternatives to create runtime instances that meet the needs of developers and system administrators and open-source, Apache-licensed[4]. The project originated from IBM developers.

4.2 Data Collection and Preprocessing

All software issues were collected from JIRA repositories[5] of the respective projects. The summary of datasets was described in Table 2, with the main, owner firm of each project, the time frame of the issues collected for analysis, the total number of issues, number of stakeholders (assigned developers and issue reporters, who collaborated with the project during this period), the total number of issues in the repository and the total number of issues that we used for our analysis.

Table 2. Issue collection from cases study

Info.\ Projects	Qt	Qpid	Geronimo
Main Firms	Qt (Nokia)	Red Hat, JP Morgan	IBM
Time Frame	11/03-12/10 (85 months)	9/06-12/10 (51 Months)	8/03-12/10 (87 Months)
Number of Stakeholders	1568	126	405
Number of issues	16818	3016	5697
Number of selected issues	9921	2278	4787

Issue resolution time was computed by using the *created time* field and the *issue resolved time* field. We excluded 3514 issues that are not possible to calculate the issue resolution time. We removed 2171 issues that have the state OPEN, DUPLICATE or INVALID. We also deleted 2838 issues that do not have *reporter* or *assignee* information (stated as *unassigned* or *unknown*), and issues with invalid

---

[2] Qt project - http://qt.nokia.com/
[3] Qpid project - http://qpid.apache.org/
[4] Geronimo project - http://geronimo.apache.org/
[5] JIRA–bug, issue and project tracking system, http://www.atlassian.com/software/jira/

stakeholder information (as described below). Twenty-two data points were also taken out by an outlier detection function implemented in the R^6 package. At the end of the data preprocessing procedure, 16986 issues were selected for further analyses, which consumes 67% of total number of issues.

The classification of stakeholder type (firm-paid or volunteer) was manually executed by searching stakeholder name and professional information in the Internet. The first information source is the list of contributor and mailing list from the project repository. We found these stakeholders with explicit company information, either as project initiators or main contributors of the open projects. With stakeholders that company information was not given in the project site, we determined the affiliation by: (1) the stakeholder's profile from social networking site such as Facebook, LinkedIn and personal blogs, and (2) the stakeholder's email with a private company domain. The stakeholder company information were extracted by the time when the stakeholder worked in the OSS project. We assumed that the group of stakeholders (more than three) come from the same company participate in the OSS project as a company representative and are paid by the company. The stakeholders without any identified company information were classified as volunteers.

After collecting stakeholder information, we synchronized the stakeholder name and alias to avoid replicated data. Table 2 describes the total number of stakeholders that involve in the OSS projects in the time period that data are collected. Collaboration information was extracted from issue tracking systems and the mailing lists of OSS projects using a Perl script. For each issue, we collected comments, edits on the issue report and issue-related messages from the project mailing list.

4.3 Descriptive Statistics

Table 3 presents the distribution of reported issues by stakeholder types in Qpid, Geronimo and Qt correspondingly. As our expectation, stakeholders from Redhat and JP Morgan in Qpid (53.6% of reported issues) and stakeholders from IBM in Geronimo (60.8% of reported issues) are the main contributors in reporting issues. However, the largest amount of reported issues in Qt comes from volunteer reporters (44.9% of reported issues). This observation can be explained by the large amount of end-users involved in the Qt project, who directly report their problem, in the issue project tracking system. Table 4 shows the distribution of resolved issues by different stakeholder types. Not surprisingly, most of the issues are resolved by developers from the main firms such as Redhat and JP Morgan (62.4% of resolved issues) in Qpid, IBM (71.6% of resolved issues) in Geronimo and Nokia (62% of resolved issues) in Qt.

Figure 4 presents box plot charts of issue centrality and issue-based messages in the three projects. In Figure 4a shows that most of issues are touched by one to three stakeholders, other than the reporter. In Figure 4b, the average number of issue-based messages is similar among three projects. We see that number of message exchanged around an issue in three projects is from none to four messages, slightly vary among projects.

---

[6] The R Project for Statistical Computing - http://www.r-project.org/

Table 3. Distribution of contribution in reporting issue

Issues from	Qpid	Geronimo	Qt
Individual	453 (19.9%)	1205 (25.0%)	4452 (44.9%)
Other company	605 (26.5%)	683 (14.2%)	1124 (11.3%)
Main Firms	1220 (53.6%)	2919 (60.8%)	4345 (43.8%)
Total	**2278 (100%)**	**4787 (100%)**	**9921 (100%)**

Table 4. Distribution of contribution in resolving issue

Issues from	Qpid	Geronimo	Qt
Individual	252 (11.1%)	401 (8.4%)	2463 (24.8%)
Other company	604 (26.5%)	956 (20.0%)	1315 (13.2%)
Main Firms	1422 (62.4%)	3420 (71.6%)	6143 (62.0%)
Total	**2278 (100%)**	**4787 (100%)**	**9921 (100%)**

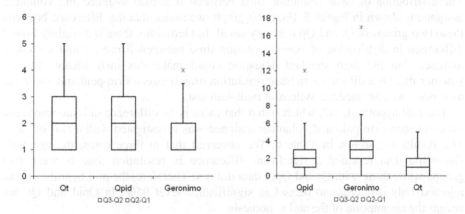

Fig. 4a, b. Descriptive of issue centrality and issue-based messages

5 Hypotheses Testing Results

5.1 H1: *The stakeholder's centrality degree of a firm-paid assignee is higher than those of a volunteer assignee.*

Due to the fact that stakeholder centrality degrees are not normally distributed as observed from histogram and descriptive statistics, we used Wilcoxon rank-sum test [13].

Table 5. Resolution time by volunteer vs. firm-paid assignees

Projects	Median centrality of Firm-paid	Median centrality of Volunteer	Significance level
Geronimo	0.0169	0.0049	p=0.0014
Qpid	0.0114	0.0057	p=0.0251
Qt	0.0131	0.0024	p=0.0014

All the tests are performed using the statistic package R with alpha = 0.05. The null hypothesis H1, which stated that there is no difference in stakeholder centrality degree between firm-paid and volunteer assignee was investigated with a one-tail test. The results are shown in Table 5. In all cases, the median values of centrality degree in the firm-paid groups are significantly higher than those in the volunteer groups. In particular, the number of issues involved by a firm-paid stakeholder is at least two times higher than ones involved by volunteer stakeholder in all projects. The p-values in all tests allow us to reject the null hypotheses in all projects. We accept the alternative hypothesis that the centrality degree of firm-paid stakeholders is higher than one of volunteer stakeholders.

5.2   H2: *There is a difference in mean issue resolution time between a firm-paid assignee and a volunteer assignee*

The distribution of issue resolution time between firm-paid assignee and volunteer assignee is shown in Figure 5. From the graph, we notice that the difference between these two groups in Qt and Qpid is very small. In Geronimo, there is a slightly higher difference in distribution of issue resolution time between firm-paid and volunteer assignee, but the high standard deviation could make this insignificant. To test whether there is a difference in issue resolution time between firm-paid and volunteer assignees, we also used the Wilcoxon rank-sum test.

The null hypothesis H2, which stated that there is no difference in issue resolution time between firm-paid and volunteer assignee was investigated with a two-tail test. The results are shown in Table 6. We observed that in three cases, the test with Geronimo data revealed a significant difference in resolution time between two groups while those with Qt and Qpid data did not. Therefore, the null hypothesis was rejected only in Geronimo dataset at significance level 95%. In Qpid and Qt, we accept the assumption of the null hypothesis.

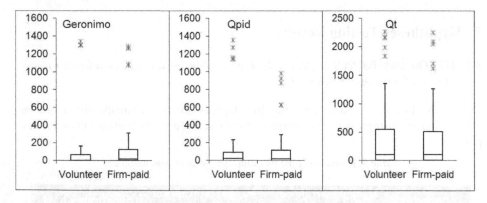

Fig. 5. Issue resolution time (days) between volunteer and firm-paid stakeholder

Table 6. Resolution time of volunteer vs. firm-paid assignees

Projects	Median resolution time by Firm-paid	Median resolution time by Volunteer	Significance level
Geronimo	10	18	p= 0.0000
Qpid	23	17	p= 0.1653
Qt	102	101	p= 0.4911

5.3   H3: *The larger number of stakeholders involve in an issue is, the longer the issue resolution time is,* **and H4:** *The larger number of exchanged message on an issue is, the longer the issue resolution time is*

We performed a pair-wise correlation analysis among number of message, issue centrality degree, sum of stakeholder centrality and issue resolution time. The correlation matrixes for Qt, Qpid and Geronimo projects are shown in Table 7, Table 8 and Table 9 respectively. The mark "**" represents a significance level at 0.01. Referring to Hopskin interpretation of value of correlation coefficient, which classify the value of correlation coefficient as trivial (<0.1), minor (0.1 – 0.3), moderate (0.3-0.5), large (0.5 – 0.7), very large (0.7 – 0.9) and almost perfect (0.9 - 1.0) [22], the correlation between number of task-based messages and issue resolution time is significant at minor level in Qt, Qpid while it is at moderate level in Geronimo. The correlation between issue centrality and its resolution time is at a minor level for Qt and at a moderate level for Qpid and Geronimo. Besides, the correlation coefficient between stakeholder accumulative centrality and resolution time is slightly higher than the one of issue centrality. All of these correlation coefficients are significant at level 0.01, which allow us reject the null hypotheses for H3, H4 and accept the alternative ones. It is noticed that among three variables, the accumulative stakeholder centrality degree has the largest correlation coefficient with issue resolution time in all projects.

Table 7. Pairwise correlation for Qt

	No of message	Issue centrality	Sum. Stak. centrality	Resolution time
Number of message	1	0.413**	0.460**	0.125**
Issue centrality		1	0.213**	0.172**
Sum. Stak. centrality			1	**0.262****
Resolution time				1

Table 8. Pairwise correlation for Qpid

	No of message	Issue centrality	Sum, Stak. centrality	Resolution time
Number of message	1	0.569**	0.423**	0.243**
Issue centrality		1	0.199**	0.310**
Sum. Stak. centrality			1	**0.331****
Resolution time				1

Table 9. Pairwise correlation for Geronimo

	No of message	Issue centrality	Sum. Stak. centrality	Resolution time
Number of message	1	0.491**	0.382**	0.416**
Issue centrality		1	0.251**	0.303**
Sum. Stak. centrality			1	**0.409****
Resolution time				1

6 Discussion of Results

Table 10 summarizes the testing results for each hypothesis. Concerning **hypothesis H1**, the statistical test results reject the null hypotheses in all cases, which show the centrality degree of an average firm-paid assignee are significantly higher than that of an average volunteer assignee. This finding infers the distribution of labor between firm-paid and volunteer assignees. It indicates that in the issue-resolving process, a firm-paid assignee involves in much more issues than a volunteer assignee does.

On testing **hypothesis H2**, the issue resolution time significantly varies between firm-paid assignees and volunteer assignees in only one out of three investigated projects. Therefore, we can conclude that the stakeholder type is unlikely an influenced factor on issue resolution time. The data suggests that while volunteer and firm-paid assignees participate in OSS projects with different motivation and working approaches, these differences do not have an impact on their issue resolution time.

In the result for **H3 and H4**, the correlation tests reveal a positive correlation between collaboration measures, such as number of message, number of involved stakeholder and issue resolution time. It implies that the high collaboration level in an issue, e.g. high number of messages exchanged or high number of involved stakeholders indicates a longer resolution time. This may be due to the complexity of the task that relates other issues or software modules; or the poor quality of the issue description leads to demands of explanation and discussion. However, we are aware that the result of correlation analysis doesn't imply cause-effect relationship due to the effect of compounding factors. To validate the provided hypothesis, a further regression analysis is necessary. From the results, we also observe that there is significant positive correlation between issue centrality and number of messages exchanged. This observation was expected as the larger number of stakeholders involved in an issue (i.e. editing the reports or commenting on the issue) clearly leads to the increasing of number of comments or report edits. Therefore, these two variables should be checked for compounding factors if they are both used in regression models.

Table 10. Results of Hypotheses testing

Hypotheses	H1	H2	H3	H4
Test	Mann Whitney U	Mann Whitney U	Spearman correlation	Spearman correlation
Geronimo	Accept	Accept	Accept	Accept
Qt	Accept	Reject	Accept	Accept
Qpid	Accept	Reject	Accept	Accept

7 Threats to Validity

First, a major threat of the study validity lies in the division of stakeholders as volunteer or firm-paid. Although a major amount of stakeholder's affiliation is identified, there are still some stakeholders with no company information. However this group of unidentified stakeholders is responsible for a very small portion of issues in general. Since the major portion of the issues comes from identified stakeholders, the comparison of resolution time between groups of stakeholders would not be significantly influenced.

Second, a main concept investigated in this study is collaboration, which is measured by the number of comments, messages and number of issue-involved stakeholders. Although collaboration between stakeholders can be done via other channels, such as IRC, Skype and face-to-face discussion, issue tracking system and mailing list are the most common discussion means. The most relevant discussion about an issue should be found here. The other concern in the data collection process is the quality of the issue report since the data can be randomly filled in and the occurrence of duplicated reports. However, the quality of report is also an included factor in this study since it might influence the issue resolution time.

Third, another threat to validity comes from the generality of the research findings. As in many empirical studies of OSS projects, few case studies are definitely not significant enough to generalize what we found to the population of OSS projects. In this study, the cases were thoroughly selected to represent for an active, medium-size and on-going OSS projects.

Last but not least, compounding factors is an unavoidable threat in a correlational study. The high correlation between number of messages, number of stakeholders and issue resolution time can be caused by a latent variable, not investigated in this study, such as complexity of the issue, or dependencies among issues. Therefore, this concern could be a subject for a future investigation.

8 Conclusion and Future Work

In this study, we investigated the impact of different stakeholder types and their collaboration on issue resolution time in three medium-size and ongoing OSS projects. The statistic test result provides some interesting findings for OSS practitioners as well as OSS researchers. First, we found that in firm-involved OSS projects, there is not only a large portion of firm-paid labor contributed to the projects, but also a higher workload on an average firm-paid assignee than on a average volunteer assignee. However, we did not find a difference in issue resolution time between volunteer and firm-paid assignees. The result contributes to the understanding of distribution of workload and resolving time between volunteer and firm-paid assignees. Second, we found a significant impact of stakeholder collaboration on issue resolution time. Particularly, the issue with fewer stakeholders is resolved faster than the one with more stakeholders. The issue with fewer comments is also resolved faster than the ones with more comments. For practitioners, these metrics can be integrated in the issue tracking system or defect repository to provide a recommendation for issue resolving process. Particularly, the collaboration information collected overtime will help developers being aware of which issue is going to take longer time to resolve. For researchers who want to integrate collaboration measures in software quality or productivity prediction models, they should consider of not only the usefulness of number of involved stakeholders, number of exchanged messages but also the compounding effect between them.

The paper contributes to fill in a gap in the literature gap by providing an empirical investigation of firm-paid stakeholders and their cooperation with others in OSS projects. The findings were supported by descriptive statistic and correlation analysis and further work should employ regression analysis to validate these findings. The study is also limited in using simple SNA metrics, such as the Freeman Centrality Degree. In future, we will explore more SNA metrics to investigate other aspects of stakeholder collaboration. Besides, the findings are based on only three projects, so the analysis should be replicated with more datasets to generalize conclusions on OSS community.

Acknowledgements. The authors would like to thank Tor Stålhane for his valuable comments and help with checking the statistical procedures.

References

[1] Abreu, R., Premraj, R.: How developer communication frequency relates to bug introducing changes. In: Proceedings of the Joint International and Annual ERCIM Workshops on Principles of Software Evolution (IWPSE) and Software Evolution (Evol) Workshops, pp. 153–158. ACM, Amsterdam (2009)

[2] Anbalagan, P., Vouk, M.: On predicting the time taken to correct bug reports in open source projects. In: Proceedings of IEEE International Conference on Software Maintenance (ICSM 2009), pp. 523–526 (2009)

[3] Ayala, C.P., Cruzes, D., Hauge, Ø., Conradi, R.: Five Facts on the Adoption of Open Source Software. IEEE Software 28, 95–99 (2011)

[4] Berdou, E.: Insiders and outsiders: paid contributors and the dynamics of cooperation in community led F/OS projects. In: Open Source Systems, pp. 201–208. Springer, Boston (2006)

[5] Bettenburg, N., Hassan, A.: Studying the Impact of Social Structures on Software Quality. In: Proceedings of IEEE 18th International Conference on Program Comprehension (ICPC 2010), pp. 124–133 (2010)

[6] Bettenburg, N., Just, S., Schroter, A., Weiss, C., Premraj, R., Zimmermann, T.: What makes a good bug report? In: Proceedings of 16th ACM SIGSOFT International Symposium on Foundations of Software Engineering (ESE 2008), pp. 308–318. ACM, Atlanta (2008)

[7] Bird, C., Nagappan, N., Gall, H., Murphy, B., Devanbu, P.: Putting It All Together: Using Socio-technical Networks to Predict Failures. In: Proceedings of 20th IEEE International Symposium on Software Reliability Engineering (ISSRE 2009), pp. 109–119 (2009)

[8] Bonaccorsi, A., Rossi, C.: Comparing motivations of individual programmers and firms to take part in the open source movement: From community to business. Knowledge, Technology & Policy 18, 40–64 (2006)

[9] Bonaccorsi, A., Lorenzi, D., Merito, M., Rossi, C.: Business firms' engagement in community projects. Empirical evidence and further developments of the research. In: Proceedings of 1st International Workshop on Emerging Trends in FLOSS Research and Development (FLOSS 2007), May 21, pp. 1–5. IEEE Computer Society, Minneapolis (2007)

[10] Capra, E., Francalanci, C., Merlo, F., Rossi-Lamastra, C.: Firms' involvement in Open Source projects: A trade-off between software structural quality and popularity. Journal of System and Software 84, 144–161 (2011)

[11] Crowston, K., Scozzi, B.: Bug Fixing Practices within Free/Libre Open Source Software Development Teams. Journal of Database Management (JDM) 19(2), 1–30 (2008)

[12] Crowston, K., Wei, K., Li, Q., Howison, J.: Core and Periphery in Free/Libre and Open Source Software Team Communications. In: Proceedings of 39th Annual Hawaii International Conference on System Sciences (HICSS 2006), vol. 06, p. 118. IEEE Computer Society, Los Alamitos (2006)

[13] Devore, J.L.: Probability and Statistics for Engineering and the Sciences. Technometrics 46(4), 497

[14] Feczak, S., Hossain, L.: Measuring Coordination Gaps of Open Source Groups through Social Networks. In: Proceedings of 11th International Conference on Enterprise Information Systems (ICEIS 2009), pp. 84–90 (2009)

[15] Freeman, L.: The Development of Social Network Analysis. Empirical Press, Vancouver (2006)

[16] Guo, P.J., Zimmermann, T., Nagappan, N., Murphy, B.: Characterizing and predicting which bugs get fixed: an empirical study of Microsoft Windows. In: Proceedings of 32nd ACM/IEEE International Conference on Software Engineering (ICSE), Cape Town, South Africa, vol. 1, pp. 495–504 (2010)

[17] Hars, A., Ou, S.: Working for Free? - Motivations of Participating in Open Source Projects. International Journal of Electronic Commerce 6, 25–39 (2002)

[18] Hauge, Ø., Ayala, C., Conradi, R.: Adoption of open source software in software-intensive organizations - A systematic literature review. Journal of Information Software and Technology 52, 1133–1154 (2010)

[19] Hauge, Ø., Sørensen, C., Conradi, R.: Adoption of Open Source in the Software Industry. In: Open Source Development, Communities and Quality, pp. 211–221. Springer, Boston (2008)

[20] Henkel, J.: Champions of revealing the role of open source developers in commercial firms. Journal of Industrial and Corporate Change 18(3), 435–471 (2009)

[21] Herbsleb, J.D.: Global Software Engineering: The Future of Socio-technical Coordination. In: Proceedings of 29th International Conference on Software Engineering- Future of Software Engineering (ICSE), pp. 188–198 (2007)

[22] Hopkins, W.G.: A scale of magnitudes for the effect statistics. A New View of Statistics (June 2002)

[23] Kagan, S.L.: United we stand: Collaboration for child care and early education services. Teachers College Press, New York (1991)

[24] Lakhani, K.R., Wolf, R.: Why Hackers Do What They Do: Understanding Motivation and Effort in Free/Open Source Software Projects. In: Perspectives on Free and Open Source Software. MIT Press, Cambridge (2005)

[25] Martinez-Romo, J., Robles, G., Gonzalez-Barahona, J., Perez, M.: Using social network analysis techniques to study collaboration between a FLOSS community and a company. Open Source Development, Communities and Quality 275, 171–186 (2008)

[26] Mattessich, P.W., Murray-Close, M., Monsey, B.R.: Collaboration: What Makes it Work: A review of Research and Literature on Factors Influencing Successful Collaboration. Amherst H. Wilder Foundation (2001)

[27] Meneely, A., Williams, L.: Secure open source collaboration: an empirical study of linus' law. In: Proceedings of 16th ACM Conference on Computer and Communications Security (CCS 2009), Illinois, USA, pp. 453–462 (2009)

[28] Meneely, A., Williams, L., Snipes, W., Osborne, J.: Predicting failures with developer networks and social network analysis. In: Proceedings of 16th ACM SIGSOFT International Symposium on Foundations of Software Engineering (FSE 2008), Atlanta, Georgia, USA, pp. 13–23 (2008)

[29] Nguyen, T., Adams, B., Hassan, A.: Studying the impact of dependency network measures on software quality. In: Proceedings of IEEE International Conference on Software Maintenance (ICSM 2010), pp. 1–10 (2010)

[30] Pinzger, M., Nagappan, N., Murphy, B.: Can developer-module networks predict failures?. In: 16th ACM SIGSOFT International Symposium on Foundations of software Engineering (FSE), Atlanta, Georgia, pp. 2–12 (2008)

[31] Rahman, M., Ruhe, G.: Resource allocation and activity scheduling: bug fixing perspective, Technical Report, Software engineering decision support laboratory, University of Calgary (2010)

[32] Wolf, T., Schroter, A., Damian, D., Nguyen, T.: Predicting build failures using social network analysis on developer communication. In: Proceeding of 31st International Conference on Software Engineering (ICSE), pp. 1–11 (2009)

Towards a Unified Definition of Open Source Quality

Claudia Ruiz and William Robinson

Georgia State University, Computer Information Systems Department,
35 Broad Street NW, Atlanta, GA 30303, USA
{cruiz5,wrobinson}@gsu.edu

Abstract. Software quality needs to be specified and evaluated in order to determine the success of a development project, but this is a challenge with Free/Libre Open Source Software (FLOSS) because of its permanently emergent state. This has not deterred the growth of the assumption that FLOSS is higher quality than traditionally developed software, despite of mixed research results. With this literature review, we found the reason for these mixed results is that that quality is being defined, measured, and evaluated differently. We report the most popular definitions, such as software structure measures, process measures, such as defect fixing, and maturity assessment models. The way researchers have built their samples has also contributed to the mixed results with different project properties being considered and ignored. Because FLOSS projects are evolving, their quality is too, and it must be measured using metrics that take into account its community's commitment to quality rather than just its software structure. Challenges exist in defining what constitutes a defect or bug, and the role of modularity in affecting FLOSS quality.

Keywords: open source, software, quality, measurement, literature review.

1 Introduction

Quality is extremely subjective, with as many definitions as there are people with opinions. It is no surprise that studies evaluating the quality of FLOSS and comparing it with traditionally developed software have produced mixed results [1], [2], [3], [4]. This is probably because of two main reasons: each study has defined quality differently and has evaluated it using different characteristics of different FLOSS projects. Defining quality differently will of course produce mixed results, but even when studies define quality in similar terms, they evaluate it using dissimilar projects and compare different project characteristics.

In order to understand what it is about certain FLOSS projects that lead them to produce high quality software, the antecedents of FLOSS quality must be found. There is, however, no current research on the antecedents of FLOSS quality [5].

This paper takes the first step towards addressing this issue by reviewing the FLOSS literature order to understand how is quality being conceptualized and to propose a unified definition of FLOSS quality.

S.A. Hissam et al. (Eds.): OSS 2011, IFIP AICT 365, pp. 17–33, 2011.
© IFIP International Federation for Information Processing 2011

The rest of the paper is organized as follows: sections two and three provide a brief background on FLOSS and software quality; section four presents the methodology followed; section five describes the findings, and section six discusses the implications of the findings., categorizing the findings according to research approach, and definition of quality.

2 FLOSS

FLOSS has grown dramatically in the 2000s and is an integral part of the IT industry. It directly supports 29% of the software that is developed in-house in the EU and 43% in the US and could reach a 32% share of all IT services by 2010 [6].

Linux, Apache, Firefox are commonly found in many computers today and were developed using open source models. Apache is a Web server used by 60% of Websites worldwide [7] and 23.2% of European and 14.5% of North American Web surfers use the Firefox Web browser [8].

This growing popularity begs the question: is FLOSS *better* than traditionally developed software? Traditionally developed software projects are considered successful if they finish on time, on budget and meet specifications. But the same standards cannot be applied to judge the success of FLOSS projects, since they usually have minimal budgets, are always in a state of development, do not have an official end time, and do not have formal specifications [9].

This lack of objective measures of success has not deterred the adoption of FLOSS products. It even has become a common assumption that FLOSS products are higher quality than traditionally developed software [10], [11] with firms entering FLOSS projects citing FLOSS's "quality and reliability" as one of the main motivating reasons for the endeavor [12].

This assumption can be traced back to Linus Torvalds, the architect of the Linux kernel, who said that "given enough eyeballs, all bugs are shallow" [13]. Torvalds believed that FLOSS's public peer review and frequent releases lead to fewer bugs because there are more people looking at the software, reporting errors, and fixing those errors. This assumption has a kernel of truth: it has been observed that most problem (bug) reports and solutions in FLOSS projects are contributed by periphery community members and not so many by the core developers [14].

A FLOSS project is one that offers its software product's license in accordance to the Open Source Definition [15] providing for free redistribution of the compiled software and the openly accessible source code.

A typical FLOSS project is composed by a community, whose structure has been described as being like an "onion" with the most actively contributing members, who are the most invested in the project and have the greatest decision power in the inner part and the least contributing members with the least amount of decision power on the outside. The project leader is at the center and radiating out are the core members, the active developers, the peripheral developers, the bug fixers, the bug reporters, the readers, and the passive users [16]. These roles are dynamic, changing as the community evolves as the system they are building evolves [16].

Although each FLOSS project is different and has different development practices and processes, The Apache project can be used as a model of a mature FLOSS development project given its success and well documented and researched development cycle stages. Its stages are below:

Identifying Work to Be Done. Core developers look at the bug reporting database and the developer forums for change and enhancement requests. The core developers need to be persuaded of the priority of the request for it to be included in the agenda status list.

Assigning and Performing Development Work. Core developers look for volunteers to perform the work. Priority is given to code owners (those who created or have been actively maintaining the particular module). The developer then identifies a solution and gets feedback from the rest of the developers.

Prerelease Testing. Each developer performs unit testing of his/her own work. There is no integrated or systems testing.

Inspections. Each developer then commits his/her changes and the code is then reviewed before it is included in a stable release, while changes to development releases are reviewed after being included in the release.

Managing Releases. A core team member volunteers to be the release manager and makes the decisions pertaining to the individual release. He or she delineates the scope of the release by making sure that all open requests and problems are resolved and restricts access to the code repository to avoid any more changes [17].

These development cycle stages draw a parallel to the Scrum Agile development methodology, where a product owner creates a backlog, a prioritized list of functional and non-functional requirements for building into the product. Development is performed in sprints which are 30 day iterations of development activities, which include only the highest priority backlog requirements that can be successfully completed in the allotted time [18].

Most FLOSS development practices are very similar to Scrum Agile development methods but less structured and more ad hoc.

3   Software Quality

The origins of software quality can be traced back to industrial engineering and operations management and their development of product quality concepts and quality management practices. For these fields, quality is adherence to process specification [19], [20] in order to produce a product that meets customer requirements with zero defects [21], [22]. In order to achieve this goal, approaches such as TQM (total quality management) [23], [24] were developed to integrate quality into all company activities and Six Sigma to measure for quality [25].

Industrial engineering and operations management's view of quality can be categorized as the manufacturing, user, and product approaches to quality as described by [26]. Table 1 summarizes these definitions of quality as well as others (transcendent and value). As well as categorizing definitions of quality, Garvin also categorized the eight dimensions of product quality (Table 2) [26].

Table 1. Garvin's quality definitions [26]

Approach	Definition of Quality
Transcendent	Innate excellence that cannot be defined, only recognized through experience.
Product	Discrete and measurable product characteristics.
User	Subjective consumer satisfaction.
Manufacturing	Conformance to specification.
Value	Conformance to specification at an acceptable cost or price.

With this legacy from industrial engineering and operations management, software quality started with the product definition of quality by defining frameworks of factors. The most popular were Boehm's model of 23 factors (dimensions of quality) [27] and McCall's with 11 factors [28], [29] which are all listed in Table 2.

Both of these frameworks of factors left out the measure (actual thing that is counted) for each factor. Each implementer and developer was left to define his or her own metrics and criteria for each factor. The ISO 9126 [30] Information technology – Software product evaluation: quality characteristics and guidelines for their use, which is part of the ISO 9000 set of standards by the ISO (The International Organization for Standardization) for quality management, was an attempt to standardize the quality factors to six main factors with three sub-factors under each one.

Table 2. Quality Factors

Model	Factors
McCall	Accessibility, Accountability, Accuracy, Augmentability, Communicativeness, Completeness, Conciseness, Consistency, Device-independence, Efficiency, Human engineering, Legibility, Maintainability, Modifiability, Portability, Reliability, Robustness, Self-containedness, Self-descriptiveness, Structuredness, Testability, Understandability, Usability
Boehm	Correctness, Efficiency, Flexibility, Integrity, Interoperability , Maintainability, Portability, Reliability, Reusability, Testability, Usability
ISO 9126	Efficiency, Functionality, Maintainability, Portability, Reliability, Usability
Garvin	Aesthetics, Conformance, Durability, Features, Perceived quality, Performance, Reliability, Serviceability

While the quality factor frameworks were used to assess the software product, process frameworks were developed to assess the quality of the process producing the software and to accommodate the manufacturing definition of quality [26]. One such framework is the CMMI (Capability Maturity Model Integration), which is a process improvement framework that can be used to drive organizational change and to judge the process maturity of another organization. CMMI has five levels, with one being the lowest. A level five organization is one where processes are defined (level 3), quantitatively managed (level 4), and are continually being optimized (level 5). This maturity model lists the processes that an organization should have in order to be considered at a certain CMMI level but it leaves the details of how to put them into place up to the organization.

Quality factor frameworks come close to Garvin's product definition of quality because they distill quality into a set of measurable characteristics while the process maturity models most closely resemble Garvin's manufacturing definition of quality because they define quality by evaluating how close an organization's processes meet a predetermined specification. Garvin's user definition of quality, on the other hand, is hard to implement for commercially offered software because a user's satisfaction or rather dissatisfaction with a software's features or performance cannot be immediately addressed. Rather, user satisfaction must be bundled with the problem resolutions and new feature requests of all other users into a new release, patch, or service pack, which are infrequently issued due to their cost. Because of this limitation, software quality has become Garvin's value definition of quality: conformance to specification at an acceptable cost.

In contrast, FLOSS software quality most closely fits Garvin's user definition of quality. Users can directly log problem reports and new functionality requests directly into the software project's issue tracking system that is used by developers. Because FLOSS has frequent releases, those requests can become part of the software much more quickly than commercially offered software, thus better satisfying users.

4 Methodology

In order to answer the research question, how is quality defined in the FLOSS literature, we performed a literature review. Since the definition of quality is very subjective, we adopted an interpretive approach [31] to this review by applying a grounded theory methodology [32].

We used the *Straussian* type of grounded theory in order to allow previous theories and our own interpretations of quality to guide the data collection and analysis [33].

4.1 Data Collection

In order to comply with the theoretical sampling necessary in grounded theory, we searched Google Scholar for journal articles and conference papers containing the terms "open source" and "quality". We retained papers that met the following criteria: explicit definition of quality and empirical validation of the quality definition. We decided these criteria would provide a relevant sample because the authors of these papers would have to explicitly define quality and operationalize it in order to empirically validate it.

This process left us with 24 papers, to which we then added 16 from the quality and defect-fixing categories in [34] that met the above stated criteria.

This left us with 40 papers that defined quality and performed some form of empirical validation of that definition.

4.2 Analysis

The papers of this literature review were analyzed using open, axial, and selective coding [32]. As the papers were read, they were coded using open coding Text segments from each of the papers were highlighted and labeled with a code to categorize and conceptualize the data.

The open coding phase produced 75 codes, which were used to label 637 text segments from the 40 papers gathered. The codes reflected how the authors defined quality, the measures used to operationalize it, the research methods used to analyze it, and the characteristics considered in the FLOSS projects that were used to validate their definition of quality.

The axial coding phase produced five codes which were categories containing the codes from the open coding phase. From the 75 codes from the open coding phase, four were discarded because they labeled few text segments and did not help explain how quality is interpreted in FLOSS research.

Table 3 show the categories produced from the axial coding phase. The categories were based on how the authors approached the research, how they analyzed the data, and how they defined and operationalized quality, with the two main categories being quality as a process and quality as a product. The final category deals with the type of data sampled by the authors to validate their models, in this case, the characteristics of the FLOSS projects they examined. These categories were chosen because they follow the research process: an approach must be chosen along with an analysis method; the phenomenon of interest must be operationalized and finally, the data sample must be chosen.

The final phase, the selective coding phase, produced and integrated category that narrated the conceptualization of quality by FLOSS researchers. This phase was complete when theoretical saturation was reached, meaning, not new conceptualizations could be obtained from the data.

Table 3. Categories from Axial Coding

Category	Description	Sample of codes within category
Research approach	Approach used to analyze quality in research study	Case study
		Survey
		Factor model
		Maturity model
Analysis method	Methodology used to analyze data	Regression
		Structural equation model
		Machine Learning
		Social Network Analysis
Quality as a process	Quality operationalized as processes that can be measured.	Defect fixing rate
		Defect fixing time
		Definition of bug
		Quality assurance procedures
		Process metrics
Quality as a product	Quality operationalized as characteristics of final software product.	Product metrics
		Number of post-release defects
		Cyclomatic complexity
		Halstead Volume
		CBO (coupling between objects)
Examined project	Characteristics considered of FLOSS projects used to validate operationalization of quality	Maturity
		Popularity
		Number of developers
		Development time examined
		Software type
		Version

5   Findings

In this section, we describe how researchers interpret quality in FLOSS publications.

5.1   Quality as a Product

These studies defined quality as structural code quality [1], [35], [36], [37], [38], [39], [40]. Metrics that are used to measure structural quality are number of statements [1], [35], cyclomatic complexity [1], [35], [36], [38], [39], [41], [42], number of nesting levels [1], [35], [38], [43], Halstead volume [1], [35], [42], coupling [35], [36], [37], [41], [43], coding style [35], statements per function, files per directory, percentage of numeric constants in operands [35], growth of LOC (lines of code)[38], modularity [2], average coupling between objects, cohesion, number of children, depth of inheritance tree, methods inheritance factor and other internal software structure metrics [1], [35], [36], [37], [38], [39], [42].

The idea behind measuring software code structure is that well-designed software is less complex, less likely to contain faults, and easier to maintain [42].

Measuring code structure left the researchers with more questions than answers. The most successful projects in terms of number of downloads and popularity were not the ones with the highest structural quality [42], [43]. Another study found that the software modules with the highest rate of change were not the ones with the highest structural complexity [38]. Even using machine learning algorithms with structural quality measures in order to predict faults did not produce clear results [39].

Comparing structural quality between open and closed projects produced mixed results with some studies finding that FLOSS projects had quality comparable to closed projects [44] while others found that open source software did not prove to have structural code quality higher than commercial software [1], [35].

Using structural quality to define, measure, and compare FLOSS quality has not proven effective with different researchers achieving different results even when using the same metrics.

5.2   Quality as a Process

5.2.1   Defect Fixing

Defect fixing [2], [3], [38], [39], [45], [46], [47], [48], [49], [50], [51], [52], [53], [54], [55], [56], [57], [58], [59], [60], [61] is by far the most popular definition of quality as a process in the FLOSS quality literature. Authors have done studies defining the what constitutes the process itself in order to determine how it works [38], [48], [49], [50], [51], [52], [53], [54], [55], [61] and developed models to test its effectiveness [45], [62].

They have approach it in terms of total bugs fixed [37], [48], [49], [58], [59], [61] and speed of bug resolution [3], [46], [47], [51], [61]. These approaches take into account the evolving nature of FLOSS, which is never truly finished, but rather, remains in a permanently emerging state. It also considers that FLOSS testing and defect reporting and fixing is a community activity where developers, users, and periphery members collaborate to create the software.

However, results using this approach have been mixed. Some open source projects resolved service requests more quickly than their closed counterparts, others did not [3]. For other studies, software type (database, financial, game, networking) made more of a difference in determining defect resolution speed along with number of developers (groups with less than 15 developers were the most efficient) [47]. The main difference with closed software is that in most FLOSS projects, bugs are only addressed after feedback is received from users. There is no way to measure the quality of a release pre hoc, only ad hoc [50]. However, this attitude is changing with projects such as GNOME, Debian, and KDE forming their own Quality Assurance teams and enforcing quality assurance tasks [57].

Even though the defect fixing approach to measure FLOSS quality considers the evolving nature of its quality, research using it has not operationalize it in an evolutionary manner. Most studies have looked at the bug databases of FLOSS projects cumulatively after a certain amount of time (i.e. after six months of activity) rather than looking at defect resolution rates per release (except for [39], [53], [58] which did compare product releases), which is the evolutionary cycle of FLOSS software. The studies that looked at defects per release found that FLOSS has lower post-feature test defects than commercial software but higher post-release defects than commercial software [58] and that release software quality is cyclical, with the Mozilla 1.2 showing a major decrease in quality, which was improved in later releases [39].

The cyclical nature of FLOSS quality is illustrated in a study that showed that bug arrival rates follow a bell curve through time between releases. Whenever drastic changes were introduced to the software, the rate would also drastically change [53]. This would suggest that defect rates and thus defect resolution rates vary across releases depending on the changes being introduced. If the release introduces new features, or makes major changes to the architecture of the software, many defects will be introduced, while those that simply introduce defect fixes and enhancements, will introduce less.

Another issue stems from the definition of bug. Bug reports in bug databases in FLOSS project management Web sites could include anything from "failures, faults, changes, new requirements, new functionalities, ideas, and tasks"[49]. Not to mention duplicate bug reports, poorly defined ones, and those that are out of scope with the product [56]. This happens because bug reporting systems are usually open to the public, and users without enough technical skills will make mistakes writing the bug reports [57].

Measuring defect-fixing effectiveness in FLOSS projects has provided mixed results because different studies have defined and thus measured defects or bugs differently. They have also calculated defect-fixing rates by looking at bug tracking databases in the cumulative, without considering that defects are introduced and fixed cyclically in FLOSS, per release.

5.2.2 Other Processes
FLOSS projects tend to rely on tools to enforce policies and standards [56]. Such tools include defect tracking systems, version control, mailing lists, automatic builds, etc. [55], [56], [57].

One quality assurance activity that is performed in traditional software development, peer review, is done differently in FLOSS projects with successful results. Peer reviews in open source were more efficient because there is no time wasted scheduling meeting since people work asynchronously and have more detailed discussions [14]. An example from the Apache project shows that it has three types of peer review procedures, depending on the experience and trustworthiness of the developer [14].

Despite the successful inclusion of peer reviews into the FLOSS development process, testing procedures have not managed to make the necessary cross over into FLOSS. Most projects do not have a baseline test suite to support testing, this means no regression testing can be performed [54].

Developers perform their unit testing and do sometimes better than commercial software [58], but it is up to the users to discover bugs and defects which could be eliminated with system or integrated testing.

5.2.3 Process Maturity Models

Maturity assessment models have been formulated to help users and integrators evaluate the quality of a FLOSS project versus another [59], [60], [63], [64], [65], [66], [67], [68]. These models provide a set of criteria to evaluate a FLOSS project. Different models concentrate on different criteria, but they all provide a way to quantify and evaluate the quality of a FLOSS product.

Maturity models use organizational trustworthiness as a proxy for product trustworthiness and thus quality—if the product is built correctly, it will then have high quality.

The assessments are mostly for the FLOSS integrator who must assess the risk of adding the FLOSS product to his or her existing architecture. The assessment models are not predictive (they do not evaluate the factors that lead to quality, nor do they provide a construct for quality) they simply provide a set of criteria with different scores that the integrator can then use to make the decision to adopt the FLOSS product.

5.3 Modularity as the Enabler to FLOSS Quality

FLOSS's paradox of having and adding more developers without compromising its productivity (in contrast to Brook's law that says that adding more developers increases coordination costs and decreases performance) is due to its approach to modularity. A FLOSS project is made up of many subprojects where only a few developers work together without ever having to interact with the developers in other subprojects or modules [69].

It is believed this is the reason that projects such as Linux and Apache are considered successful. They have been able to scale because of their modularity. Because of modularity, defects in one module, do not affect the rest [58].

However, there is no single definition of modularity. The studies that have defined and measured it do so differently. One study used "correlation between functions added and functions modified" to measure modularity. It then compared modularity across a set of open and closed projects. The open projects did not prove to be more modular than the closed ones [2].

Another study used average component size, which was measured as program length (sum of the number of unique operands and operators) divided by number of statements. The study found that applications with smaller average component size received better user satisfaction scores [1].

In terms of influencing quality, modularity has produced mixed results. A study found that higher modularity does not lead to higher quality. This study defined modularity as the distance of each package in a release from the main sequence. Because higher modularity is associated with reduced software complexity, it should result in higher structural code quality, but the authors found that the projects with higher modularity contained the greater number of defects [37].

Yet another study contests that it is small component design that leads to low defect density, higher user satisfaction, and easier maintenance and evolution [55].

Work distribution is another way of conceptualizing modularity. In another study, the authors found that a lower concentration of developers making changes to a module led to higher quality for the module. The authors speculated that this could explain the FLOSS paradox of many developers and high productivity: at the project level there could be many developers, but within the project, they should be organized into small teams; this would keep the concentration of authors to code low, thus fostering simpler code, higher quality and better maintainability [41].

The literature has defined and operationalized modularity differently using either software structure measures (component size, distance from main sequence, etc.) or development organizational measures such as author concentration per class, number of authors per module, etc.

The development organizational measures have proven to be more effective at finding a correlation between quality and modularity, but these measures are still vaguely defined and more research needs to be performed to optimally define them and operationalize them in order to produce a universal measure of modularity.

5.4 Characteristics of Samples

The researchers seem to have made very arbitrary choices when it came to choosing FLOSS project to make up their samples.

By far, the most popular place to obtain data for FLOSS quality studies is the SourceForge repository maintained by Notre Dame University. But it was not the only place to find data; some case studies concentrated on popular projects that are not hosted by SourceForge such as Apache [14], [58], Linux [42], and Eclipse [46].

Some projects considered software type a defining factor and only looked at projects of the same type [40], [41], [42], [62], while the rest did not consider it a factor. However one study did consider it and found that project category affects the bug resolution time [47].

The development time of the projects examined by the authors was extremely variable. There were studies that examined FLOSS project data that covered development time for one week [48], 105 weeks [51], four months [56], six months [46], etc. with one project capturing data from initial commit until the last commit before the stable version was released [41].

Another factor that varied across the studies was the number of projects examined. From four [48], to 52 [49], all the way to 140 [61], and beyond.

Another factor used to choose candidate was the success of the project measured in popularity terms such as number of downloads [61] and SourceForge rank [43], [68], which uses number of downloads and recommendations. Researchers refrained from including *failed* projects and only looked at those that high success measures.

The way researchers are choosing their samples is definitely a reason why there are mixed results in the FLOSS quality literature. They are looking at different types of projects, examining them for different amounts of time, and only considering popular projects.

5.5 Summary

Quality is very subjective and hard to define absolutely. With this challenge, FLOSS researchers have used many ways to define quality. They have used product and process metrics and have found mixed results. FLOSS software is always evolving and one version might produce more defects than a pervious one because of some major change in the software or the community structure.

Successful projects are those that have adopted a modular organization of their code and their community, allowing them to grow and isolate defects. They have also implemented tools to automate policy enforcement and adapted traditional software development practices to their context.

There is a need to evaluate the quality of FLOSS projects, and maturity assessment models have emerged to meet this need. However, they are hard to automate, and their scores are hard to interpret.

An important reason as to why researchers have obtained mixed results in researching FLOSS quality is that their samples have different characteristics in terms of number of projects examined, software type, time evaluated, and popularity of projects examined.

6 Discussion

The reviewed papers show that there is a need to define and quantify quality in FLOSS development projects in order to compare them among each other and to traditionally developed software. Identifying projects that produce high quality products will lead to further research into understanding the factors that lead to higher quality and the interaction of those factors in FLOSS development projects.

The development of assessment models to ascertain quality comes from the position in traditionally developed software that established and repeatable processes lead to the development of quality products.

6.1 FLOSS Quality as Evolving

With each release, the FLOSS software and its community change. Quality is not linear: the tenth release of a software product might not have fewer defects than its first. It all depends on what type of release it is; whether it is adding new features, restructuring the entire product, restructuring the way it is developed, or simply posting defect fixes. Which type of release of the product is a more important determinant of its quality than its software structure, or its number of developers.

This explains the mixed results obtained from research that only used product measures as a measure of quality – the modules with the highest change rate and the highest number of defects were not those with the lowest design quality or complexity [38].

6.2 Quality as Defect Resolution Rate

Number of defects added by a release divided by the number of lines of code added by the release would seem a good measure of software product quality [58] that would allow comparison between open and closed software products.

But this assumption is wrong because the release of a commercial closed software product is not the same as the one from an open source project. An open source product release resembles the commercial software after its feature test, since there is neither system nor regression testing in open source projects [58].

These measures do not take into account FLOSS projects' community development. That is why a better measure of FLOSS quality is defect resolution rate, in terms of number of bugs resolved and average time of bug resolution. These measures not only show the quality of the code but also the community's effectiveness at achieving quality.

A key issue is to define bugs as defects in the software product. Bug databases, which are used to calculate the defect resolution rates, are riddled with non-bugs, which must not be taken into account when calculating these rates.

6.3 Modularity as Driver of Quality

The "many eyeballs" looking at the bugs include core developers, periphery developers, sometime contributors, and users, who can easily find their way to the project's publicly available bug tracking system. This group has activity rates, contribution amounts, contributions included per release, problem reports contributed, problem reports resolved, and download statistics. These are all metrics of the community's quality efforts.

But making sure that these community members can work effectively with each other is very necessary. A modular architecture of the code and the community allows a project to grow and attract new developers without having the defects of one group affect another group.

That is why modularity needs to be defined in terms of technical modularity (the coupling of the modules) and organizational modularity (the coupling of the module managers/owners and the core project manager) [70].

However, a downside of modularity is that if a member leaves, his or her module might become orphaned. That is why projects such as Debian are developing their own "quality assurance" groups (http://qa.debian.org), where anyone interested can join and help with mass bug filing and transitions, track orphan code, etc.

It seems that too much modularity might be bad for quality in the long run. It is important to do more research to understand what the right amount of modularity looks like.

6.4 Process and Product as Drivers of Quality

Product and process requirements, the traditional specific quality requirements [71] are still relevant in driving quality in FLOSS products: they are universally understood and any project community still needs to reach for them. But because of this, they are sometimes taken for granted; their inclusion needs to be aided by automation tools, such as testing tools included in the automatic build software that calculates and posts correctness and reliability metrics and compares them with benchmark numbers, alerting the community members if their software product falls below the thresholds.

Relying on the "many eyeballs" to report and fix defects has helped FLOSS achieve quality, but there is something to be said for automating the process in order to produce a higher quality product before it is released.

6.5 FLOSS Requires Its Own Maturity Model for Quality

The development of maturity models such as QualOSS, QSOS, OpenBRR, shows the need for a process evaluation model like CMMI but for FLOSS.

This means that quality could also be defined in terms of this process maturity model, but for this approach to reach maturity (so that one day we might have *level 5* FLOSS projects) more research needs to occur to define, if not the ideal, the most effect FLOSS development processes.

7 Conclusion

Just like in traditionally developed software, there is little consensus in the FLOSS literature when it comes to defining quality.

Linux and Apache are by far the most studied projects in FLOSS literature. All the reviewed papers studied projects that they considered successful: they had released several versions, and had high popularity rating, and download numbers. However, failed projects also need to be studied in order to determine what led to their downfall.

FLOSS communities and their software product are emergent and need to a measure of quality that will reflect their nature. Defect resolution rates (amount of defects resolved, speed of resolution) are the best way to measure a community's commitment to quality, because they recognize that FLOSS is not a static product, but ever evolving. These rates should be calculated per release, and not cumulatively, because the cycle of FLOSS evolution is the release. Researchers should be careful to only include defects and not new feature requests, duplicates, or poorly reported bugs into their calculations.

Modularity is being touted as the main driver of FLOSS quality success, but it needs to be further defined and studied in order to understand how it works.

References

1. Stamelos, I., Angelis, L., Oikonomou, A., Bleris, G.L.: Code Quality Analysis in Open Source Software Development. Information Systems Journal 12, 43–60 (2002)
2. Paulson, J.W., Succi, G., Eberlein, A.: An empirical study of open-source and closed-source software products. IEEE Transactions on Software Engineering 30, 246–256 (2004)

3. Kuan, J.: Open Source Software as Lead-User's Make or Buy Decision: A Study of Open and Closed Source Quality. In: Second Conference on The Economics of the Software and Internet Industries (2003)
4. Raghunathan, S., Prasad, A., Mishra, B.K., Chang, H.: Open source versus closed source: software quality in monopoly and competitive markets. IEEE Transactions on Systems, Man and Cybernetics, Part A 35, 903–918 (2005)
5. Crowston, K., Wei, K., Howison, J., Wiggins, A.: Free/Libre Open Source Software Development: What We Know and What We Do Not Know. ACM Computing Surveys 44 (2012)
6. Ghosh, R.A.: Economic Impact of Open Source Software on Innovation and the Competitiveness of the Information and Communication Technologies Sector in the E. U (2006)
7. von Hippel, E., von Krogh, G.: Open Source Software and the "Private-Collective" Innovation Model: Issues for Organization Science. Organization Science 14, 209–223 (2003)
8. Hales, P.: Firefox use continues to rise in Europe. The Inquirer (2006)
9. Scacchi, W.: Understanding Requirements for Open Source Software. In: Lyytinen, K., Loucopoulos, P., Mylopoulos, J., Robinson, B. (eds.) Design Requirements Engineering. LNBIP, vol. 14, pp. 467–494. Springer, Heidelberg (2009)
10. Stewart, K.J., Gosain, S.: The Impact of Ideology on Effectiveness in Open Source Software Development Teams. MIS Quarterly 30, 291–314 (2006)
11. Ajila, S.A., Wu, D.: Empirical study of the effects of open source adoption on software development economics. Journal of Systems and Software 80, 1517–1529 (2007)
12. Bonaccorsi, A., Rossi, C.: Comparing Motivations of Individual Programmers and Firms to Take Part in the Open Source Movement. Knowledge, Technology and Policy 18, 40–64 (2006)
13. Raymond, E.: The cathedral and the bazaar. Knowledge, Technology, and Policy 12, 23–49 (1999)
14. Rigby, P.C., German, D.M., Storey, M.-A.: Open source software peer review practices: a case study of the apache server. In: 0th International Conference on Software Engineering (ICSE 2008), Leipzig, Germany, pp. 541–550 (2008)
15. The Open Source Initiative, http://www.opensource.org/docs/osd
16. Ye, Y., Kishida, K.: Toward an Understanding of the Motivation of Open Source Software Developers. In: Proceedings of the 25th International Conference on Software Engineering, pp. 419–429 (2003)
17. Mockus, A., Fielding, R.T., Herbsleb, J.D.: Two case studies of open source software development: Apache and Mozilla. ACM Trans. Softw. Eng. Methodol. 11, 309–346 (2002)
18. Schwaber, K.: Agile Project Management with Scrum. Microsoft Press (2004)
19. Deming, W.E.: Quality, Productivity, and Competitive Position. MIT Center for Advanced Engineering Study, Cambridge (1982)
20. Deming, W.E.: Out of the Crisis. MIT Center for Advanced Engineering Study, Cambridge (1986)
21. Juran, J.M.: Planning for Quality. Collier Macmillan, London (1988)
22. Crosby, P.B.: Quality is Free: The Art of Making Quality Certain. McGraw-Hill, New York (1979)
23. Feigenbaum, A.: Total quality control: engineering and management: the technical and managerial field for improving product quality, including its reliability, and for reducing operating costs and losses. McGraw-Hill, New York (1961)

24. Ishikawa, K.: What is total quality control? The Japanese way. Prentice-Hall, Englewood Cliffs (1985)
25. Tennant, G.: Six Sigma: SPC and TQM in manufacturing and services. Gower Publishing (2001)
26. Garvin, D.A.: What does 'Product Quality' really mean? Sloan Management Review 1, 25–43 (1984)
27. Boehm, B.W., Brown, J.R., Lipow, M.: Quantitative evaluation of software quality. In: Proceedings of the 2nd International Conference on Software Engineering. IEEE Computer Society Press, San Francisco (1976)
28. Cavano, J.P., McCall, J.A.: A Framework for the Measurement of Software Quality. In: Proceedings of the ACM Software Quality Workshop, pp. 133–139. ACM, New York (1978)
29. McCall, J.A., Richards, P.K., Walters, G.F.: Factors in Software Quality. National Technology Information Service 1, 2, 3 (1977)
30. ISO: ISO 9126-1:2001, Software engineering - Product quality, Part 1: Quality model (2001)
31. Walsham, G.: The Emergence of Interpretivism in IS Research. Information Systems Research 6, 376–394 (1995)
32. Strauss, A.L., Corbin, J.M.: Basics of Qualitative Research: Grounded Theory Procedures and Techniques. Sage Publications, Newbury Park (1990)
33. Strauss, A., Corbin, J.: Grounded Theory Methodology - An Overview. In: Denzin, N.K., Lincoln, Y.S. (eds.) Handbook of Qualitative Research, pp. 273–285. Sage Publications, Thousand Oaks (1994)
34. Aksulu, A., Wade, M.: A Comprehensive Review and Synthesis of Open Source Research. Journal of the Association for Information Systems 11, 576–656 (2010)
35. Spinellis, D.: A Tale of Four Kernels. In: 30th International Conference on Software Engineering, ICSE 2008, pp. 381–390. ACM/IEEE, Leipzig, Germany (2008)
36. Capra, E., Francalanci, C., Merlo, F.: An Empirical Study on the Relationship among Software Design Quality, Development Effort, and Governance in Open Source Projects. IEEE Transactions on Software Engineering 34, 765–782 (2008)
37. Conley, C.A.: Design for quality: The case of Open Source Software Development. Stern Graduate School of Business Administration, vol. PhD, pp. 43. New York University, New York (2008)
38. Koru, A.G., Tian, J.: Comparing high-change modules and modules with the highest measurement values in two large-scale open-source products. IEEE Transactions on Software Engineering 31, 625–642 (2005)
39. Gyimothy, T., Ferenc, R., Siket, I.: Empirical validation of object-oriented metrics on open source software for fault prediction. IEEE Transactions on Software Engineering 31, 897–910 (2005)
40. Koru, A.G., Liu, H.: Identifying and characterizing change-prone classes in two large-scale open-source products. Journal of Systems and Software 80, 63–73 (2007)
41. Koch, S., Neumann, C.: Exploring the Effects of Process Characteristics on Product Quality in Open Source Software Development. Journal of Database Management 19, 31–57 (2008)
42. Yu, L., Schach, S.R., Chen, K., Heller, G.Z., Offutt, J.: Maintainability of the kernels of open-source operating systems: A comparison of Linux with FreeBSD, NetBSD, and OpenBSD. Journal of Systems and Software 79, 807–815 (2006)

43. Barbagallo, D., Francalenei, C., Merlo, F.: The Impact of Social Networking on Software Design Quality and Development Effort in Open Source Projects. In: Proceedings of the International Conference on Information Systems (2008)
44. Samoladas, I., Stamelos, I., Angelis, L., Oikonomou, A.: Open source software development should strive for even greater code maintainability. Commun. ACM 47, 83–87 (2004)
45. Ghapanchi, A.H., Aurum, A.: Measuring the Effectiveness of the Defect-Fixing Process in Open Source Software Projects. In: Proceedings of the 44th Hawaii International Conference on System Sciences, Hawaii, USA (2011)
46. Kidane, Y., Gloor, P.: Correlating temporal communication patterns of the Eclipse open source community with performance and creativity. Computational & Mathematical Organization Theory 13, 17–27 (2007)
47. Au, Y.A., Carpenter, D., Chen, X., Clark, J.G.: Virtual organizational learning in open source software development projects. Information & Management 46, 9–15 (2009)
48. Crowston, K., Scozzi, B.: Bug fixing practices within free/libre open source software development teams. Journal of Database Management 19, 1–30 (2008)
49. Koru, A.G., Tian, J.: Defect handling in medium and large open source projects. IEEE Software 21, 54–61 (2004)
50. Glance, D.G.: Release Criteria for the Linux Kernel. First Monday 9 (2004)
51. Huntley, C.L.: Organizational learning in open-source software projects: an analysis of debugging data. IEEE Transactions on Engineering Management 50, 485–493 (2003)
52. Sohn, S.Y., Mok, M.S.: A strategic analysis for successful open source software utilization based on a structural equation model. Journal of Systems and Software 81, 1014–1024 (2008)
53. Zhou, Y., Davis, J.: Open source software reliability model: an empirical approach. In: Proceedings of the Fifth Workshop on Open Source Software Engineering, pp. 1–6. ACM, St. Louis (2005)
54. Zhao, L., Elbaum, S.: Quality assurance under the open source development model. Journal of Systems and Software 66, 65–75 (2003)
55. Aberdour, M.: Achieving Quality in Open Source Software. IEEE Software 24, 58–64 (2007)
56. Halloran, T.J., Scherlis, W.L.: High Quality and Open Source Software Practices. In: Proceedings of the 2nd Workshop on Open Source Software Engineering (ICSE 2002), Orlando, FL, USA (2002)
57. Michlmayr, M., Hunt, F., Probert, D.: Quality Practices and Problems in Free Software Projects. In: Scotto, M., Succi, G. (eds.) Proceedings of the First International Conference on Open Source Systems, Genova, Italy, pp. 24–28 (2005)
58. Mockus, A., Fielding, R.T., Herbsleb, J.: A Case Study of Open Source Software Development: The Apache Server. In: Proceedings of the 22nd International Conference on Software Engineering, ICSE (2000)
59. Samoladas, I., Gousios, G., Spinellis, D., Stamelos, I.: The SQO-OSS Quality Model: Measurement Based Open Source Software Evaluation. In: 4th International Conference on Open Source Systems (OSS 2008), Milan, Italy, pp. 237–248 (2008)
60. Deprez, J.-C., Alexandre, S.: Comparing Assessment Methodologies for Free/Open Source Software: OpenBRR and QSOS. In: Jedlitschka, A., Salo, O. (eds.) PROFES 2008. LNCS, vol. 5089, pp. 189–203. Springer, Heidelberg (2008)
61. Crowston, K., Howison, J., Annabi, H.: Information Systems Success in Free and Open Source Software Development: Theory and Measures. Software Process: Improvement and Practice 11, 123–148 (2006)

62. Wray, B., Mathieu, R.: Evaluating the performance of open source software projects using data envelopment analysis. Information Management & Computer Security 16, 449 (2008)

63. del Bianco, V., Lavazza, L., Morasca, S., Taibi, D., Tosi, D.: The QualiSPo approach to OSS product quality evaluation. In: Proceedings of the 3rd International Workshop on Emerging Trends in Free/Libre/Open Source Software Research and Development (FLOSS 2010), pp. 23–28. ACM, Cape Town (2010)

64. Soto, M., Ciolkowski, M.: The QualOSS open source assessment model measuring the performance of open source communities. In: Proceedings of the 2009 3rd International Symposium on Empirical Software Engineering and Measurement, pp. 498–501 (2009)

65. Deprez, J.-c., Monfils, F.F., Ciolkowski, M., Soto, M.: Defining Software Evolvability from a Free/Open-Source Software Perspective. In: Third International IEEE Workshop on Software Evolvability, pp. 29–35. IEEE, Paris (2007)

66. Glott, R., Groven, A.-K., Haaland, K., Tannenberg, A.: Quality Models for Free/Libre Open Source Software–Towards the "Silver Bullet"? In: 36th EUROMICRO Conference on Software Engineering and Advanced Applications, Lille, France, pp. 439–446 (2010)

67. Groven, A.-K., Haaland, K., Glott, R., Tannenberg, A.: Security measurements within the framework of quality assessment models for free/libre open source software. In: Proceedings of the Fourth European Conference on Software Architecture: Companion Volume, pp. 229–235. ACM, Copenhagen (2010)

68. Michlmayr, M.: Software Process Maturity and the Success of Free Software Projects. In: Proceeding of the 2005 Conference on Software Engineering: Evolution and Emerging Technologies, pp. 3–14 (2005)

69. Schweik, C.M., English, R.C., Kitsing, M., Haire, S.: Brooks' Versus Linus' Law: An Empirical Test of Open Source Projects. In: Proceedings of the 2008 International Conference on Digital Government Research, Montreal, Canada, pp. 423–424 (2008)

70. Tiwana, A.: The Influence of Software Platform Modularity on Platform Abandonment: An Empirical Study of Firefox Extension Developers. University of Georgia, Terry School of Business (2010)

71. Glinz, M.: On Non-Functional Requirements. In: Proceedings of the 15th IEEE International Requirements Engineering Conference, Delhi, India, pp. 21–26 (2007)

Ginga-J - An Open Java-Based Application Environment for Interactive Digital Television Services

Raoni Kulesza[1,2], Jefferson F.A. Lima, Álan L. Guedes, Lucenildo L.A. Junior[1], Silvio R.L. Meira[2], and Guido L.S. Filho[1]

[1] Laboratory of Digital Video Application (LAVID)
Federal University of Paraiba (UFPB)
Cidade Universitaria - 58059-900,
João Pessoa/PB, Brazil
{raoni,jefferson,alan,lucenildo,
guido}@lavid.ufpb.br
[2] Informatic Center (CIn)
Federal University of Pernambuco (UFPB)
Cidade Universitaria - 50740-560,
Recife/PE, Brazil
srlm@cin.ufpe.br

Abstract. This paper aims to present a Ginga-J's reference implementation. Although based on a particular platform, the implementation not only works as a proof of concept, but also raised several issues and difficulties on the software architecture project that should be taken into account to ease extensibility and porting to other platforms. Ginga is the standard middleware for the Brazilian DTV System. Its imperative environment (Ginga-J) is based on new JavaDTV specification and mandatory for fixed terrestrial receptors.

1 Introduction

With arise of the Digital TV a new set of functionalities were incorporated to the shows offered by stations as well as to the DTV's receptors. Therefore, the TV environment has become more interactive, as TV systems (or middleware) now offer an environment for the execution of interactive applications. These applications can be transmitted and executed along with multimedia content such as audio, video, image, text, etc, thus enabling the increase of interactivity between viewers and the television through applications such as games, polls, etc. [1]. All those services or applications could not be possible without the support of an intermediary software layer, called middleware, installed on each access terminal.

The execution of a same application in different devices with distinct processing capacities and hardware architectures is achieved through the specification of well-defined software architecture. The main role of a Digital TV middleware is to act as an intermediary software layer between the operating system and the interactive applications, abstracting specific characteristics of the platform and providing a series

S.A. Hissam et al. (Eds.): OSS 2011, IFIP AICT 365, pp. 34–49, 2011.

of specific services to the above layers. Thus, it is possible the developing of portable applications to many distinct receptors.

The main DTV middleware open specifications offer support to the execution of interactive applications in two environments: a declarative and an imperative [2]. On the Brazilian Digital TV System, the declarative environment is represented by the Ginga-NCL [3][4], which supports applications based on NCL language (Nested Context Language) and the imperative environment is represented by Ginga-J [5], which provides support to the execution of applications written in Java language.

This paper main goal is to present the first implementation of Ginga-J, highlighting its singularities when compared to other middleware's implementations. On [5] is described all the information about the Ginga-J's functionalities specification. This article describes the reference implementation on the GingaCDN's[1] project context in order to serve as basis to future Ginga-J's implementations from different manufacturers and its platforms. Another point discussed in this article is the evolution and validation of Ginga-J architecture's reference implementation, since it is based on preceding work on middleware developing, realized by the same research group from LAVID at UFPB [6] . The main results from the work shown here were: (i) definition of a flexible architecture that allows reuse and software extensibility and; (ii) developing of a reference implementation in conformity with the new JavaDTV API recently adopted by the Ginga middleware.

This article is organized as follows: section 2 describes the Ginga middleware. The section 3 presents Ginga-J's specification history. The section 4 talks about the implementation architecture proposed for the Ginga-J. The Section 5 details the developed implementation. The section 6 discusses the main existing DTV's middleware projects for fixed terminals, performing a brief comparison between these middlewares and the Ginga-J. And, lastly, section 7 presents the final considerations, future and current works.

2 Ginga Middleware

Ginga is the SBTVD's middleware specification, it resulted from the fusion of FlexTV [6] and MAESTRO[7] middlewares, developed through a consortium led, respectively, by UFPB and PUC-Rio on the SBTVD [8] project.

The FlexTV, procedural middleware proposed by SBTVD's project, included an API set compatible with other standards along with novel functionalities such as the possibility of communication with multiple devices, allowing different viewers to interact with the same interactive application using remote devices. The MAESTRO was the declarative middleware proposal of SBTVD's project. Focusing on space-time synchronization between multimedia objects using the NCL (Nested Context Language) declarative language combined with the functionalities of the scripting language Lua.

---

[1] GingaCDN Project. available on http://www.openginga.net

Ginga combined these two solutions, now called Ginga-J and Ginga-NCL, considering the ITU's international recommendations [11]. Thus the Ginga middleware is divided in two main interconnected subsystems (Figure 1), also known as Execution Machine (Ginga-J) and Presentation Machine (Ginga-NCL). The imperative content execution is possible through the Java Virtual Machine (JVM). Depending on the application requirements, one programming paradigm can be more appropriate than other.

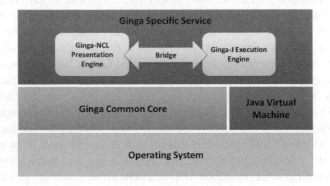

Fig. 1. Overview of Ginga middleware

Another important aspect is that the two environments, for the execution of interactive applications, are not necessarily independent, since that ITU's recommendation includes a "bridge", which provides mechanisms for the communication between them. This bridge API allows imperative applications to use available services on declarative applications, and vice versa. Therefore the execution of hybrid applications one level above the layer of execution and presentation environments is made possible, allowing to combine the NCL language facilities of multimedia elements presentation and synchronization with the power of the object oriented Java language.

Ginga Common Core is the Ginga subsystem responsible for providing specific functionalities of Digital TV common to the imperative and declarative environments, abstracting the specific characteristics of platform and hardware for the above layers. Some of its main functions we can mention are: media exhibition and control, system resources control, return channel management, storage devices, access to service information, channel tuning, among others.

3   Ginga-J Specification

The Ginga-J (Figure 2) is composed by a set of APIs, defined to provide all the necessary functionalities for the developing of DTV applications, from the multimedia data manipulation, to access protocols. Its specification is formed by an adaptation of the information access API of the Japanese standard service (ISDB ARIB B.23), the Java DTV [12] specification (which includes the JavaTV API), besides an additional set of extensions or innovation APIs.

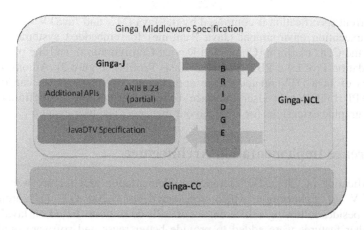

Fig. 2. Ginga-J overview

The additional APIs include a set of available classes for the bridge between applications written in NCL and Java language, additional functionalities for tuning channels, sending asynchronous messages through the interactivity channel and integration of external devices, enabling the support to multimedia resources and simultaneous interaction of multiple users on DTV [13] applications.

The Java DTV [12] specification is an open and interoperable platform that allows the implementation of interactive services in Java language, which has been recently adopted to the Ginga-J's APIs set. Functionally, the JavaDTV replaces the APIs collection that was previously used and defined by the GEM standard (Globally Executable MHP), such as DAVIC (Digital Audio Video Council) and HAVi (Home Audio Video Interoperability). The goal was to provide royalties free solution for device manufacturers and application developers, allowing the production of TV sets and/or set-top-boxes at an affordable cost.

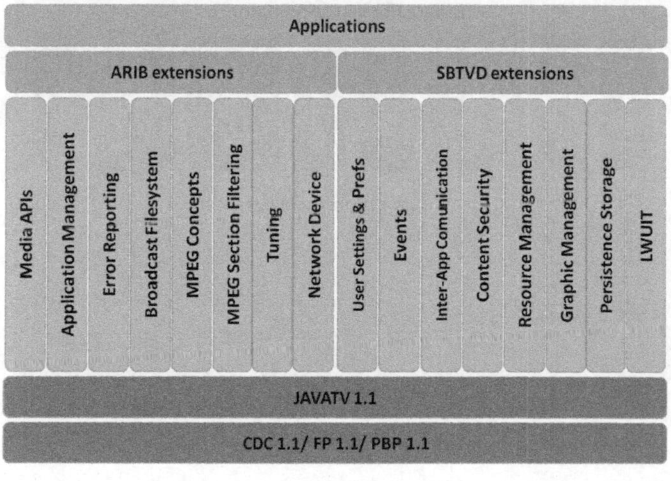

Fig. 3. Ginga-J's APIs set

The current specification is composed by the Java DTV and JavaTV APIs, added to the Java execution environment (Java Runtime) for embedded systems (JavaMe), including the CDC platform (Connected Device Configuration), and the profile APIs: FP (Foundation Profile) e PBP (Personal Basis Profile) (Figure 3). Among the main differences of Java DTV related to the application development, we can quote the LWUIT API (LightWeight User Interface Toolkit), responsible for defining graphic elements, graphic extensions for DTV, layout managers and user events.

4   Reference Implementation Architecture

The specification of Ginga-J's reference implementation architecture was based on the FlexTV [6] architecture, which considered the J.200 ITU [11] architecture. However, besides following a different set of APIs definitions (based on JavaDTV, not GEM), other features were added to provide better reuse and software quality. The Figure 4 illustrates the modularized conceptual architecture: (i) operating system, (ii) common core layer and; (iii) Ginga-J's execution machine. Following are described the three stages which were adopted to define the architecture solution.

The first step for the architecture's definition was to choose a suitable execution platform for the characteristics and differential limitations of a Digital TV fixed terminal. With that in mind, we chose the Linux operating system for personal computers (x86) and the PhoneMe Java [15] virtual machine, which is an official implementation of JavaME/CDC's environment. The main reason of this choice was the Linux and PhoneME availability as open platforms, and also the offer of many development tools without additional cost. Besides, Linux supports heterogeneous systems [16]. The aim was to allow the implementation's development on an environment closer to an access terminal, but that could also be available to as many developers as possible. In this case, a personal computer, without the need to buy any specific hardware.

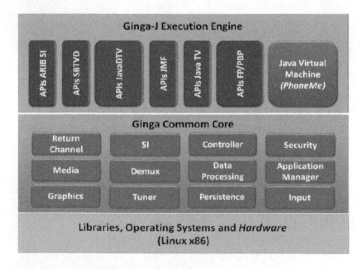

Fig. 4. Conceptual Architecture

The second stage was to develop and refine FlexTV's common core architecture. Nearly no change was performed in the conceptual definition of these subsystem's modules, there was merely a refactoring in order to attain better functionalities cohesion. The main change was to specify the common core using a component-based approach, adopting a component model and an execution environment: FlexCM[17].

The goal was to emphasize the software modeling by decomposing the system in functional components with well-defined interaction interfaces. In this context, a component model defines the instantiation scheme, composition, life cycle of the system components and an environment of software execution responsible for managing the components ensuring the specifications defined by the respective components' model.

The FlexCM model follows a declarative approach, in which the components define its dependencies explicitly (required interfaces) and the execution environment loads and provides the dependencies through a dependencies' injection standard. The FlexCM model allows its components to know only the interfaces; the implementations are treated through the execution environment. Besides the required interfaces, the components can also declare configuration parameters which values are also injected through the execution environment allowing the developer to easily configure the component in the final product where it will be installed. The FlexCM's execution environment is capable of loading the entire system from an architectural description file in which the connections and configurations are specified.

The adoption of the FlexCM's components model offered a series of specific advantages for Ginga-J's implementation besides the commonly known advantages for a component based development, like modularity, maintainability and reuse we can quote: (1) knowing the architecture in model level; (2) facilities on the configuration of individual components and; (3) on the system configuration as a whole. Lastly, these characteristics bring the possibility of managing different architectures also easing the execution of unit tests and integration of different portions of the architecture. A test process proposal for the Ginga-J based on FlexCM can be found at [18].

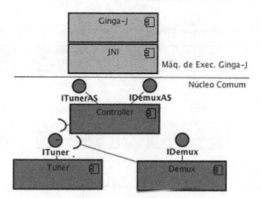

Fig. 5. Ginga-J Execution Layer and Common Core integration

The third and last stage was to define an integration model of the Common Core layer with the Ginga Execution Machine. As mentioned, the Common Core is responsible for offering services for the Ginga-J execution machine. Consequently, it contains native code (in C or C++ languages) and it depends on the platform's execution libraries. It was then important to define a communication model in order to reduce the coupling and the dependency between these two subsystems. The adopted solution was based on the *Proxy, Facade* and *Adapter* [19] design patterns. The idea here was to centralize all Java execution machine use on a Controller module, which exposes the services for the applications (Application Services). Figure 5 illustrates the module Controller with two AS interfaces: ITunerAS and IDemuxAS. These services are offered for the Ginga-J's applications through JNI (Java Native Interface) callings implemented internally through Java's packages. The Controller calls by delegation the component that implements the required functionality. For example, ITuner and IDemux calls (shown on Figure 5). If a Common Core component needs another Common Core component functionality, it can call it directly. The main advantage of this approach is to isolate the layer(s) above the Common Core, in such way to prevent platform dependencies, as well to decrease the coupling between Java API's specifications and the implementation in C/C++ code. For example, the port of a Ginga-NCL's presentation machine or a Java's execution machine from another Digital TV (GEM) system to the Common Core used in this work would be facilitated.

5   Implementation

In this section the Ginga-J's implementation is described focusing on its Common Core components. Figure 6 displays this subsystem overview, which contains the following components:

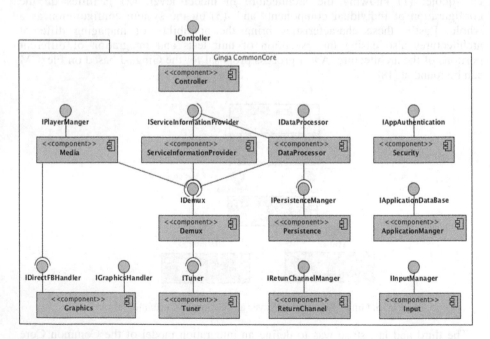

Fig. 6. GingaJ's Common Core implementation (in this case, Controller is not a component, just a facade)

(1) **Tuner** - tunes and controls the access to the multiple network transport streams; (2) **SI** – obtains service information from the transport stream, in other words, which elementary streams (semantics) of audio, video and data has been transmitted, besides information as parental rating, synopses and time scheduling; (3) **Demux** – provides specific filters to select streams; (4) **Media** - Provide access to media decoders (hardware and software) to manage and display the presentation of video and audio elementary streams; (5) **Data Processor** – processes and separates transmitted data (e.g. applications) in multiplexed MPEG-2 transport streams; (6) **Graphics** – provides graphical user interface handling; (7) **Input Manager** – handles user key events through the remote control, STB's panel keys, keyboard, or another input device; (8) **Return Channel** – provides interfaces for the return channel's usage, for example, through dial-up, ADSL, Ethernet, WiMax or 3G; (9) **Application Manager**: loads, instantiates, configures and runs applications; (10) **Persistence** – manages non-volatile storage resources; (11) **Security** – verifies an interactive application's authenticity and permission; (12) **Middleware Manager** – responsible for the middleware's functional management.

As previously mentioned, for Ginga-J's execution machine reference implementation the RC2 version of Linux's PhoneMe Advanced was used [15]. A native port of the Java AWT graphic API for the DirectFB^2 was performed in the virtual machine. The generated code was based on the PhoneME built-in native implementation in Qt. Then, it was possible to implement the JavaDTV APIs, using the Java's base classes, which are present in PhoneME, for example, the graphical interface API and user events handling API. These functionalities were encapsulated in Graphics and Input components, respectively. For functionalities not present on the Java environment it was necessary integration with the Common Core. To allow Java applications management, it was also necessary to integrate the JVM with the Common Core through the `Controller` and `ApplicationManager` components. The `Controller` component implemented a new proxy element, which enabled the execution of *Xlets* through the `ansiJavaMain()` fuction (available on JVM's code). This function starts the JVM and runs a Java class that initiates all graphical layers available (as specified on SBTVD's standard), and also loads the interactive applications' data (*Xlets*) which are started as separated *Threads,* since Ginga allows the execution of more than one application at the receptor.

The `Tuner` implements required services of the `com.sun.dtv.tuner` package, using a scanning process for identifying non-blocking channels based on events and on the Observer pattern[19].

The `Demux` component contains functionalities from the `com.sun.dtv.filtering` package, allowing the selection of different types of elementary streams. Internally it uses a "circular queue buffer" with different start pointers, one to feed each user, trying to avoid that users lose their contents consumed by others.

The `SI` component obeys the APIs' requirements which deals with service information (ARIB, JavaTV and JavaDTV, besides allowing the component user to

---

^{2} DirectFB is an open source project which provides graphic acceleration, input events treatment, graphic layers management and reproduction of several medias through multimedia providers: Available on: http:// www.directfb.org

obtain final information about the stream, without the need of another processing, since it implements a *cache* mechanism. All the abstractions for Service Information provided by SBTVD's standards [20] (Table, Descriptors and Events) can be generated from an object factory, which uses the *Factory Method* pattern [19]. This component also warns the `DataProcessing` to perform the signalization, execution of applications and data carousel.

The `Media` component is responsible for the middleware's processing of continuous media (audio and video) received from `Demux` using the `vlc` infrastructure to present the media over a DirectFB surface. Acquiring validation of the implementation with a performance analysis [27]. This component was designed considering the requirements of the JMF API, since it provides basic reproduction functionalities for the Java API through the JNI calls.

The `ApplicationManager` offer interface abstraction for applications in your database, this abstraction is called `ApplicationProxy`, witch offer the control of the applications lifecycle (start, stop, pause, resume and destroy). A example, `JavaProxy` is a child class of `ApplicationProxy`, that has the capabilities to call the `ansiJavaMain()` function to start the a xlet. As the same, also exist the `NCLProxy` that has de capabilities to start a NCL Presentation Engine [10] to start a NCL document.

Considering the execution of the applications in deferents process, the `ApplicationProxy` must use `IPC`(Inter Process Communications) strategy to send commands to your engine execution, example send events received through the `InputManager` or a control command.

Beside the lifecycle of the applications, the each `Proxy`'s interface contains functionalities to offer communication inter applications. In Java Engine, this happens through `javax.microedition.xlet.ixc` for interaction with another xlet, and `br.org.sbtvd.bridge` for control NCL documents.

The `Persistence` and `Security` components work together to strictly follows the JavaDTV[12] model to pack, authenticate and authorize the applications and file storage. That consist in persist the jar file of the application and study the application access permissions in platform.

The `Persistence` component has important interaction with the `ApplicationManger` component, given that the last send destroy events when a execution of a application is finished, this provides the trigger to `Persistence` deallocate the finished application resources.

The `Return Channel` component implements the TCP/IP communication for different network technologies, offering abstraction about the orientation to connection in two types `ConnectionReturnChannel`(dial-up, ADSL, 3G) and `ReturnChannel`(Ethernet, WiMax). The Return Channel and the Persistence component acquired validation of your implementation by used in LARISSA project[26]. The Figure 7 below illustrates 4 (four) use scenarios of Ginga-J's implementation.

The Figure 7(a) displays an Java (*Xlet*) application using the access APIs for Service Information (JavaTV SI and ARIB SI) and Ginga-J's graphic elements (LWUIT) APIs. The Figure 7(b) shows an application displaying 3 video streams

(2 locals and 1 live) as a validation scenario for the implementation of the media's execution API (JMF). Now the Figure 7(c) and the Figure 7(d) illustrate the possibility to execute a Java application from a local file (for example, USB device) or from a transport stream, respectively. So, as on a TV set, many middleware configurations can be modified through an OSD resident application (*On Screen Device*). The two last examples supported the APIs' validation for the lifecycle control of the application (JavaTV), data carousel, persistence and security (JavaDTV). The tests were conducted using a personal computer with the following specifications: Core 2 Duo 2.16GHz processor; 1GB RAM; operating system Ubuntu 9.10 Kernel 2.6.31-14, and; a 100 GB hard drive.

Fig. 7. Ginga-J's use Scenarios

6 Related Works

The main existing middleware's implementations on the Digital TV context might be divided in two categories: (1) declarative environments (2) imperative environment. The first group is represented by: (i) LASeR(*Lightweight Application Scene Representation*) [21]; (ii) BML (*Broadcast Markup Language*)[22]; (iii) GingaNCL for portable devices[23] and; (iv) Ginga-NCL for fixed devices[24]. However, for the second, we can quote: i) FlexTV[6] and (ii) OCAP-RI (*OpenCable Application Platform – Reference Implementation*)[25].

On [23] it was presented a comparative analysis between the solutions LASeR, BML and Ginga-NCL for portable devices. The main difference of these solutions regarding the implementation proposed here (Ginga-J) is at the architecture project. None of these solutions uses a component-oriented approach, not defining a model and execution environment for the system modules. Besides, we can observe that these environments seek to implement the following functionalities: medias' synchronization, adaptability, support of multiple devices, supports on air edition, and also supports reuse. The Ginga-NCL for portable devices is the only solution that supports multiple devices and meets reuse support standards. The solution proposed here also attends all the requirements presented by the declarative environments, but uses an imperative approach, through the object-oriented language Java. The use of this kind of language is much harder and susceptible to errors and also requires a bigger *footprint* from the application. Nonetheless, it carries a power of expression larger than that offered by declarative languages. The goal is to offer more advanced applications that need to use, for example, access and security mechanisms, finer control to information and audio-visual content.

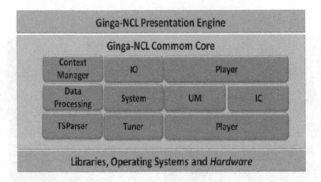

Fig. 8. Overview of Ginga-NCL's fixed devices

Figure 8 displays an overview of the implementation for Ginga-NCL's fixed devices [24]. Ginga-NCL's Presentation Machine is also a logic subsystem capable of starting and controlling NCL applications. Ginga-NCL's Common Core is responsible for offering the previously mentioned services for Ginga-NCL's Presentation Machine. This solution, although attending a different set of applications, displays further similarity on the definition of Ginga-J's Common Core functionalities. One of the differences is on the absence of security functionalities and a lower set of informations about the offered service. The Tuner, DataProcessing, ContextManager, InputOutput (IO) e InteractiveChannel (IC) components of Ginga-NCL, are equivalent, respectively, to Tuner, DataProcessing, ApplicationManager, Input and ReturnChannel of Ginga-J. Media and Player of Ginga-J represent functionalities of Ginga-NCL's Player module. Demux and SI Ginga-J modules represent the Ginga-NCL's TSParser. The module System of Ginga-NCL is implemented internally on GingaJ. The main reason for representing Ginga-J's functionalities with more modules is to allow better cohesion and, consequently, larger extension flexibility and code maintenance. Another important difference concerns the implementation on the modules' management mechanism. On Ginga-NCL this is implemented by ContextManager and UpdateManager (UM) and on Ginga-J a model and execution environment of software components (FlexCM) are defined.

Table 1. Comparison between Ginga-J and Ginga-NCL for fixed devices

	Ginga-J	Ginga-NCL for fixed devices
Creation model	Dependency injection	Factory
Architecture's knowledge	On model's level	On source's level
Life cycle	Creation, initialization, pause and destruction	Creation and destruction
Components configuration	Standard model	Absent

Table 1 displays a comparison of the solutions. On the criteria for evaluation, we observed that the Ginga-NCL model uses an approach of object factory, imposing that the architecture knowledge is spread through the system's source code. This characteristic limits the flexibility in which the architecture may be instantiated. Besides, the lack of standardization in order to configure the components prevents an effective management of the system modules. So it is believed that the model used on Ginga-J's implementation best meets the requirements for modularity, maintainability and reuse of the project and implementation of the Common Core's code.

The FlexTV implementation was realized by Ginga-J's same group and the current proposal is an evolution of the same in two points: (1) functionalities (new set of APIs Java based on JavaDTV) and (2) architecture (adoption of a model and environment of components execution).

OCAP-RI Moreno, F. M. A Declarative Middleware for Digital TV Systems. (Master Thesis); PUC-Rio, DI, 2006

[25] is a proposal of imperative middleware implementation based on the American standard of Digital Cable TV. One of the differences is on the set of offered functionalities, fewer than Ginga-J, since OCAP's Java APIs do not support multiple devices nor users, management of the multimedia streams and asynchronous messages. Another important point is related to the architecture project (Figure 9) which is divided into: (I) OCAP Java – set of Java APIs available for applications and defined by the TVD American standard; (ii) JVM and OCAP Native – Java's virtual machine and set of specific native code to implement OCAP Java's functionalities; (iii) MPE *(Multimedia Platform Extensions)* – layer that abstracts the execution platform for the JVM and the OCAP Native code; (iv) MPEOS *(Multimedia Platform Extensions Operating System)* -implements   platform dependent code offering services for the MPE, which means, MPEOS is the code that needs to be ported for each platform and; (v) *RI Platform* – represents the operating system and the hardware that runs the middleware. The MPE and MPEOS layers from OCAP-RI are equivalent to the set of components of the Ginga-J's Common Core, where MPE is represented by the interfaces of *Controller* and MPEOS by internal implementations of each component. As already quoted, such characteristic facilitates the port of the Java execution machine for different platforms. However, MPE and MPEOS are implemented using C language and do not use any model and components execution environment. Therefore the OCAP-RI architecture does not offer any modulate division of functionalities, making reuse and code flexibility more difficult.

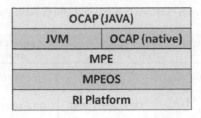

Fig. 9. OCAP-RI's Architecture

Based on the points discussed in this section the main differences between Ginga-J's implementation and other proposals can be understood. The first is related to the programming model and the set of different functionalities offered by an imperative environment in relation to declarative environments or imperative environment based on GEM. The second refers to the architecture project, which tried to attend reuse, maintenance and code flexibility the best way possible. Such aspects are important for the implementation of reference, since itself offers a starting point and can be adapted to many platforms by manufacturers and other developers.

7 Development Process

The GingaCDN (Ginga Code Development Network) was idealized as a group of developers and contributors (scattered across the globe) of Brazil's Digital TV middleware the Ginga. Among the various projects being carried out by this network is Ginga-J reference implementation. Nowadays, the number of registered developers reached 570 from 15 different countries on GingaCDN community site.

In order to become a distributed software development team and gain from its benefits like reducing costs and time spent, while improving the software's quality, it was necessary to thwart it's known drawbacks such as inefficient communication, loss in coordination and providing a full vision of the project. The solution came through a well-defined development process with roles and a proper tool (based on Redmine[3] tool) to support it all.

The collaborative development process was designed to standardize agile and objective practices as to attend the deployment of distributed component for digital TV's middleware. Thus, we defined roles for the collaborative team members in order for users to know their responsibilities and have the freedom to attend their assigned tasks independent and simultaneously. Such roles are distributed in the five phases defined in the collaborative development process. Each phase of the process is mainly conducted by a specific role (except for the Review phase), they are: 1) Conception, where the Manager creates a new subproject; 2) Elaboration, when the Leader specifies tasks to accomplish the subproject; 3) Construction, is carried out by the Developers undertaking the tasks; 4) Review, the Reviewers review the component and Integrators check if they integrate with the whole project; 5) Transition, the

---

[3] Redmine Project: Available at: http://www.redmine.org

Manager once again comes along to check if the component is in accordance to what was out lined initially.

8 Final Remarks

This work describes the Ginga-J's reference implementation, SBTVD's imperative middleware. The development was based on JavaDTV's specification, an architecture based on software components. As a form of proposal validation, the architecture was instantiated for the Linux environment on a personal computer. The main Java packages of Ginga-J's standard were implemented through the integration of basic Java's environment (*PhoneME*) functionalities as well as implementations of specific functionalities for Digital TV (Ginga-J Common Core).

The project and implementation of a development based architecture using components brought a series of benefits, such as: (i) knowing the architecture at the model level; (ii) ease when configuring individual components; (iii) configuration of the system as a whole and (iv) the possibility of managing different architectures making the execution of unity tests and integration of different architecture portions easier. Such aspects are very important for the implementation of reference.

As a result of this experience, many works are already being accomplished, such as (i) port of PUC-Rio's Ginga-NCL's presentation machine to the Common Core; (ii) the development of management tools for the architecture and conception of different middleware's versions; (iii) proposal of a conformance validation model for Digital TV's middleware.

Acknowledges. We thank the support of our institutions, the Laboratory of Digital Video Application of the Federal University of Paraiba and of the Federal University of Pernambuco, as well as the funding provided by Brazilian research agencies: National Education and Research Network (RNP) and Science and Technology Ministry (MCT) under the CTIC program[4].

References

[1] Peng, C.: Digital Television Applications (PhD Thesis) – Helsinki University of Technology, Espoo (2002)
[2] Morris, S., Smith-Chaigneau, A.: Interactive TV Standards: A Guide to MHP, OCAP, and JavaTV. Focal Press (2005)
[3] ABNT NBR 15606-2 Digital terrestrial television – Data coding and transmission specification for digital broadcasting – Part 2: Ginga-NCL for fixed and mobile receivers – XML application language for application coding (2007)
[4] ABNT NBR 15606-5 Digital terrestrial television – Data coding and transmission specification for digital broadcasting Part 5: Ginga-NCL for portable receivers – XML application language for application coding (2008)

---

[4] CTIC Program: Available at: http://www.ctic.rnp.br/

[5] ABNT NBR 15606-4 Digital terrestrial television — Data coding and transmission specification for digital broadcasting Part 4: Ginga-J — The environment for the execution of procedural applications (2010)

[6] Leite, L.E.C., et al.: FlexTV – Towards a Middleware Architecture to Brazilian Digital TV System. Journal of Computer Engineering and Digital Systems 2, 29–50 (2005)

[7] Soares, L.F.G.: MAESTRO: The Declarative Middleware Proposal for the SBTVD. In: Proceedings of the 4th European Interactive TV Conference (EUROITV 2006), Athens (2006)

[8] SBTVD. Brazilian Digital TV System Project, http://sbtvd.cpqd.com.br

[9] de Souza Filho, G.L., Leite, L.E.C., Batista, C.E.C.F.: Ginga-J: The Procedural Middleware for the Brazilian Digital TV System. Journal of the Brazilian Computer Society 12, 47–56 (2007)

[10] Soares, L.F.G., Rodrigues, R.F., Moreno, M.F.: Ginga-NCL: the Declarative Environment of the Brazilian Digital TV System. Journal of the Brazilian Computer Society 12, 37–46 (2007)

[11] ITU J.200. ITU-T Recommendation J.200: Worldwide common core – Application environment for digital interactive television services (2001)

[12] JavaDTV API. Java DTV API 1.3 Specification, Sun Microsystems (2009), http://www.oracle.com/technetwork/java/javatv/overview/index.html

[13] Silva, L.D.N., et al.: Digital TV Multiuser and Multidevices Application Development Support with Ginga. Amazonia Magazine (12), 75–84 (2007)

[14] ETSI TS 102 819: Globally Executable MHP (GEM). ETSI Standard (May 2004)

[15] Projeto PhoneME, http://phoneme.dev.java.net/

[16] Yaghmour, K.: Building Embedded Linux Systems. O'Reilly Media, Inc., Sebastopol (2003)

[17] Miranda Filho, S., et al.: Flexcm - A Component Model for Adaptive Embedded Systems. In: COMPSAC IEEE International Computer Software and Applications Conference, Beijing, pp. 119–126 (2007)

[18] Caroca, C., Tavares, T.A.: Test Process Model to Ginga Common Core Components. In: Proceedings of the 15th Brazilian Symposium on Multimedia and the Web (WebMedia 2009), Fortaleza (2009)

[19] Gamma, E., Helm, R., Johnson, R., Vlissides, J.: Design Patterns: Elements of Reusable Object-Oriented Software. Addison-Wesley, Reading (1994)

[20] ABNT NBR 15603-2 Digital terrestrial television — Multiplexing and service information (SI) Part 2: Data structure and definitions of basic information of SI (August 2008)

[21] ISO 14496-20. Lightweight Application Scene Representation (LASeR) and Simple Aggregation Format (SAF) (2006)

[22] B24 Appendix 5 – Operational Guidelines for Implementing Extended Services for Mobile Receiving System (2004)

[23] Cruz, V.M., Moreno, M.F., Soares, L.F.: Ginga- NCL: Reference implementation for portable devices. In: Proceedings of the 14th Brazilian Symposium on Multimedia and the Web (WebMedia 2008), pp. 67–74. ACM, New York (2008)

[24] Moreno, F. M.: A Declarative Middleware for Digital TV Systems. (Master Thesis); PUC-Rio, DI (2006)

[25] OCAP – Reference Implementation, http://ocap-ri.dev.java.net

[26] Oliveira, M., Cunha, P.R.F., da Silva Santos, M.E., Bezerra, J.C.C.: Implementing home care application in Brazilian Digital TV. In: Global Information Infrastructure Symposium (GIIS 2009), Hammamet (2009)

[27] Trojahn, T.H., Gonçalves, J.L., Mattos, J.C.B., Da Rosa, L.S., Agostini, L.V.: A Media Processing Implementation Using Libvlc for the Ginga Middleware. In: Proceedings of the 5th International Conference on Future Information Technology (FutureTech) (2010)

[28] Cabral, P.A., et al.: GingaCDN A Code Development Network to DTV Brazilian Middleware. In: Proceedings of the 16th Brazilian Symposium on Multimedia and the Web (WebMedia 2010), 1st Workshop of Interactive Digital TV, Belo Horizonte, vol. 2 (2010)

Developing Architectural Documentation
for the Hadoop Distributed File System

Len Bass, Rick Kazman, and Ipek Ozkaya

Software Engineering Institute, Carnegie Mellon University
Pittsburgh, Pa 15213 USA
lenbass@cmu.edu,{kazman,ozkaya}@sei.cmu.edu

Abstract. Many open source projects are lacking architectural documentation that describes the major pieces of the system, how they are structured, and how they interact. We have produced architectural documentation for the Hadoop Distributed File System (HDFS), a major open source project. This paper describes our process and experiences in developing this documentation. We illustrate the documentation we have produced and how it differs from existing documentation by describing the redundancy mechanisms used in HDFS for reliability.

1 Introduction

The Hadoop project is one of the Apache Foundation's projects. Hadoop is widely used by many major companies such as Yahoo!, E-Bay, Facebook, and others. (See http://wiki.apache.org/hadoop/PoweredBy for a list of Hadoop users.) The lowest level of the Hadoop stack is the Hadoop Distributed File System [2]. This is a file system modeled on the Google File System [1] that is designed for high volume and highly reliable storage. Clusters of 3000 servers and over 4 petabytes of storage are not uncommon within the HDFS user community.

The amount and extent of documentation of the architecture [3] that should be produced for any given project is a matter of contention. There are undeniable costs associated with the production of architectural documentations and undeniable benefits. The open source community tends to emphasize the costs and downplay the benefits. As evidence of this claim, there is no substantive architectural documentation for a the vast majority of open source projects, even the very largest ones. The existing description of the architecture of most widely used open source systems tend to be general descriptions rather than systematic architectural documentation targeted for the system's stakeholders [4].

This paper describes the process we used to produce architectural documentation with emphasis on what is different about producing documentation for open source projects. This production was the first step in a more ambitious project that will analyze the community for evidence as to the value of the documentation but we have nothing to report on that front as yet.

HDFS makes two assumptions that take it out of the realm of a standard file system: it assumes high volumes of data in primarily a write-once, read-many-times

S.A. Hissam et al. (Eds.): OSS 2011, IFIP AICT 365, pp. 50–61, 2011.

environment. The only block size that HDFS supports is 64Mbytes. There is very little synchronization supported since the kinds of applications for which it is designed are primarily batch – collect data and process it later. The second assumption that HDFS makes is that it will run primarily on commodity hardware. With 3000 servers, hardware failures, even with all RAID devices, become a normal occurrence. As a consequence the software was designed to handle failure smoothly. Since the software must handle failure in any case, use of commodity hardware makes the use of a multi-thousand server cluster much more economical.

The structure of this paper is that we will first describe our idealized process for producing more detailed architectural documentation. We then discuss what we actually did and how it differed from the idealized process. Throughout the paper we use the description of the HDFS availability strategy as illustrative of both the existing documentation and our additions to it.

2 Our Process for Developing the Documentation

When writing architectural documentation it is necessary to have an overview of what the system components are and how they interact. When there is a single architect for the system, the easiest route is to simply talk to this person. Most open source projects, however, do not have a single identifiable architect—the architecture is typically the shared responsibility of the group of committers.

The first step of our documentation process is to gain this overview. Subsequent steps include elaborating the documentation and validating and refining it. To do this we needed to turn first to published sources.

2.1 Gaining the Overview

HDFS is based on the Google File System and there are papers describing each of these systems 1, 2. Both of these papers cover more or less the same territory. They describe the main run-time components and the algorithms used to manage the availability functions. The main components in HDFS are the NameNode that manages the HDFS namespace and a collection of DataNodes that store the actual data in HDFS files. Availability is managed by maintaining multiple replicas of each block in an HDFS file, recognizing failure in a DataNode or corruption of a block, and having mechanisms to replace a failed DataNode or a corrupt block.

In addition to these two papers, there is an eight page "Architectural Documentation" segment on the Apache Hadoop web site [5]. This segment provides somewhat more detail than the two academic papers about the concepts used in HDFS and provides an architectural diagram, as shown in Figure 1.

Code level documentation (JavaDoc) is also available on the HDFS web site. What currently exists, then, are descriptions of the major concepts and algorithms used in HDFS as well as code-level JavaDoc API documentation.

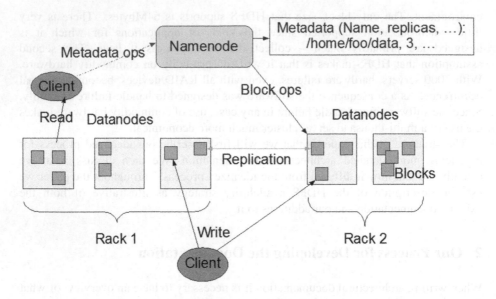

Fig. 1. HDFS Architecture Diagram from 5

What is missing from the existing documentation can be seen by considering how architectural documentation is used. Architectural documentation serves three purposes: 1) a means of introducing new project members to the system, 2) a vehicle for communication among stakeholders, and 3) the basis for system analysis and construction [3,6]. These uses of architectural documentation include descriptions of the concepts and, where important, the algorithms. But architectural documentation, to be truly useful for those who wish to modify the system, must also connect the concepts to the code. This connection is currently missing in the HDFS documentation. A person who desires to become a contributor or committer needs to know which modules to modify and which are affected by a modification. Communication among stakeholders over a particular contribution or restructuring is also going to be couched in terms of the relation of the proposed contributions to various code units. Finally, for system construction, maintenance, and evolution to proceed, the code units and their responsibilities must be unambiguously identified. Existence of such focused architecture documentation can assist contributors become committers faster. It could also assist addressing many current open major issues. As of April 12, 2011 out of the 834 total issues in HDFS Jira 628 of the issues are major issues.

Architectural documentation occupies the middle ground between concepts and code and it connects the two. Creating this explicit connection is what we saw as our most important task in producing the architectural documentation for HDFS.

2.2 Expert Interview

Early in the process of gaining an overall understanding of HDFS, we interviewed Dhruba Borthakur of Facebook, a committer of the HDFS project and also the author of the existing architectural documentation posted on the HDFS web site [5]. He was also one of the people who suggested that we develop more detailed architectural documentation for HDFS. We conducted a three hour face to face interview where we explored the technical, historical, and political aspects of HDFS. Understanding the history and politics of a project is important because when writing any document you need to know who your intended audience is to describe views that are most relevant to their purposes [3].

In the interview, we elicited and documented a module description of HDFS as well as a description of the interactions among the main modules. The discussion helped us to link the pre-existing architectural concepts—exemplified by Figure 1—to the various code modules. The interview also gave us an overview of the evolutionary path that HDFS is following. This was useful to us since determining the anticipated difficulty of projected changes provides a good test of the utility, and driver for the focus, of the documentation. Figure 2 shows a snippet from our interview and board discussions where Dhruba Borthakur described to us the three DataNode replicas in relationship to the NameNode.

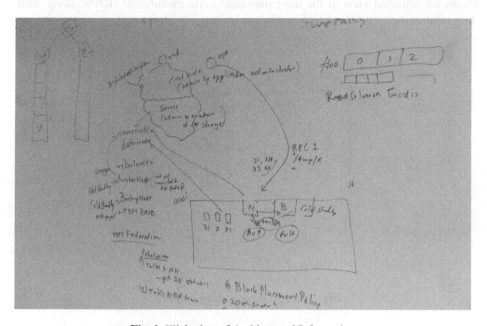

Fig. 2. Elicitation of Architectural Information

2.3 Directory Structure

A final item that proved very helpful is the directory structure of HDFS. The code is divided cleanly into the following pieces:

- The library used by the client to communicate with the NameNode and the DataNodes.
- The protocols used for the client communication

- The NameNode code
- The DataNode code
- The protocols used for communication between the NameNode and the DataNodes.

In addition, there are a few other important directories containing functionality that the HDFS code uses, such as Hadoop Common.

2.4 Tool Support

An obvious first step in attempting to create the architectural documentation was to apply automated reverse engineering tools. We employed SonarJ [7] and Lattix [8], both of which purport to automatically create architectural representations of a software product by relying on static analysis of the code dependencies. However, neither of these tools provided useful representations although they did reveal the complexity of the dependencies between Hadoop elements. For example, Figure 3 shows an extracted view of the most important code modules of HDFS, along with their relationships, produced by SonarJ.

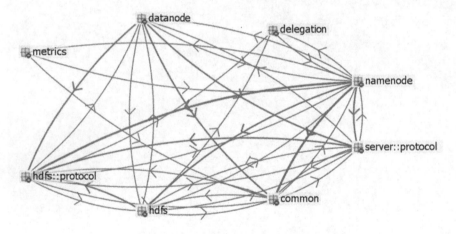

Fig. 3. Module Relationships in HDFS

What are we to make of this representation? It appears to be close to a fully connected graph. Is the code a "big ball of mud" [9]? The answer lies in the purpose and goals of the architecture. The main quality attribute foci of HDFS are performance and availability. These concerns dominate the architectural decisions and the discussions amongst the project's committers. Of significant, but decidedly lesser concerns, are qualities such as modifiability and portability. The process Hadoop follows in handling modification is a planned evolutionary processes where a committer suggests alternative design, it is vetted among the key committers, and then planned for an agreed upon future release cycle. The goals of the project should be aligned with the focus of the architecture. Since performance and availability were the top goals of HDFS, it is not surprising that these concerns shaped the architectural decisions. Since modifiability and portability were of lesser concern, it is also not surprising that these qualities were not strongly reflected in the architectural structures chosen.

The reverse engineering tools SonarJ and Lattix are primarily focused on these latter concerns—modifiability and portability. They aid the reverse engineer in determining the modular and layered structures in the architecture by allowing the definition of design rules to detect violations for such architectural structures. We thus see a mismatch between the goals of the tools and the goals of HDFS. For this reason, the structures that these tools were able to automatically extract were not particularly interesting ones, since they did not match the goals of the project and the important structures in the architecture. HDFS does not have any interesting layering, for example, since its portability concerns are, by and large, addressed by the technique of "implement in Java". The governing architectural pattern in HDFS is a master-slave style, which is a run-time structure. And modifiability, while important, has been addressed simply by keeping the code base at a relatively modest size and by having a significant number of committers spending considerable time learning and mastering this code base.

The modest code size, along with the existing publications on the availability and performance strategies of HDFS allows us to document the architecture by tracing the key use cases through the code. While this is not an easily repeatable process for larger open source projects, it proved to be the most accurate and fit for purpose strategy for creating the architecture documentation of HDFS.

This lack of attention to specific architectural structures aimed at managing modifiability is a potential risk for the project as it grows, since it makes it difficult to add new committers—the learning curve is currently quite steep. Our architectural documentation is one step in addressing this risk. Another step that the HDFS committers could take is to simplify the "big ball of mud".

3 Elaboration

This project began Nov. 1, 2010 and the interview with Dhruba Borthakur took place on Nov. 22, 2010. Thus, it took a month to gain an overview of HDFS. December was devoted to exploring the architecture using the reverse engineering tools and the month of January was devoted to writing a first version of the architectural documentation. It is in the elaboration phase that value is added to the existing materials.

The elaboration phase of the architectural documentation is when the connections between the concepts and the algorithms are made manifest. Consider the following section from the existing architectural documentation (found at http://hadoop.apache.org/common/docs/r0.20.0/hdfs_design.html) about one of the key mechanisms for maintaining high data availability: the heartbeat.

Data Disk Failure, Heartbeats and Re-Replication

Each DataNode sends a Heartbeat message to the NameNode periodically. A network partition can cause a subset of DataNodes to lose connectivity with the NameNode. The NameNode detects this condition by the absence of a Heartbeat message. The NameNode marks DataNodes without recent Heartbeats as dead and does not forward any new IO requests to them. Any data that was registered to a dead DataNode is not available to HDFS any more. DataNode death may cause the replication factor of some blocks to fall below their specified value. The NameNode constantly tracks which blocks need to be replicated and initiates replication whenever necessary. The necessity for re-replication may arise due to many reasons: a DataNode may become unavailable, a replica may become corrupted, a hard disk on a DataNode may fail, or the replication factor of a file may be increased.

Our elaboration of this concept, as we have produced it in the architectural documentation, is:

Heartbeats are the mechanism by which the NameNode determines which DataNodes are currently active.   There is a heartbeat monitor thread in NameNode that controls the management.

   NameNode   maintains   information   about   the   DataNodes   in DataNodeDescriptor.java. DataNodeDescriptor tracks statistics on a given DataNode, such as available storage capacity, last update time, etc., and maintains a set of blocks stored on the DataNode. This data structure is internal to the NameNode. It is not sent over-the-wire to the Client or the DataNodes. Neither is it stored persistently in the fsImage (namespace image) file.

A DataNode communicates with the NameNode in response to one of four events.

   1. *Initial registration. When a DataNode is started or restarted it registers with the NameNode. It also registers with the NameNode if NameNode is restarted. In response to a registration, NameNode creates a DataNodeDescriptor for the DataNode. The list of the DataNodeDescriptors is checkpointed in fsimage (the namespace image file). Only the DataNodeInfo part is persistent, the list of blocks is restored from the DataNode block reports.*
   2. *In response to a heartbeat request from the NameNode. If NameNode has not heard from a DataNode for some period of time, it sends a request for a Heartbeat. If this request does not generate a response, the DataNode is considered to have failed and each of the replicas it maintains must be created on a different DataNode.   When the DataNode has reported, NameNode:*
       *   *Records the heartbeat, so the DataNode isn't timed out*
       *   *Adjusts usage stats for future block allocation*
       If a substantial amount of time passed since the last DataNode heartbeat then NameNode requests an immediate block report.

3. `In response to a blockReport() request from the NameNode. NameNode may request the DataNode to report all of the replicas that it is currently maintaining. NameNode does this when it has reason to believe that its list of blocks in the DataNode is not up to date. i.e., on start up or if it has not heard from the DataNode for some period of time.*

4. Completion of a replica write. When the DataNode has successfully written a replica, it reports this event through a blockReport().*

The differences between the original documentation [5] and the new version that we have produced are as follows:

- *Much more detail.* Rather than giving a general description of the concepts, the specific interactions between a DataNode and the NameNode are described.
- *Code is explicitly named.* The classes that contain the code that provides the heartbeat responsibility are identified.

- *Subtle optimizations are identified.* For example, a blockReport() sent by the DataNode to the NameNode indicates that the DataNode is alive and there is no need for a heartbeat query to that DataNode.

At the time of writing of this paper we are also working on adding sequence diagrams of major use cases to further highlight the architectural details mapping the architecture documentation more explicitly to the code.

4 Validation and Refinement

The final phase of the production of the architectural documentation is to validate it. We have now received comments on our draft from two committers of HDFS - Dhruba Borthakur of Facebook and Sanjay Radia of Yahoo!. Based on their comments we have modified the draft architectural documentation. At the time of writing this paper, we are preparing the documentation for publication on the HDFS web-site and appropriate blogs. Before doing so, we are creating baseline metrics on the existing state of the basic metrics that we can track to architecture documentation such as number of committers, contributors, and surveys that we will conduct with them to establish a baseline impression of the state of the architecture for HDFS (like actual versus perceived architecture).

5 Structure of the Documentation

The documentation that we have produced has 6 major sections. These are

1. *Introduction*
2. *HDFS Assumptions and Goals.* This section talks about the design rationale and major quality attribute concerns for HDFS.
3. *Overview of HDFS Architecture.* This section introduces the three types of processes within HDFS – the application, the NameNode, and the DataNode.

4. *Communication among HDFS elements.* This section describes the four canonical runtime interactions between the three types of HDFS processes. These interactions, as shown in Figure 1, are: Application code <-> Client, Client <-> NameNode, and Client <-> DataNode, NameNode <-> DataNode.

5. *Decomposition and Basic Concepts of HDFS elements.* This section describes how each of the basic elements reacts to client requests to create, write, read, and close files. It also describes the modes and thread structure within NameNode and how these modes and threads are used to manage the file systems and provide high availability.

6. *Use Cases.* The basic use cases of create, write, read, and close are described in terms of sequence diagrams.

6 Discussion

We will now discuss three aspects of creating architectural documentation that are lessons learned from this process: where to start, how to evolve the documentation, and the use of tools. We will also discuss how the production of architectural documentation for an open source project differs from the production of architectural documentation for a closed project.

6.1 Where to Start

There were two documents that helped us get started in documenting HDFS: the Google File System paper (the Google file system was the original model for HDFS) and the existing architectural documentation on the web-site. These two documents provided a good start on our gaining an early understanding of HDFS. What would we have done if this level of documentation had not existed?

Whether or not there is existing documentation, our process calls for interviewing experts. The documentation that exists is invariably out of date (if it were not, we would not be doing this job) and much of the information that we require typically resides in the heads of just a few individuals. These individuals are usually very busy and without sufficient time or interest to produce the architectural documentation. For HDFS, one interview was sufficient. If the pre-existing level of documentation is not sufficient to gain an overview level of understanding, then more interviews may have been necessary. The interview was quite lengthy, involved multiple drawings on a whiteboard and, although it did not work out, we had hopes of arranging another interview in the same trip. Although face to face interviews are difficult and expensive to arrange, it is hard for us to image the same results from a video conferencing meeting.

Our notes from the interview contain several photos of drawings from a white board. It is possible that additional information could have been gained from either e-mail or telephone conferences, but we did not feel the need for that for this case, In an open source project, the committers are usually easy to find, although possibly difficult to arrange time with.

To summarize, the techniques for gaining an understanding of the architecture from dealing with the committers include face to face meetings, off line communication, and telephone conferences.

6.2 Evolution

Systems evolve and (hopefully) more slowly, architectures evolve. This means that the architecture documentation in an evolving system may quickly become out of date. Using HDFS as an example, this fear seems to be overblown. The fundamental structures of HDFS—for example, the separation and relationships between client, DataNode, and NameNode—have not changed since its inception. Of course, the details of each of these elements and their interactions have evolved, but at the architectural level there was considerable stability.

A recent major change is the addition of the ability to append data to an existing file. In architectural terms, this involved multiple classes and the addition of a major new functionality. Yet in terms of the documentation, a search of the documentation we produced finds several one-word references to *append* plus an 11 line paragraph describing how *append* is different than *write*. Generating the additional architectural documentation associated with *append* would have been the work of just a few hours.

Major changes currently being considered for HDFS are a refactoring of the code and several different proposals being considered to break the current design of an HDFS deployment being limited to one cluster. These two types of changes raise different issues in terms of the evolution of the architectural documentation.

- *Refactoring to simplify the code structure.* Refactoring the code would not change the concepts or the algorithms used, but it would have an impact on the mapping of the important concepts to classes. Yet the changes to the architectural documentation can be kept to a minimum. A refactoring will add new classes, modify existing classes, or deleting classes. Any major new class will be constructed from portions of existing classes. We can match a list of classes being modified or deleted with the classes mentioned in the documentation. For each match, the changes to the documentation will consist of adding a class name or removing a class name. These are minimally intrusive changes.
- *Breaking the current limitation of a single cluster per deployment.* This type of change is more far reaching and will have more impact on the documentation. Yet this type of change is not made radically or quickly. In fact, the discussion of the proposed changes can be found in the Jira bug-tracking system prior to the change actually taking place.

The discussion in Jira provides exactly the type of information that needs to be captured in the architectural documentation. Consider the following example, open issue from HDFS Jira, HDFS 1783 created March 24, 2011 by Dhruba Borthakur:

The current implementation of HDFS pipelines the writes to the three replicas. This introduces some latency for realtime latency sensitive applications. An alternate implementation that allows the client to write all replicas in parallel gives much better response times to these applications.

Although architectural documentation is created once and then needs to be maintained and evolved, we argued here that if one considers the type of evolution that a system like HDFS undergoes once it has become successful, the concomitant evolution to the architectural documentation is relatively minor and painless.

6.3 The Use of Tools

Tools are most useful in the elaboration stage of the documentation. As discussed above, tools are not much help in gaining an initial understanding of the concepts and algorithms, and may be of limited use in understanding the module structure. But tools can be very useful in tracking the effects of a method call or the use of a particular class.

One tool that is particularly useful is the call graph. A call graph enables tracking how a call to "write" by the client goes through NameNode. It is not the best way to understand that NameNode is not involved with the data transfer, per se but it will provide a track through the classes that allocate blocks.

As discussed above, we originally tried to create a module view with tool support but that effort was unsuccessful. The tools that we used that support the construction of a module view require some initial guesses as to the decomposition of the modules. Beyond the decomposition of HDFS into the client, NameNode, and DataNode, finding further decompositions proved unsuccessful for us. As a result, the architectural documentation that we produced only has those three major components identified.

6.4 Open Source Specifics

One distinction between producing architectural documentation for an open source project and a proprietary project comes from the openness and availability of discussions about issues. In an open source project, Jira, mailing lists and bulletin boards become the repository of these discussions and they can be mined for rationale information. In a closed source project we must rely upon the availability, good will, and good memory of individuals. Although in principle there is nothing to stop closed source projects from adopting these practices, in our experience, we have rarely seen evidence of their existence.

There are other factors that are traditionally cited as distinctive to open source activities—multiple eyes, requirements arising from the contributors rather than from an explicit elicitation process, and so forth. But none of these other distinctive characteristics of open source appears to substantially affect the process of creating architectural documentation.

7 Next Steps

The production of the architectural documentation is the first step in a more ambitious research project to measure the *impact* of architectural documentation. The current group of committers of HDFS is stressed because of their HDFS-related workload and would like to grow the HDFS committer community. Currently there are 221 contributors as opposed to 28 committers. Our conjecture is that the existence of architectural documentation will shorten the learning curve for potential contributors and committers, thus lowering the bar to entry.

To test this conjecture we have created improved architectural documentation for HDFS, and begin disseminating it on May 19, 2011 via http://kazman. shidler. hawaii.edu/ArchDoc.html. We announced the availability of the documentation to the Hadoop community through HDFS Jira (issue number HDFS-1961)

In addition, we are collecting a number of project measures. We will measure the usefulness of the documentation by tracking how often it is downloaded and how often it is mentioned in discussion groups. We are tracking project health measures, such as the growth of the committer group, and the time lag between someone's appearance as a contributor and their acceptance as a committer and other measures. And we are tracking product health measures, such as the number of bugs per unit time and bug resolution time.

This study is much more long-term than the production of the architectural documentation, although it crucially depends on the documentation as a first step. We will report on our results in due course.

Acknowledgements. The work was supported by the U.S. Department of Defense. We would also like to thank Dhruba Borthakur and Sanjay Radia for their assistance.

References

1. Ghemawat, S., Gobioff, H., Leung, S.-T.: The Google file system. In: ACM SIGOPS Operating Systems Review - SOSP 2003, vol. 37(5), pp. 23–43 (2003)
2. Shvachko, K., Kuang, H., Radia, S., Chansler, R.: The Hadoop Distributed File System. In: IEEE 26th Symposium on Mass Storage Systems and Technologies, Incline Village, NV, pp. 1–10 (2010)
3. Clements, P., Bachmann, F., Bass, L., Garlan, D., Ivers, J., Little, R., Merson, P., Nord, R., Stafford, J.: Documenting Software Architectures: Views and Beyond, 2nd edn. Addison-Wesley, Reading (2010)
4. Brown, A., Wilson, G.: The Architecture of Open Source Applications (2010), http://www.aosabook.org/en/index.html (accessed June 6, 2011)
5. Apache Hadoop, HDFS Architecture, http://hadoop.apache.org/common/docs/r0.19.2/hdfs_design.html (accessed April 7, 2011)
6. ISO/IEC 42010:2007 – Recommended Practice for Architectural Description of Software-intensive Systems (2007), http://www.iso-architecture.org/ieee-1471/ (accessed June 6, 2011)
7. Sonar, J.: http://www.hello2morrow.com/products/sonarj (accessed April 9, 2011)
8. Lattix, http://www.lattix.com (accessed April 9, 2011)
9. Foote, B., Yoder, J.: Big Ball of Mud. In: Fourth Conference on Patterns Languages of Programs (PLoP 1997/EuroPLoP 1997), Monticello, Illinois (1997)

Modding as an Open Source Approach to Extending Computer Game Systems

Walt Scacchi

Institute for Software Research and
Center for Computer Games and Virtual Worlds
University of California, Irvine
wscacchi@ics.uci.edu

Abstract. This paper examines what is known so far about the role of open source software development within the world of game mods and modding practices. Game modding has become a leading method for developing games by customizing or creating OSS extensions to game software in general, and to proprietary closed source software games in particular. What, why, and how OSS and CSS come together within an application system is the subject for this study. The research method is observational and qualitative, so as to highlight current practices and issues that can be associated with software engineering and game studies foundations. Numerous examples of different game mods and modding practices are identified throughout.

1 Introduction

User modified computer games, hereafter *game mods*, are a leading form of user-led innovation in game design and game play experience. But modded games are not standalone systems, as they require the user to have an originally acquired or licensed copy of the unmodded game software.

Modding, the practice and process of developing game mods, is an approach to end-user game software engineering [4] that establishes both social and technical knowledge for how to innovate by resting control over game design from their original developers. At least four types of game mods can be observed: user interface customization; game conversions; machinima; and hacking closed game systems. Each supports different kinds of open source software (OSS) extension to the base game or game run-time environment. Game modding tools and support environments that support the creation of such extensions also merit attention. Furthermore, OSS game extensions are commonly applied to either proprietary, closed source software (CSS) games, or to OSS games, but generally more so to CSS games. Why this is so also merits attention. Subsequently, we conceive of game mods as covering customizations, tailorings, remixes, or reconfigurations of game embodiments, whether in the form of game content, software, or hardware denoting our space of interest.

The most direct way to become a game mod developer (a game *modder*) is through self-tutoring and self-organizing practices. Modding is a form of learning – learning how to mod, learning to be a game developer, learning to become a game content/software developer, learning computer game science outside or inside an

S.A. Hissam et al. (Eds.): OSS 2011, IFIP AICT 365, pp. 62–74, 2011.
© IFIP International Federation for Information Processing 2011

academic setting, and more [5,20]. Modding is also a practice for learning how to work with others, especially on large, complex games/mods. Mod team efforts may also self organize around emergent software development project leaders or "want to be" (W.T.B.) leaders, as seen for example in the *Planeshift* (http://www.planeshift.it/) OSS massively multiplayer online role-playing game (MMORPG) development and modding project [20].

Game mods, modding practices, and modders are in many ways quite similar to their counterparts in the world of OSS development, even though they often seemingly isolated to those unaware of game software development. Modding is increasingly a part of mainstream technology development culture and practice, and especially so for games, but also for hardware-centered activities like automobile or personal computer   customization. Modders are players of the games they reconfigure, just as OSS developers are also users of the systems they develop. There is no systematic distinction between developers and users in these communities, other than there are many users/players that may contribute little beyond their usage, word of mouth they share with others, and their demand for more such systems. At OSS portals like SourceForge.net, the domain of "Games" is the second most popular project category with nearly 42K active projects, or 20% of all projects[1]. These projects   develop either OSS-based games, game engines, or game development tools/SDKs, and all of the top 50 projects have each logged more than 1M downloads. So the intersection of games and OSS covers a substantial socio-technical plane, as game modding and traditional OSS development are participatory, user-led modes of system development that rely on continual replenishment of new participants joining and migrating through project efforts, as well as new additions or modifications of content, functionality and end-user experience [19,20,21]. Modding and OSS projects are in many ways experiments to prototype alternative visions of what innovative systems might be in the near future, and so both are widely embraced and practiced primarily as a means for learning about new technologies, new system capabilities, new working relationships with potentially unfamiliar teammates from other cultures, and more [cf. 21].

Consequently, game modding appears to be (a) emerging as a leading method for developing or customizing game software; (b) primarily reliant of the development and use of OSS extensions the ways and means for game modding; and (c) overlapping a large community of OSS projects that develop computer game software and tools that has had comparatively little study. As such, the research questions that follow then are why do these conditions exist, how have they emerged, and how are they put into practice in different game modding efforts.

This paper seeks to examine what is known so far about game mods and modding practices. The research method in this study is observational and qualitative. It seeks to snapshot and highlight current practices that can be associated with software engineering and game studies, as well as how these practice may be applied in CSS

---

[1] See http://www.sourceforge.net/softwaremap/index.php, accessed 15 April 2011. The number one category of projects is for "Development" with more than 65K OSS projects, out of 210K projects. So OSS Development and OSS Games together represent half of the projects currently hosted on SourceForge.

versus OSS game modding. Numerous examples of different game mods and modding practices are identified throughout to help establish an empirically grounded baseline of observations, from which further studies can build or refute. Furthermore, the four types of game mods and modding practices identified in this paper have been employed first-hand in game development projects led or produced by the author. Such observation can subsequently serve as a basis for further empirical study and technology development that ties together computer games, OSSD, software engineering, and game studies [19,20,21,22].

2   Related Research

Two domains of research inform the study here: software extension within the field of software engineering, and modding as cultural practice within game studies. Each is addressed in turn.

2.1   Software Extension

Game mods embody different techniques and mechanisms for software extension. However, the description of game mods and modding is often absent of its logical roots or connections back to software engineering. As suggested, mods are extensions to existing game software systems, so it is appropriate to review what we already know about software extensions and extensibility.

Parnas [15] provides an early notion of software extension as an expression of modular software design. Accordingly, modular systems are those whose components can be added, removed, or updated while satisfying the original system functional requirements. Such concepts in turn were integrated into software architectural design language descriptions and configuration management tools [14]. But reliance on explicit software architecture descriptions is not readily found in either conventional game or mod development. Hentonnen and colleagues [8] examine how software plug-ins support architectural extension, while Leveque, et al. [11] investigate how extension mechanisms like views and model-based systems support extension, also at the architectural level. Last, the modern Web architecture is itself designed according to principles of extensibility through open APIs, migration across software versions, network data content/hypertext transfer protocols, and representational state transfer [6]. Mod-friendly networked multi-player games often take advantage of these capabilities.

Elsewhere, Batory and associates [3] describe how domain-specific languages (for scripting) and software product lines support software extension, and now such techniques are used in games that are open for modding. Next, OSS development as a complementary approach to software engineering, relies on OSS code and associated online artifacts that are open for extension through modification and redistribution of their source representations [21]. Finally, other techniques to extend the functionality or operation of an existing CSS system may include unauthorized modifications that might go beyond what the end-user license agreement might allow, and so appear to fall outside of what software engineering might anticipate or encourage. These include extensions via hacking methods like code injection or hooking, whose purpose is to gain/redirect control of normal program flow through overloading or intercepting system function calls, or provide a hidden layer of interpretation, which allow for "man in the middle" interventions. So software extensions and extensibility is a foundational concept in software engineering, as well as foundational to the development of game mods. However, the logical connections and common/uncommon legacy of game modding, OSS development, and software engineering remain under specified, which this paper begins to address.

2.2 Modding as Cultural Practice

Game modding is a practice for user content creation that creates/networks not only game mods but game modders. Within anthropological, behavioral, and sociological studies of computer game play, modding has been studied as an emerging cultural practice that mediates both game play and player interaction with other players (including the game's developers). In some early studies, modding has been designated as a form of "playbour" whereby player actions to create game extensions for use by other players is observed as a form of unpaid (or underpaid) labor that primarily benefits the financial and property interests of game development corporations or hegemonic publishers [10,16,26].

Game modding also modifies or transforms game play experience, since what is play and what is experience(d) are culturally situated. Examples of this may include single player games being modded into multi-player games. So the experience of single player versus the game environment is transformed into other situations including player versus player, multi-player group play, or team versus team play. Similarly, the modding of games to enable experiences other than expected game play, like using a modded game for storytelling or film-making experiences is also a practice of growing interest, with the emergence of a distinguishable community of gamer-filmmakers who produce *machinima* (described later) as either a literary medium, or an art form [9,12,13].

Other studies have observed that user/modders also benefit from modding as a way to achieve a sense of creative ownership and meaning in the modded games they share and play with others [17,19,20,23], and that game mods and modding practices become central elements in what constitutes play with and through games [24]. Finally, as already observed, OSS project portals like SourceForge host thousands of OSS game development projects that develop and deploy role-playing games (4.3K projects), simulation-based games (2.6K), board games (2.3K), side-scrolling/arcade games (2K), turn-taking strategy games (1.7K), multi-user dungeons or text-based adventure/virtual worlds (1.6K), first-person shooters (1.6K), MMORPG (0.6K) and more. So development of OSS games and related game development tools can be recognized as a central element in the cultural world of computer games and game development, as well as the world of OSS development [19,20,21].

3 Four Types of Game Mods

At least four types of game mods are realized through OSS development practices. These include (i) user interface customizations and agents, (ii) game conversions, (iii) machinima, and (iv) hacking closed source game systems. Each is examined in turned, and each is facilitated (or prohibited) according to its copyright license.

3.1 User Interface Customizations and Agents

User interfaces to games embody the practice and experience of interfacing users (game players) to both the game system and the play experience designed by the game's developers. Game developers act to constrain and govern what users can do, and what kinds of experience they can realize. Some users in turn seek to achieve a form of competitive advantage during game play by modding the user interface software for their game, when so enabled by game developers. These mods acquire or reveal additional information that users believe will help their play performance and experience. User interface add-ons subsequently act as the medium through which game development studios support game product customization, which is a strategy for increasing end-user satisfaction and thus the likelihood of product success [4].

Three kinds of user interface customizations can be observed. First and most common, is the player's ability to select, attire or accessorize a *player's in-game identity*. Second, is for players to customize *the color palette and representational framing borders* of the their game display within the human-computer interface, much like what can also be done with Web browsers (e.g, Firefox 4 "personas" and "themes") and other end-user software applications. Third, are *user interface add-on modules* that modify the player's in-game information management dashboard, but do not modify the underlying game play rules or functions. These add-ons provide additional information about game play state that may enhance the game play experience, as well as increasing a player's sense of immersion or omniscience within the game world through perceptual expansion. This in turn enables awareness of game events not visible in the player's pre-existing in-game view. Furthermore, some add-on facilities (e.g., those available with the proprietary *World of Warcraft* MMORPG, scripted in the LUA language) accommodate the creation of automated agent scripts that can read/parse data streamed to the UI within an existing or other add-on dashboard component, and then provide some additional value-added play experience, such as sending out messages or status reports to other players automatically. Such add-on agents thus modify or reconfigure the end-user play experience, rather than the core functionality or play mechanics available to all other of the game's players. Consequently, the first two kinds of customizations result from meta-data selections within parametric system functions, while the third represents a traditional kind of user-created modular extension; one that does not affect the pre-existing game's functional requirements, nor one included in the operational source code base during subsequent system builds or releases, unless they do alter the software's requirements (e.g., by introducing a new security vulnerability or exploit that must be subsequently prevented).

3.2 Game Conversions

Game conversion mods are perhaps the most common form of game mods. Most such conversions are partial, in that they add or modify: (a) in-game characters including user-controlled character appearance or capabilities, opponent bots, cheat bots, and non-player characters; (b) play objects like weapons, potions, spells, and other resources; (c) play levels, zones, maps, terrains, or landscapes; (d) game rules; or (e) play mechanics. Some more ambitious modders go as far as to accomplish (f) total conversions that create entirely new games from existing games of a kind not easily determined from the originalgame. For example, one of the most widely distributed and played total game conversions is the *Counter-Strike* (CS) mod of the *Half-Life* (HL) first-person action game from Valve Software. As the success of the CS mod gave rise to millions of players preferring to play the mod over the original HL game, then other modders began to access the CS mod to further convert in part or full, to the point that Valve Software modified its game development and distribution business model to embrace game modding as part of the game play experience that is available to players who acquire a licensed copy of the HL product family. Valve has since marketed a number of CS variants that have sold over 10M copies as of 2008, thus denoting the most successful game conversion mod, as well as the most lucrative in terms of subsequent retail sales derived from a game mod.

Another example is found in games converted to serve a purpose other than entertainment, such as the development and use of games for science, technology, and engineering applications. For instance, the *FabLab* game [22] is a conversion of the *Unreal Tournament 2007* retail game, from a first-person shooter to a simulator for training semiconductor manufacturing technicians in diagnosing and treating potentially hazardous materials spills in a cleanroom environment. This conversion is not readily anticipated by knowledge of the Unreal games or underlying game engine, though it maintains operational compatibility with the Unreal game itself. So game conversions can re-purpose the look, feel, and intent of a game across application domains, while maintaining a common software product line [cf. 3].

Finally, it is common practice that the underlying game engine has one set of license terms and conditions to protect original work (e.g., no redistribution), while game mod can have a different set of terms and conditions as a derived work (e.g., redistribution allowed only for a game mod, but not for sale). In this regard, software licenses embody the business model that the game development studio or publisher seeks to embrace, rather than just a set of property rights and constraints. For example, in *Aion*, an MMORPG from South Korean game studio NCSoft, no user created mods or user interface add-ons are allowed. Attempting to incorporate such changes would conflict with its EULA and subsequently put such user-modders at risk of losing their access to networked *Aion* multi-player game play. In contrast, the MMORPG *World of Warcraft* allows for UI customization mods and add-ons only, but no other game conversions, no reverse engineering of the game engine, and no activity intended to bypass WoW's encryption mechanisms. And, in one more variation, for games like *Unreal Tournament, Half-Life, NeverWinterNights, Civilization* and many others, the EULAs encourage modding and the free redistribution of mods without fee to others who must have a licensed copy of the proprietary CSS game, but not allowing reverse engineering or redistribution of the CSS game engine required to run the OSS mods. This restriction in turns helps game companies realize the benefit of increased game sales by players who want to play with known mods, rather than with the un-modded game as sold at retail. Mods thus help improve games software sales, revenue, and profits for the game development studio, publisher, and retailer, as well as enable new modes of game play, learning, and skill development for game modders.

3.3 Machinima

Machinima can be viewed as the product of modding efforts that intend to modify the visual replay of game usage sessions. Machinima employ computer games as their creative media, such that these new media are mobilized for some other purpose (e.g., creating online cinema or interactive art exhibition). Machinima focuses attention to playing and replaying a game for the purpose of story telling, movie making, or retelling of daunting or high efficiency game play/usage experience [12,13]. Machinima is a form of modding the experience of playing a specific game, by recording its visual play session history, so as to achieve some other ends beyond the enjoyment (or frustration) of game play. These play-session histories can then be further modded via video editing or remixing with other media (e.g., adding music) to better enable cinematic storytelling or creative performance documentation. Machinima is a kind of play/usage history process re-enactment [cf. 18] whose purpose may be documentary (replaying what the player saw or experienced during a play session) or cinematic (creatively steering a play session so as to manifest observable play process enactments that can be edited and remixed off-line to visually tell a story). Machinima mods are thus a kind of extension of game software use experience that is not bound to the architecture of the underlying game software system, except for how the game facilitates a user's ability to structure and manipulate emergent game play to realize a desired play process enactment history.

3.4 Hacking Closed Game Systems

Hacking a closed game system is a practice whose purpose oftentimes seems to be in direct challenge to the authority of commercial game developers that represent large, global corporate interests. Hacking proprietary game software is often focused not so much on how to improve competitive advantage in multi-player game play, but instead is focused on expanding the range of experiences that users may encounter through use of alternative technologies [7,20]. For example, Huang's [7] study instructs readers in the practice of "reverse engineering" as a hacking strategy to understand both how a game platform was designed and how it operates in fine detail. This in turn enables reconfiguration of new innovative modifications or original platform designs, such as installing and running a Linux operating system (instead of Microsoft's proprietary CSS offering). While many game developers seek to protect their intellectual property (IP) from reverse engineering through end-user license agreements (EULAs) whose terms attempt to prohibit such action under threat of legal action, reverse engineering is not legally prohibited. Consequently, the practice of modding closed game consoles/systems is often less focused on enabling players to achieve competitive advantage when playing retail computer games, but instead may encourage those few so inclined for how to understand and ultimately create computing innovations through reverse engineering or other modifications.

Closed game system modding is a style of software extension by game modders who are willing to forego the "protections" and quality assurances that closed game system developers provide, in order to experience the liberty, skill, knowledge acquisition, conceptual appropriation ("pwned"), and potential to innovate, that mastery of reverse engineering affords. Consequently, players/modders who are willing to take responsibility for their actions (and not seek to defraud game producers due to false product warranty claims or copyright infringement), can enjoy the freedom to learn how their gaming systems work in intimate detail and to potentially learn about game system innovation through discovery and reinvention with the support of others like-minded [cf. 20]. Proprietary game development studios may sometimes allow for such mod-based infringement of their games. For example, the team of modders behind the hacking and conversion of the single-player CSS game, *Grand Theft Auto*, have produced an OSS (now GPL'd) game mod using code injection and hooking cheating methods to realize a networked multi-player variant called *Multi Theft Auto,* that Rockstar Games has chosen not to prosecute for potential EULA violation, but instead to embrace as GTA fan culture [25]. Nonetheless, large corporate interests may assert that their IP rights allow them to install CSS rootkits that collect potentially private information, or that prevent the reactivation of previously available OSS (e.g., the Linux Kernel on the Sony PS3 game console[2]) that game system hackers seek to undo.

Finally, games are one of the most commonly modified types of proprietary CSS that are transformed into "pirated games" that are "illegally downloaded." Such game modding practice is focused on engaging a kind of meta-game that involves hacking into and modding game IP from closed to (more) open. Game piracy has thus become recognized as a collective, decentralized and placeless endeavor (i.e., not a physical organization) that relies on torrent servers as its underground distribution venue for pirated game software. As recent surveys of torrent-based downloads reveals, in 2008 the top 10 pirated games represented about 9M downloads, while in 2009 the top 5 pirated games represent more than 13M downloads, and in 2010 the top 5 pirated games approached 20M, all suggesting a substantial growth in interest in and access to such modded game products[3]. Thus, we should not be surprised by the recent efforts of game system hackers that continue to demonstrate the vulnerabilities of different hardware and software-based techniques to encrypt and secure closed game systems from would be crackers. However, it is also very instructive to learn from these exploits how difficult it is to engineer truly secure software systems, whether such systems are games or some other type of application or package.

4 Game Modding Software Tools and Support

Games are most often modded with tools providing access to unencrypted representations of game software or game platform. Such a representation is accessed and extended via a domain-specific (scripting) language. While it might seem the case

---

[2] For details, see http://en.wikipedia.org/wiki/George_Hotz#Hacking_the_PlayStation_3
[3] For 2008, see http://torrentfreak.com/top-10-most-pirated-games-of-2008-081204/ For 2009, http://torrentfreak.com/the-most-pirated-games-of-2009-091227/ For 2010, http://torrentfreak.com/call-of-duty-black-ops-most-pirated-game-of-2010-101228/

that game vendors would seek to discourage users from acquiring such tools, a widespread contrary pattern is observed.

Game system developers are increasingly offering software tools for modifying the games they create or distribute, as a way to increase game sales and market share. Game/domain-specific Software Development Kits (SDKs) provided to users by game development studios represent a contemporary business strategy for engaging users to help lead product innovation from outside the studio. Once Id Software, maker of the *DOOM* and *Quake* game software product line, and also Epic Games, maker of the *Unreal* software game product line, started to provide prospective game players/modders with software tools that would allow them to edit game content, play mechanics, rules, or other functionality, other competing game development studios were pressured to make similar offerings or face a possible competitive disadvantage in the marketplace. However, the CSS versions of these tools do not provide access to the underlying source code that embodies the proprietary game engine—a large software program infrastructure that coordinates computer graphics, user interface controls, networking, game audio, access to middleware libraries for game physics, and so forth. But the complexity and capabilities of such a tool suite mean that any one person, or better said, any game development or modding team, can now access modding tools or SDKs to build commercial quality CSS games through OSS extensions. But mastering these tools appears to be an undertaking likely to be only of interest to highly committed game developers who are self-supported or self-organized.

In contrast to game modding platforms provided by game development studios, there are also alternatives provided by the end-user community. One approach can be seen with facilities provided in meta-mods like *Garry's Mod* or the *AMX Mod X* mod-making package. Modders can use these packages to construct a variety of plug-ins that provide for development of in-game contraptions as game UI agents or user created art works, or to otherwise create comic books, program game conversions, and produce other kinds of user created content. But both packages require that you own a licensed CSS game like *Counter-Strike: Source*, *Half-Life2* or *Day of Defeat: Source* from Valve Software.

A different approach to end-user game development platforms can be found arising from OSS games and game engines. The *DOOM* and *Quake* games and game engines were released as free software subject to the GPL, once they were seen by Id Software as having reached the end of their retail product cycle. Thousands of games/engines, as already observed, have been developed and released for download. Some started from the OSS that was previously the CSS platform of the original games. However, the content assets (e.g., in-game artwork) for many of these CSS- then-OSS games are not covered by the GPL, and so user-developers must still acquire a licensed copy of the original CSS game if its content is to be reused in some way[4]. Nonetheless, some variants of the user-created GPL'd games now feature their own content that is limited/protected by Creative Commons licenses.

---

[4]   For example, see http://assault.cubers.net/docs/license.html, accessed 13 April 2011.

5 Opportunities and Constraints for Modding

Game modding demonstrates the practical value of software extension as a user-friendly approach to customizing software. Such software can extend games open to modding into diverse product lines that flourish through reliance on domain-specific game scripting languages, and integrated SDKs. Modding also demonstrates the success of end-users learning how to extend software to create custom user interface add-ons, system conversions, replayable system usage videos, as well as to discover security vulnerabilities. Game modding therefore represents a viable form of end-user engineering of complex software that may be transferable to other domains.

Modding is a form of OSS-enabled collaboration. It is collaboration at a distance where the collaborators, including the game developers and game users, are distant in space and time from each other, yet they can interact in an open but implicitly coordinated manner through software extensions. Comparatively little explicit coordination arises, except when CSS game developers seek to embrace and encourage the creation of OSS game mods that rely on the proprietary CSS game engine (and also SDK), as a way to grow market share and mid share for the proprietary engine as a viable strategy to entry into the game industry.

However, mods are vulnerable to evolutionary system version updates that can break the functionality or interface on which the mod depends. This can be viewed as the result of inadequate software system design practice, such that existing system modularization did not adequately account for software extensions that end-users seek, or else the original developer wanted to explicitly prohibit end-users from making modifications that transform game play mechanics/rules or unintentionally allow for modification or misappropriation of copy protected code or media assets.

Last, one the key constraints on game modding in particular, and software extension in general, are the rights and obligations that are expressed in the original software EULA. Mods tend to be licensed using OSS or freeware licenses that allow for access, study, modification, and redistribution, rather than using free software licenses (e.g., GPLv2 or GPLv3). Software extensions that might be subject to a reciprocal GPL style license require that the base/original software system incorporate an explicit software architectural design that requires the propagation of reciprocal rights across an open interface, except through an LGPL software shim [1]. Otherwise, the scope of effectiveness and copyright protections of either free or non-free software (or related media assets) cannot be readily determined, and thus may be subject to copyright infringement or licenses non-compliance allegations. They may also be treated as social transgressions within a community of modders whose perceived ownership of the game mods demands respect and honor of a virtual license that may or may not be legally valid [2]. As the OSS community has long recognized, software rights and freedoms are expressed through IP licenses that insure whether or not a person has the right to access, study, modify, and redistribute the modified software, as long as the obligation to include a free software license is included that restates these rights in unalterable form, is included with the OSS code and its modified distributions.

6 Conclusions

Modding is emerging as a viable approach for mixing proprietary CSS systems with OSS extensions. The result is modded systems that provide the benefits of OSSD to developers of proprietary CSS systems, and to end-users who want additional functionality of their own creation, or from others they trust and seek to interact with through game play.

In contrast, modding is not so good for protecting software and media/content copyrights. Modding tests the limits of software/IP copyright practices. Some modders want to self-determine what copy/modding rights they have or not, and sometimes they act in ways that treat non-free software and related media as if it were free software. Who owns what, and which copy rights or obligations apply to that which is modded, are core socio-technical issues when engaging in modding.

This study helps to demonstrate that game modding is becoming a leading method for developing or customizing game software, whether based on proprietary CSS or OSS game systems. OSS-based software extensions are the leading ways and means for modding game-based user interfaces, converting games from one style/genre to another, for recording game play sessions for cinematic production and replay, and for hacking closed source game systems. Finally, the development of computer game software and tools itself represents a large community of OSS projects that has had comparatively little study, and thus merits further attention as its own cultural world as well as one for OSS development. This last consideration may be important as other empirical studies of OSS development that rely on data from SourceForge will increasingly include OSS game projects within large project samples. This study has therefore begun to address why and how these conditions have they emerged, and how are they put into practice in different game modding efforts. Future study should also consider whether and how modding might be applied and adopted in other application domains where CSS can be extended through OSS mods.

Acknowledgments. The research described in this paper has been supported by grants #0808783 and #1041918 from the National Science Foundation, and grant #N00244-10-1-0077 from the Naval Postgraduate School. No review, approval or endorsement implied. The anonymous reviewers also provided helpful suggestions for improving this paper.

References

1. Alspaugh, T.A., Asuncion, H.A., Scacchi, W.: Intellectual Property Rights Requirements for Heterogeneously Licensed Systems. In: Proc. 17th. Intern. Conf. Requirements Engineering (RE 2009), Atlanta, GA, September 24-33 (2009)
2. Alspaugh, T.A., Scacchi, W., Asuncion, H.A.: Software Licenses in Context: The Challenge of Heterogeneously Licensed Systems. J. Assoc. Information Systems 11(11), 730–755 (2010)
3. Batory, D., Johnson, C., MacDonald, B., von Heeder, D.: Achieving extensibility through product lines and domain specific languages: a case study. ACM Trans. Software Engineering and Methodology 11(2), 191–214 (2002)

4. Burnett, M., Cook, C., Rothermel, G.: End-User Software Engineering. Communications ACM 47(9), 53–58 (2004)
5. El-Nasr, M.S., Smith, B.K.: Learning Through Game Modding. ACM Computers in Entertainment 4(1), Article 3B (2006)
6. Fielding, R.T., Taylor, R.N.: Principled Design of the Modern Web Architecture. ACM Trans. Internet Technology 2(2), 115–150 (2002)
7. Huang, A.: Hacking the Xbox: An Introduction to Reverse Engineering. No Starch Press, San Francisco (2003)
8. Henttonen, K., Matinlassi, M., Niemela, E., Kanstren, T.: Integrability and Extensibility Evaluation in Software Architectural Models—A case study. The Open Software Engineering Journal 1(1), 1–20 (2007)
9. Kelland, M.: From Game Mod to Low-Budget Film: The Evolution of Machinima. In: Lowood, H., Nitsche, M. (eds.) The Machinima Reader, pp. 23–36. MIT Press, Cambridge (2011)
10. Kücklich, J.: Precarious playbour: Modders and the digital games industry. Fiberculture (5) (2005), http://journal.fibreculture.org/issue5/kucklich.html (accessed April 13, 2011)
11. Leveque, T., Estublier, J., Vega, G.: Extensibility and Modularity for Model-Driven Engineering Environments. In: 16th IEEE Conf. On Engineering Computer-Based Systems (ECBS 2009), pp. 305–314 (2009)
12. Lowood, H., Nitsche, M. (eds.): The Machinima Reader. MIT Press, Cambridge (2011)
13. Marino, P.: 3D Game-Based Filmmaking: The Art of Machinima. Paraglyph Press, Scottsdale (2004)
14. Narayanaswamy, K., Scacchi, W.: Maintaining Evolving Configurations of Large Software Systems. IEEE Trans. Software Engineering SE-13(3), 324–334 (1987)
15. Parnas, D.L.: Designing Software for Ease of Extension and Contraction. IEEE Trans. Software Engineering SE-5(2), 128–138 (1979)
16. Postigo, H.: Of mods and modders: Chasing down the value of fan–based digital game modifications. Games and Culture 2(4), 300–313 (2007)
17. Postigo, H.: Video Game Appropriation through Modifications: Attitudes Concerning Intellectual Property among Modders and Fans. Convergence 14(1), 59–74 (2008)
18. Scacchi, W.: Modeling, Integrating, and Enacting Complex Organizational Processes. In: Carley, K., Gasser, L., Prietula, M. (eds.) Simulating Organizations: Computational Models of Institutions and Groups, pp. 153–168. MIT Press, Cambridge (1998)
19. Scacchi, W.: Understanding the Requirements for Developing Open Source Software. IEE Proceedings—Software Engineering 149(1), 24–39 (2002); Revised version in Lyytinen, K., Loucopoulos, P., Mylopoulos, J., Robinson, W., (Eds.), Design Requirements Engineering: A Ten-Year Perspective. LNBIP, vol. 14, pp. 467–494. Springer, Heidelberg (2009)
20. Scacchi, W.: Free/Open Source Software Development Practices in the Game Community. IEEE Software 21(1), 59–67 (2004)
21. Scacchi, W.: Free/Open Source Software Development: Recent Research Results and Emerging Opportunities. In: Proc. European Software Engineering Conference and ACM SIGSOFT Symposium on the Foundations of Software Engineering, Dubrovnik, Croatia, pp. 459–468 (September 2007)
22. Scacchi, W.: Game-Based Virtual Worlds as Decentralized Virtual Activity Systems. In: Bainbridge, W.S. (ed.) Online Worlds: Convergence of the Real and the Virtual, pp. 225–236. Springer, New York (2010)

23. Sotamaa, O.: When the Game Is Not Enough: Motivations and Practices Among Computer Game Modding Culture. Games and Culture 5(3), 239–255 (2010)
24. Taylor, T.L.: The Assemblage of Play. Games and Culture 4(4), 331–339 (2009)
25. Wen, H.: Multi Theft Auto: Hacking Multi-Player Into Grand Theft Auto With Open Source, OSDir (May 25, 2005), http://osdir.com/Article4775.phtml Also see, http://www.mtavc.com/andhttp://en.wikipedia.org/wiki/MultiTheft_Auto (all accessed June 1, 2011)
26. Yee, N.: The Labor of Fun: How Video Games Blur the Boundaries of Work and Play. Games and Culture 1(1), 68–71 (2006)

Preparing FLOSS for Future Network Paradigms:

A Survey on Linux Network Management

Alfredo Matos, John Thomson, and Paulo Trezentos

Caixa Mágica Software
Edificio Espanha - Rua Soeiro Pereira Gomes
Lote 1 - 8 F, 1600-196 Lisboa
{alfredo.matos,john.thomson,paulo.trezentos}@caixamagica.pt

Abstract. Operating system tools must fulfil the requirements generated by the advances in networking paradigms. To understand the current state of the Free, Libre and Open Source Software (FLOSS) ecosystem, we present a survey on the main tools used to manage and interact with the network, and how they are organized in Linux-based operating systems. Based on the survey results, we present a reference Linux network stack that can serve as the basis for future heterogeneous network environments, contributing towards a standardized approach in Linux. Using this stack, and focusing on dynamic and spontaneous network interactions, we present an evolution path for network related technologies, contributing to Linux as a network research operating system and to FLOSS as a whole.

1 Introduction

Free, Libre and Open Source Software (FLOSS) is often characterized by a distributed and even fractured development model. This can lead to different applications with similar purposes, where the consequence is often effort duplication. While this can simultaneously be characterized as an advantage or handicap of the FLOSS world, these characteristics are also observable in the networking aspects of Linux. The Linux Kernel is very rich in terms of networking functionality, with a modern stack that makes it a reliable network Operating System (OS). However, most network management[1] operations are usually executed in user-space, by distribution specific tools, which can vary across distributions.

This dichotomy is further exemplified by the networking paradigms: on one hand, they require supporting multiple heterogeneous networks as specified by Next Generation Networks (NGN), relating to different concurrent technologies on the device, such as WiFi, 3G, WiMax or even Bluetooth; while on the other

---

[1] When we consider management, we are in fact referring to bringing up devices, selection of network attachment points, performing dynamic configurations and the associated integration that cannot be configured statically or hard coded into applications.

S.A. Hissam et al. (Eds.): OSS 2011, IFIP AICT 365, pp. 75–89, 2011.

hand, dynamic and spontaneous configuration require supporting peer-to-peer interactions that operate without infrastructure support such as Zeroconf [41] technologies and even Ad-Hoc or Mesh Networks. Therefore, as we move towards NGN environments, where wireless connectivity is the norm rather than the exception, powered by WiFi or 3G connections, it becomes increasingly necessary to handle all the different connectivity scenarios and technologies, without compromising the user experience. This creates added complexity for the OS and network stack that must allow a seamless user experience with competing requirements.

To handle these concurrent vectors, we need a consistent FLOSS network stack that aligns the different tools and approaches, to match the needs of the evolving network environments. To achieve this, we present a survey that looks at the current Linux network model and existing technologies. Based on the survey of both FLOSS tools and Linux distributions, we open the door to an aligned network view by proposing a reference network stack that promotes standardized approaches. This reference stack, which takes into account both heterogeneous networks and spontaneous dynamic environments, can contribute to the evolution of FLOSS, and especially Linux, by helping to prepare for new and forthcoming NGN network scenarios that are being promoted in different venues, such as the ULOOP [38] IST Project.

By promoting a common vision based on current research scenarios, it is possible to promote Linux as a leading research platform, and simultaneously contribute to less effort-duplication and avoiding fractures in the development model, thus contributing to FLOSS as a whole, as is discussed in Sect. 5. This can be achieved using the reference stack presented in Sect. 4 as a starting point. The remainder of the paper is organized as follows: Section 2 highlights the importance and structure of the survey, while the tools are presented in Sect. 3 and the choices of network strategy for Linux-based distributions are presented in Sect. 4. We conclude the paper in Sect. 6 focusing on future work.

2   Tools for Evolving Paradigms

The first step towards tackling the complexity of FLOSS user-centric network management tools is conducting a survey that reflects the current state of Linux network management. We focus on two different aspects: heterogeneous network support and dynamic configurations.

From the Linux operating system perspective, it is important to see how the different technologies are handled, like WiFi, 3G or WiMax. But, heterogeneous networking goes beyond the support of multiple technologies, and implies a seamless experience, where the different technologies are integrated from the user's perspective. It is becoming increasingly important that these technologies work together, and managing how network selection is performed. Therefore, before implementing complex network solutions [20], it is necessary to determine the current state of the art, especially in FLOSS.

As a complementary aspect to network selection, we also focus on spontaneous and dynamic configurations. It is important to analyse how dynamic configurations occur, along with the benefits that they provide, especially considering wireless environments and user-centric technologies [34], regardless of whether infrastructure support exists. In this domain, we highlight two complementary approaches: IPv6 and Zeroconf.

IPv6 has built-in mechanisms that allow automatic address configuration and peer discovery on the local link. Stateless Address Auto Configuration (SLAAC) [35] allows a node to generate an address for local communication in the fe80::/10 range through an EUI64 expansion of its MAC address. Peer discovery can be performed using special IPv6 multicast groups (e.g. *all-nodes*).

In IPv4, peer discovery can be achieved through Zeroconf [41], which is a protocol suite that aims to provide a fully functional IPv4 stack without the need for special configuration servers. It focuses on network address configuration and local name resolution, without resorting to DHCP or DNS servers. Address configuration is done through IPv4 Link-Local Addresses [9], which is a mechanism that enables the configuration of local addresses in a special address range (169.254.0.0/16), similar to IPv6 local link addresses. For local name resolution, Zeroconf defines the usage of Multicast DNS (mDNS) [11]. mDNS requires that each host stores its own DNS records (A,MX,SRV) locally, answering queries sent to a specific Multicast address. Whoever knows the answer, i.e holds the record, should respond to queries (resolve the address). This establishes a simple protocol for DNS supported communication without a central server. Using mDNS it is possible to provide service discovery on the local link through DNS Service Discover (DNS-SD) [10], also part of Zeroconf. Using DNS-SD, a node can join the proper mDNS multicast group and query for well known DNS records (SRV, TXT and PTR) that have service instances names, according to a *dns-sd.org* list [19].

In the lower layers, Ad-Hoc (802.11 Independent Basic Service Set mode, IBSS) and Mesh networking (802.11s) can provide access without centralized infrastructure, but have limited support for dynamic configurations, which usually depends on higher layer technologies. WiFi Direct [39] is a WiFi Alliance proposed certification that extends the Ad-Hoc support in 802.11 with better security and simultaneous WiFi network connections. It provides the means for establishing dynamic connections between 802.11-enabled peers and also, according to preliminary findings, supports peer discovery on the link layer.

By focusing on heterogeneous networking and dynamic configuration technologies it is possible to evaluate the FLOSS tools, and how they are integrated in the different distributions, which is presented over the next sections.

3   Linux Tools

The main objective of this survey is to catalogue and analyse the most important network tools in Linux-based operating systems. To provide a thorough survey that covers the different FLOSS tools and technological aspects, we must look at the configuration and management tools currently available in Linux distributions. We focus on those that gather information from user input (through

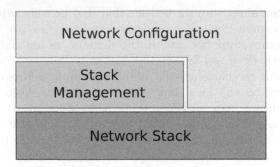

Fig. 1. Three-part network management stack

configuration interfaces) and translate it into the necessary commands and operations that are understandable by the lower level daemons and applications that interact with the network, and with the Linux Kernel.

Therefore we follow a top-down approach, as conceptually reflected in Fig. 1, focusing on the tools and processes existing at each level. We start with the connection managers in current FLOSS systems, which define how a user configures and interacts with the *Network Configuration*. We then explore the tools required to translate the configuration towards the system, providing *Stack Management*. Finally we focus on the *Network Stack*, defining how tools interact with the network, mostly within the Linux Kernel.

3.1   Network Manager

In recent years, Network Manager [29] (NM), a GPLv2 project by Red Hat and Novell, has emerged as the primary network configuration application for the Linux desktop. Its main purpose is to provide a hassle free networking experience, without compromising usability. This means that the focus is on reducing the amount of manual configuration exposed to the end-user, aiming at connectivity that "just-works".

By integrating network configuration and management, it creates a central control point across the entire desktop that is tightly integrated with the operating system and applications. Its modular design, shown in Fig. 2, includes several supported technologies, managing both wired and wireless connections. NM is split into two components: a system daemon that controls the networking infrastructure and a management application (usually graphical, e.g. network-manager-applet [30]) that handles user interactions. In fact, the daemon is controlled through a D-Bus [14] interface, a FreeDesktop [17] standard, allowing a flexible integration with different clients. NM architecture supports both IPv4 and IPv6, along with several access technologies, such as WiFi, WiMax, GSM/CDMA, Mesh and even Bluetooth, as shown by its architecture. This is achieved through different sub-systems that communicate with the main daemon through various interfaces, such as D-Bus, Netlink Sockets, Unix sockets or system call wrappers. The WiFi interactions are handled through the Supplicant

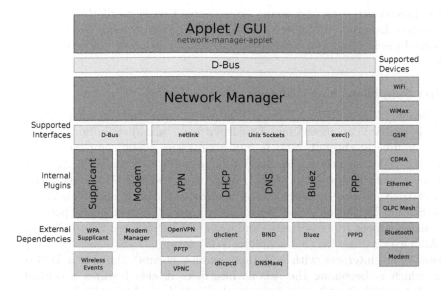

Fig. 2. Network Manager internal architecture

Manager, a D-Bus interface module for WPA Supplicant (Sect. 3.3), complemented by the Linux Wireless Extensions (WEXT, 3.4). 3G support is provided by Modem Manager [28], which supports most modern 3G devices.

Concerning network selection in Network Manager, it has a static preference list based on device types, enabling first Ethernet, WiFi, GSM, CDMA, Bluetooth, Mesh and finally WiMax. The connections, however, are timestamped and network manager will always prefer the last known active connection, when two connections of the same type exist (e.g. two wireless networks). For the actual selection of an Access Point (AP) within an Extended Service Set Identifier (ESSID), network manager relies on wpa_supplicant.

3.2 Connman

Connection Manager (ConnMan) [13] is a small and lightweight daemon designed for managing network connections on Linux embedded devices. It has a plug-in based design in order to build with as few components as possible, thus supporting customized configurations. Its main target is the inclusion in the MeeGo project [27], where it is the default network connection manager, unlike NM, which is general purpose for every Linux-based OS. It supports most technologies through plug-ins, namely, Ethernet, Bluetooth, WiFi, UMTS and even WiMax. It also supports network protocols through plug-ins, such as DNS, DHCP an VPN connections. The WiFi subsystem is composed of the main daemon and the WiFi plug-in, which connects to wpa_supplicant through a D-Bus interface (the preferred interface type). 3G support is achieved through oFono [31].

For managing connection preferences it uses a connection list, with both dynamic and static preferences. Previously used networks have a favourite status that takes precedence over new connection points. However, all things being equal it prefers Ethernet, Bluetooth, GSM, UMTS, WiMax and WiFi. These two combined mechanisms can be seen as a semi-static list that guides network selection.

3.3 Wpa_Supplicant and Hostapd

wpa_supplicant [25] is a GPLv2 licensed WPA supplicant that supports WPA/ WPA2. It is available on most Linux-based platforms and distributions, either directly or indirectly through NM. It implements the client component (the supplicant) of WPA, negotiating the encryption keys towards the WPA Authenticator (the server counterpart in WPA), supporting the 802.11i standard and also EAP/802.1X. It also supports several wireless extensions, such as 802.11r, Fast Base Station Transition (smoother roaming process between access points), 802.11w (management frame security) and even WiFi Protected Setup (WPS), a WiFI Alliance certification that simplifies WiFi setup.

wpa_supplicant interacts with NM (and similar clients) through a D-Bus interface, which is becoming the default interface. It also features a control (Unix) socket, which is still used by several clients (e.g. Android). In Linux, wpa_supplicant supports all drivers that use the recent mac80211 [23] stack (Sect. 3.4, drivers which support WEXT (v19+) and several older drivers/ chipsets.

Besides the security functions, it can also control roaming between Base-stations with the same BSSID, given the requirements for wpa_supplicant to perform scanning and associating procedures. When active, roaming decisions follow a specific priority list: WPA/WPA2 support, privacy capability support (a beacon bit that mandates encryption), transmission rate (if signal level is similar) and finally signal level.

While wpa_supplicant implements the supplicant in WPA, hostapd [24] (which shares the same author and codebase, featuring similar functionality) implements the authenticator, as well as being the most common software for running an 802.11 AP in Linux.

3.4 Wireless Communication Linux Kernel

The Linux Kernel supports all of the previous tools through device drivers and protocol implementations. The focus on wireless technologies dictates that we look at the Linux Kernel Wireless subsystem [2], composed of several building blocks. Currently, the most important component of this wireless stack is the mac80211 [23] framework. It provides a SoftMAC driver approach, i.e. most 802.11 protocol implementation (frame management) is done in software (inside the Kernel) rather than on every driver or card individually. While there are several advantages to this approach, the most important is that drivers share a common 802.11 implementation, only implementing device-specific callbacks, resulting in much simpler drivers. The main features of mac80211 include support for 802.11a/b/g/n, 802.11d, 802.11s (Mesh) and 802.11r. Interestingly, roaming is outsourced to user-land applications, like wpa_supplicant.

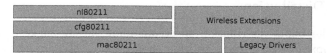

Fig. 3. Linux mac80211-based Wireless stack

As shown in Fig. 3, based on [6], mac80211 is composed of three subsystems: the mac80211 main block implements the 802.11 protocol, while cfg80211 implements 802.11 configuration and nl80211 implements the user-land communication through netlink sockets. However, as highlighted in Fig. 3, the mac80211 system also supports Wireless Extensions (WEXT) [36], a legacy configuration interface that either interacts with cfg80211 or directly with the mac80211 core. As mentioned, WEXT is a legacy wireless configuration interface (only maintained, not being developed) running over IOCTL (Input/Ouput Control) calls. IOCTL have been steadily removed in favour of other transport mechanisms, such as netlink, for user-space/kernel-space communication. However, it is still used in different places (e.g. Android).

Recently, the Linux Kernel picked up initial WiFi Direct (or WiFi P2P) support (also supported in wpa_supplicant). A key issue that has surfaced in the process of proposing the P2P extensions is the need for a standardized API between connection managers and wpa_supplicant, which in turn interacts with mac80211 through nl80211.

3.5   Avahi

As discussed in Sect. 2, one of the most important protocols in the context of local networking configuration is the Zeroconf suite. In Linux, Zeroconf is implemented by Avahi [42], which is a daemon that provides service discovery on the network through mDNS/DNS-SD and IPv4 address auto configuration through IPv4LL. IPv4 address configuration is done on demand, in most cases requested by NM, through a D-Bus interface. It is integrated into most Linux distributions, including embedded efforts such as OpenWRT, as presented in the next section.

4   Linux Network Stack

To understand how the tools are organized inside Linux, we must evaluate different Linux-based platforms. By examining the major distributions, it is possible to establish how most tools are organized in the Linux network stack and to determine the major trends concerning network management. The identified trends can provide insight into the best available tools, given that distributions spend a considerable integration effort and expertise towards building the appropriate network management stack and also consequently brings us closer to the goal of defining a reference Linux architecture.

4.1   Linux Distributions

Looking at the Linux distribution spectrum immediately suggests that there are several approaches towards network management. Distributions use different management tools, either scripts or applications, resulting in a uneven landscape. Here, we evaluate a select set of distributions, based on perceived importance [1]: we focus on those with most derivatives, from where tools are reused in each derivative distribution. We also focus on those with most user adoption, which helps determine the main ways in which users interact with Linux-based systems.

Fedora. Fedora [16] is a user oriented distribution, a development effort sponsored by Red Hat [3]. We analysed the latest release - Fedora 14. It uses a custom tool for the most static and standard network configurations, *system-network-config*, which is part of the control panel options and provides scripts and (python) tools for static system configurations. However, NM (v0.8.1) is also included, superseding most of the functionality provided by *system-network-config*. As expected, NM is accompanied by the required wpa_supplicant for wireless management and security. Also, avahi is used and takes over all the Zeroconf aspects.

CentOS. For a Red Hat Enterprise Linux (RHEL) [18] based distribution, which is a popular yet paid-for Linux distribution, we analysed CentOS [8]. CentOS is a free RHEL-based distribution, presenting an internal organization similar to Fedora, except that NM is disabled by default even though it is installed. However, NM is recommended for laptop usage [7]. In both cases, wpa_supplicant is used and avahi is running, controlling Zeroconf protocols.

Debian/Ubuntu. Debian [15] is one of the major available distributions, generating many derivatives. For static configurations it uses ifupdown, a tool that implements scripts and configuration files to easily manage network interfaces. The remaining setup is similar to Fedora, where the main tool is NM (v0.6.6 in Debian Lenny and v0.8.1 in Squeeze), complemented by wpa_supplicant for wireless support. This setup is seen both in Debian and Ubuntu [37] (Maverick 10.10), and in all versions, avahi runs by default, handling Zeroconf functionality.

OpenSUSE. OpenSUSE [32] and earlier SUSE systems, have historically relied on YAST for all system configurations. YAST handles most network configurations, using scripts to manage the different interfaces. In the initial interface configuration it is possible to activate NM, consequently becoming similar to the previous approaches, relying on NM and wpa_supplicant for most of the wireless interactions, and on avahi for Zeroconf.

Mandriva and Caixa Mágica. Mandriva [26] is the Linux distribution upon which Caixa Mágica [22] is built. We reviewed Caixa Mágica 15, as well as Mandriva 10.1 and 10.2, which share the same base. Mandriva, and consequently Caixa Mágica, do not follow the same pattern as other distributions using mostly custom tools, as seen in Fig. 4 where the Mandriva specific tools are highlighted

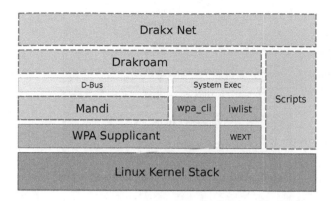

Fig. 4. Madriva and Caixa Mágica Network Management stack

via dashed stroke. The main networking configuration tool is Drakx-net, which is part of the Drak configuration toolset, a custom Mandriva system configuration tool. Drakx-net provides a configuration manager for networking settings, covering network interfaces and VPN.

Looking at the roaming/wireless subsystem, it is handled by Drakroam. This Mandriva developed tool is an application composed of scripts that interact with the OS, a graphical configuration interface, and an applet that provides a shortcut for network configuration with special emphasis on wireless. Drakroam uses mandi, a custom built D-Bus daemon that provides support for network configurations. It features a plug-in system, where the wireless part is an interface to the wpa_supplicant control interface. As a fall-back, Drakroam can support wpa_cli, a command line interface application provided by wpa_supplicant, and alternatively it falls back to iwlist (using WEXT). Beyond this, Caixa Mgica and Mandriva deploy Zeroconf mechanisms through the avahi daemon.

Other Distributions and Platforms. While we mostly explored desktop-like distributions, it is worth considering other platforms, especially embedded devices. We analyse Android [5], which targets mobile devices, and OpenWRT, which targets embedded routers.

Android, aimed at mobile phones and embedded devices, has an approach to network management that is different from the previously discussed distributions. As shown in Fig. 5, it uses Connectivity Manager [12], a Java connection manager, for controlling network interfaces and providing an API for applications interacting with the Android network management infrastructure. Similarly, WiFi is controlled through WiFi Manager [40], which has limited ca pabilities constrained by the Java exposed interface.

WiFi functionality is supported by a modified version of wpa_supplicant that supports additional control commands specific to Android mobile devices. The middleware interactions with wpa_supplicant are done through the socket interface, given that there is no D-Bus support. However, because Android devices do not support the mac80211 stack, wpa_supplicant is limited to WEXT.

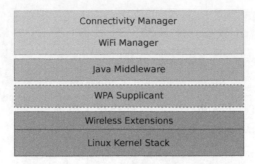

Fig. 5. Android network management stack

Furthermore, the Android approach does not support Ad-Hoc networks [4], and lacks a fully compliant Zeroconf tool (only a Java library for mDNS [21] exists).

OpenWRT [33] uses slimmed down versions of the applications used by most distributions. Given that the main purpose of the distribution is acting as an router and AP, it supports hostapd (Sect. 3.3, assuming the role of AP and WPA Authenticator with WPA/WPA2/802.11i capabilities. It supports several deployments, depending mostly on the hardware drivers to determine functionality, using the mac80211 stack as well as legacy drivers. Zeroconf can also be supported, by installing the provided packages for the avahi daemon, which can run on OpenWRT.

The static nature of the target deployment, implies that most network configurations are achieved statically through (BASH) scripts, using a flat database, the Unified Configuration Interface (UCI) module within OpenWRT, for storage.

4.2   Reference Architecture

After analysing all the different tools and distributions, an obvious pattern emerges, as shown in Table 1. The Linux network stack, especially considering the wireless subset, is centered mostly around Network Manager, both for the graphical interface, as well as the system daemon. While there are some notable alternatives in the form of ConnMan, and some distribution specific efforts, there is a convergence within Linux towards the widespread use of Network Manager on the desktop and laptop platforms. The only noteworthy exception is Android, which uses a custom connectivity manager. Network Manager is tightly integrated in the Linux operating system, providing not only means to configure network settings, but also means for applications to determine whether an active network connection exists, as highlighted in the top-most part of Fig. 6, which shows the reference Linux network architecture. However, as we follow down the proposed consolidated network stack, we observe a much clearer convergence across all platforms: WPA Supplicant. The WPA Supplicant daemon has become an expected presence on all Linux based operating systems, such as desktop, laptop and even handheld devices, also making an appearance on Android phones.

Table 1. Tools summary per distribution

Distribution	Network Management	Wireless Management	Wireless Stack	Zeroconf Support
Fedora	system-network-config and Network Manager	wpa_supplicant	mac80211 WEXT	avahi
CentOS	system-network-config[1]	wpa_supplicant	mac80211 WEXT	avahi
Debian Ubuntu	Network Manager	wpa_supplicant	mac80211 WEXT	avahi
OpenSUSE	YAST and Network Manager[2]	wpa_supplicant	mac80211 WEXT	avahi
Mandriva Caixa Mágica	Drakx-net and Drakroam[3]	wpa_supplicant	mac80211 WEXT	avahi
Android	Connectivity Mgr. and WiFI Mgr.	wpa_supplicant	WEXT	-
OpenWRT	UCI/Scripts	wpa_supplicant	mac80211 WEXT	avahi[4]

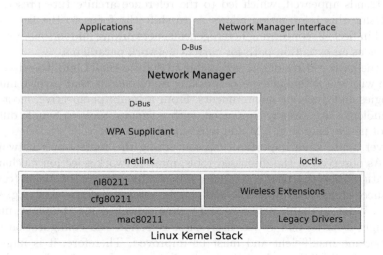

Fig. 6. Linux Network Management Reference Architecture

WPA supplicant started out by handling the security aspects of WiFi network connections, but also covers roaming between access points. It integrates with all the network connection managers, such as NM, ConnMan, and even Android's Connectivity Manager.

---

[1] Network Manager is recommended but not installed.
[2] Network Manager is installed by default but is optional.
[3] Requires more helper applications such as mandi, wpa_cli and iwlist.
[4] Not installed by default, but available for the platform.

When stepping into the actual network protocol implementations, we venture into the Linux Kernel, as depicted by the bottom-most part of Fig. 6. The main presented focus is on WiFi, which is handled through the new mac80211 wireless subsystem. This provides support for most modern wireless cards, but also supports the legacy WEXT interfaces, kept for legacy support (of both user land tools and Kernel drivers).

Beyond the presented blocks, every distribution and system complements the network stack tools with auxiliary scripts that handle the static aspects of network configurations. While this where most differences exist, it does not represent a major divergence given that most scripts tend to be distribution specific. We also omit from the illustration the Zeroconf tools but those which are shown in Table 1, given that the only real alternative in Linux is avahi, which already enjoys widespread deployment. When coupled with adequate Link Layer tools, it can provide a interesting effort in the self-managed network environments.

5   Overview and Future Directions

By looking at the previous sections, which are mostly summarized by Table 1, obvious trends appeared, which led to the reference architecture presented in Fig. 6. Using this information allows us to shift the focus to the key aspects presented in Sect. 2, concerning heterogeneous networking and autonomous configurations as an evolution path for current network paradigms.

From the gathered results, it is possible to deduce that Linux-based systems can cope with the heterogeneous networking requirements that involve multiple technologies and dynamic environments. From a Linux perspective, most wireless technologies are already supported, with adequate tools to handle different aspects of networking (e.g. NM and wpa_supplicant).

However, there is an important gap concerning the control over network selection. As observed in the discussed tools, most network selection mechanisms imply static technology-based preference lists, simple pattern repetition (connect to last successfully used network) and simple network information (e.g. signal strength). When available networks and technologies become abundant, most of which might even be new to the user, these selection and configuration mechanisms become insufficient and must be improved. Therefore, it is important to increase the flexibility of existing control structure for network attachment. This can be achieved using two complementary approaches: 1) provide a flexible interactive API, exposed by the modules that directly control the network selection (e.g. wpa_supplicant or NM); and 2) introduce a component that collects the options coming from the different technologies, and provides consistent and reliable network selection decisions, which could be a part of NM, or even an on-demand external dependency that depends on the deployment scenario.

Concerning the autonomous configuration mechanisms and technologies that can work without infrastructure support, we observed that Linux already has a strong Zeroconf support, along with IPv6 and even WiFi Direct. This places FLOSS as a front-runner when considering these types of technologies. Avahi is already distributed with most Linux-based systems, and the Linux Kernel already support most modern technologies. Therefore, we can conclude that most tools are in place, leaving FLOSS in a good position to increase the integration of these technologies in the OS. What is missing now is the widespread adoption of these techniques along with extensions that enable us to integrate them in different applications. This is where FLOSS has the upper hand: through open interfaces and a collaborative model, it is possible to develop and integrate adoptable interfaces. This provides applications with a potential agility to quickly take advantage of the discussed autonomous configuration mechanisms, in different scenarios.

Following the open source model, this allows for an uncomplicated API for information sharing across different applications. The consequence is that, instead of being bound by standards, FLOSS can use them as a launchpad towards innovative efforts, taking advantage of local loop technologies and promoting research through extensions/tools that benefit the end user.

Lastly, it is worth mentioning that as convergence on the network management architecture occurs, the disadvantages of FLOSS development model get diluted, which is what we have observed and potentially contributed to. Right now, most fragmentation occurs only in the static configuration scripts, which are a matter of preference, style and legacy for each distribution.

6   Conclusions

Throughout the presented survey, we attempted to explore FLOSS technologies in the light of current and future technologies, by undertaking the effort of investigating the current state of the art Linux tools and distributions. The result was the proposed reference network architecture, that defines the baseline for the Linux network stack, highlights the strengths and gaps of current approaches.

Using this reference stack it is possible to outline different approaches that enable FLOSS to tackle the new networking environments, and also understand what can be can expected from the current Linux networking landscape. More importantly, by relating the current state with emerging technologies we have uncovered an opportunity for proposing future developments, contributing to the usefulness of open source operating systems in light of new technologies.

The two most important conclusions revolved around the need to improve Linux network selection mechanisms, in order to tackle dynamic and mobile heterogeneous environments, and also identified that it is necessary to place a strong emphasis on providing new and innovative services that use autonomous technologies, which are already available in Linux.

Nevertheless, the main contribution of a reference architecture and future evolution path, is that it enables the reduction of divergence and repeated work in FLOSS. This can increase the traction of existing technologies, highlighting the potential advantages of the FLOSS model in light of future research and development activities.

Acknowledgements. The work described in this paper was done in the scope of IST FP7 ULOOP STREP Project. ULOOP receives funding from the European Community's Seventh Framework Programme, under contract agreement n°257418. The views expressed in this publication are those of the authors and do not necessarily reflect the project or the European Commission's view on the subject.

References

1. Distrowatch, http://distrowatch.com/ (last checked: June 2011)
2. Official linux wireless wiki, http://wireless.kernel.org/ (last checked: April 2011)
3. Red hat, http://www.redhat.com/ (last checked: April 2011)
4. Wifi: support ad hoc networking - support ticket, http://code.google.com/p/android/issues/detail?id=82 (last checked: April 2011)
5. Android. Android mobile plaftorm, http://www.android.com/ (last checked: April 2011)
6. Berg, J.M.: Wifi control plane overview, http://wireless.kernel.org/en/developers/Documentation/mac80211?action=AttachFile&do=get&target=mac80211.pdf (last checked: April 2011)
7. CentOS. Centos network manager configuration, http://wiki.centos.org/HowTos/Laptops/NetworkManager (last checked: April 2011)
8. CentOS. Rhel based linux operating system, http://www.centos.org/ (last checked: April 2011)
9. Cheshire, S., Aboba, B., Guttman, E.: Dynamic Configuration of IPv4 Link-Local Addresses. In: RFC 3927 (Proposed Standard) (May 2005)
10. Cheshire, S., Krochmal, M.: DNS based Service Discovery. Internet-Draft (February 2011)
11. Cheshire, S., Krochmal, M.: Multicast DNS. Internet-Draft (February 2011)
12. Connectivity Manager. Android connectivity manager developer information, http://developer.android.com/reference/android/net/ConnectivityManager.html (last checked: April 2011)
13. ConnMan. Network connection management daemon, http://www.connman.net (last checked: April 2011)
14. D-Bus. D-bus system message bus, http://www.freedesktop.org/wiki/Software/dbus (last checked: April 2011)
15. Debian. Linux-based operating system, http://www.debian.org/ (last checked: April 2011)
16. Fedora. Linux-based operating system, http://fedoraproject.org/ (last checked: April 2011)
17. freedesktop.org. Open source software projects working on interoperability for x window system desktops, http://www.freedesktop.org/ (last checked: April 2011)
18. Hat, R.: Red hat enterprise linux, http://www.redhat.com/rhel (last checked: April 2011)
19. Informal DNS-SD Service types list. Dns srv (rfc 2782) service types, http://www.dns-sd.org/ServiceTypes.html (last checked: April 2011)

20. Jesus, V., Sargento, S., Corujo, D., Senica, N., Almeida, M., Aguiar, R.: Mobility with qos support for multi-interface terminals: Combined user and network approach. In: IEEE Symposium on Computers and Communications (ISCC 2007), pp. 325–332 (July 2007)

21. jmDNS. Java mdns multicast implementation, http://jmdns.sourceforge.net (last checked: April 2011)

22. Caixa Mágica Linux. Mandriva based linux operating system, http://www.caixamagica.pt/ (last checked: April 2011)

23. mac80211. mac80211 development documentation, http://wireless.kernel.org/en/developers/Documentation/mac80211 (last checked: April 2011)

24. Malinen, J.: Hostapd: Ieee 802.11 ap, ieee 802.1x/wpa/wpa2/eap/radius authenticator, http://w1.fi/wpa_supplicant/ (last checked: April 2011)

25. Malinen, J.: Linux wpa/wpa2/ieee 802.1x supplicant. http://w1.fi/wpa_supplicant/ (last checked: April 2011)

26. Mandriva. Linux-based operating system, http://www.mandriva.com/ (last checked: April 2011)

27. Meego. Meego mobile platform - linux foundation, http://meego.com (last checked: April 2011)

28. ModemManager. Modem manager for linux, http://cgit.freedesktop.org/ModemManager/ (last checked: April 2011)

29. Network Manager. Linux network manager, http://projects.gnome.org/NetworkManager/ (last checked: April 2011)

30. nm applet. Network manager gnome applet, http://ftp.gnome.org/pub/GNOME/sources/network-manager-applet/0.8/ (last checked: April 2011)

31. oFono. Open source telefony (gsm/umts), http://ofono.org (last checked: April 2011)

32. OpenSuse. Linux-based operating system, http://www.opensuse.com/ (last checked: April 2011)

33. OpenWRT. Openwrt embedded linux. http://openwrt.org (last checked: April 2011)

34. Sofia, R., Mendes, P.: User-centricity impact on future internet architectures. In: FIA: Future Internet Workshop 2010 (June 2010)

35. Thomson, S., Narten, T., Jimei, T.: IPv6 Stateless Address Autoconfiguration. RFC 4862 (Proposed Standard) (September 2007)

36. Tourrilhes, J.: Linux wireless extensions, http://www.hpl.hp.com/personal/Jean_Tourrilhes/Linux/Linux.Wireless.Extensions.html (last checked: April 2011)

37. Ubuntu. Debian based linux operating system, http://www.ubuntu.com/ (last checked: April 2011)

38. ULoop. User-centric wireless local loop, http://www.uloop.eu (last checked: April 2011)

39. Wi-Fi Alliance. Wi-fi certified wi-fi direct: Personal, portable wi-fi technology (2010), http://www.wi-fi.org/knowledge_center_overview.php?docid=4685

40. WiFi Manager. Android wifi manager developer information, http://developer.android.com/reference/android/net/wifi/WifiManager.html (last checked: April 2011)

41. Zeroconf. Zero configuration networking (zeroconf) working group, http://www.zeroconf.org (last checked: April 2011)

42. Zeroconf. Zeroconf open source implementation, http://www.zeroconf.org (last checked: April 2011)

A Review of Tool Support for User-Related Communication in FLOSS Development

Aapo Rantalainen, Henrik Hedberg, and Netta Iivari

Department of Information Processing Science, University of Oulu

Abstract. Free/Libre/Open Source Software (FLOSS) projects rely on Internet tools for communication and in coordinating their work. Communication between developers is well supported in FLOSS projects, but user-developer communication has proven out to be challenging. This paper examines the following questions: "What kinds of means for communication exist in FLOSS projects for user-developer communication? What kinds of means should there be?" We have carried out a literature review addressing communication in FLOSS projects, and contrasted the findings with Human-Computer Interaction (HCI) literature on user-developer communication. HCI literature indicates that user-developer communication is needed during requirements construction, design and evaluation tasks, and HCI specialists are needed for orchestrating the communication and the user related tasks. Communication during the evaluation task is somewhat supported in FLOSS projects, but design and requirements construction are badly in need for support, even though ideas have already been presented. In addition, HCI specialists are in need of different kinds of communication support in FLOSS projects.

Keywords: Free/Libre/Open Source Software, Human-Computer Interaction, User-Developer Communication, Tool Support.

1 Introduction

The term FLOSS means Free/Libre/Open Source Software. It refers to free software, defined by Free Software Foundation [28] and open source programs defined by Open Source Initiative [51]. This software is determined by licenses. If a license allows users to modify and share software, it is considered as free software. Practically this means users must have access to the source code, thus the name open source. But open source is more than an available source code, it is a philosophy of studying the code, improving it, customizing it, and sharing it [16]. However, free software is also an ideology and a lifestyle [65, 68].

FLOSS projects are usually described by concentric circles [19]. The most inner circle represents coders who have direct write access to the official code. The next circle represents coders who send patches. Last circles are users, first active ones and then passive. Working is very meritocratic [64]. Contributors who have distinguished themselves by the quality of their work are invited to join the inner circle and gain more responsibility in the project [21]. FLOSS development

S.A. Hissam et al. (Eds.): OSS 2011, IFIP AICT 365, pp. 90–105, 2011.

originates in the hacker culture with highly talented developers [41, 42, 63], but recently also non-technical users have found FLOSS solutions; e.g., Linux operating system, OpenOffice.org application suite and Mozilla Firefox web browser are widely known and used by non-technical users. Therefore, in FLOSS projects nowadays there are typically different stakeholders with different perspectives, emphasizing, e.g., end user-oriented aspects or technical implementation-specific details. The onion model describes only coders, not e.g. translators, manual writers, HCI experts nor any domain specialist [33]. Communication between these different stakeholder groups and perspectives is a challenge [33]. Especially it has been noticed that HCI specialists tend to remain outsiders and they do not have any real decision making power in FLOSS development [1, 9, 12, 15, 49, 61, 62].

This paper is focused on software targeted to the end users (i.e. not system libraries) and will examine: "What kinds of means for communication exist in FLOSS projects for user-developer communication? What kinds of means should there be?" Communication, generally, refers to the successful exchange of information so that the sender and receiver understand each other [14]. We will examine user-developer communication in FLOSS development by relying on ideas from HCI research on user-developer communication. This literature is contrasted with the existing research on communication in FLOSS development. We will critically review the means already suggested for user-developer communication.

The article is structured as follows. The next two sections cover a general introduction to the communication means used in FLOSS projects as well as the existing means used for user-developer communication. The fourth section presents guidelines from HCI research on user-developer communication, and the fifth section contrasts those with the findings related to communication in FLOSS projects. A number of areas for improvement and associated paths for future work are identified.

2   Existing Communication Means in FLOSS Projects

Next, few common characteristics of communication in FLOSS projects are described, even though it is acknowledged that there is no particular, explicitly defined FLOSS development methodology or model [46]. Although the open source development process is not well defined, some general characteristics can still be identified. Open-source projects attempt to ship out minimally working prototypes at the earliest possible time [13]. The most well known principle is "release early and release often" [54].

Almost always FLOSS projects begin by one developer scratching a personal itch [25, 54]. Thus software might not be even planned to be used by anybody else [58]. Typically a FLOSS project is directed and managed by the initial creator and the requirements come from the programmers themselves. Massey gives additional two sources of requirements: users and standards [43].

FLOSS projects have some common practices for communication, but they have also unique practices [5, 21], as well as unique structures [20]. They are project oriented, not organization oriented like closed source projects [5]. Because FLOSS projects are often distributed [64], there is a need for methods which support distributed development. Actually, there is a long list of existing communication means used in FLOSS projects that indeed support distributed development. These means are 1) email and mailing lists, 2) web sites, 3) version control systems, 4) bug trackers, 5) real time chats, 6) wikis and 7) web forums. [5, 26].

Often projects use only emails and mailing lists [21]. The most important benefit of them is time independency. Developers can live different continents and different timezones [21]. Email communication is sufficient, so it is used. Fogel claims that a mailinglist is the most used communication channel in FLOSS projects [26]. This kind of an open mailinglist is not scalable, however. If software has millions of users and even a small part of them are posting, the input rate is still bigger than any reader can handle. Therefore there must be dedicated lists. [26].

Web sites of FLOSS projects can be one-way channels for bigger audiences [26]. Not all projects even have other web sites than the bug tracker. Web portals are virtual workspaces that merge several channels to a web site [31]. Ankolekar and colleagues claim that project's websites often have relatively little to offer to non-technical users of the software [2].

Wiki is fast, but it isn't a realtime communication channel. The challenges of Wiki include cohesion of totality and duplicate working [26]. Challenges can be minimized and quality will be better, if there are both implicit (e.g. rules and guidelines) and explicit (e.g. discussion pages) coordination [39].

A web forum can be just an interface to a mailinglist, or it can be meant only for a browser. Web forums are categorized and threaded by default so they can handle more traffic than mailinglists. Forums are typically focused for users. [26].

Bug trackers, bug repositories or issue trackers are places to store found issues and problems. They can be part of the broader web site. They are not meant for discussions [26], but can be used for coordinating and planning releases [26]. Bach and colleagues criticized that bug tracker is a hidden place to do development [5]. The amount of tickets in a bug tracker is of no value in itself. One empirical study showed that 36 % of Eclipse bugs were invalid or duplicates [3].

Mockup pictures are easier to understand than source code, due to which there can be more people to discuss and then also unwanted or unhelpful postings from people who do not understand the big picture [5, 26]. As Parkinson said in 1957, which has now come to be known as Parkinson's Law of Triviality, "the more trivial the topic the more conversation" [52].

3   User-Developer Communication Means in FLOSS Projects

"Writing code isn't the problem, understanding the problem is the problem." [22]. One of the main problems in software projects is the lack of domain knowledge [22]. This source of error is eliminated in self-driven FLOSS projects where domain experts write software for their own needs [45]. But when developers are not users of the software, they need domain expertise [45]. Complex software project's design problems can't be solved by individuals or by homogeneous groups [24]. HCI experts are needed to steer user research studies such as surveys and interviews [53]. Communication is critical to these processes [50]. Nevertheless, "the role of the average user in FLOSS is not clear" [6].

Existing literature has indicated that there is a multitude of tools available in FLOSS projects for user-developer communication (see section 2): users can deliver feature requests and bug reports through these means as well as ask questions, while developers can answer the questions as well as provide user support [29, 40, 60, 69].

However, tools should be chosen based of on the skills of users, not the skills of developers [57]. Despite the existence of these means, it might be that the users are unable to communicate with the developers. They might be unable, unwilling or scared to use these means (i.e. the mailing lists and bug trackers) [9, 15, 48, 70]. "Non-technical users may not have the technical vocabulary valued by developers" [5]. Numerous problems with bug trackers have been listed in connection to reporting usability issues. They enable no recording, uploading, showing, managing or commenting videos, audio or images. Minor usability issues are hard to describe only verbally, and not all users have painting or drawing tools. The existing bug trackers are also too complex and coder centric. They ask many questions that user (or non-technical domain specialist) don't know how to answer. Because of these points not all users are able to use trackers at all. [15].

The skill level of reporters may also vary and they don't know or understand the bug fixing process [38]. Users don't know what is a good bug report either, i.e. what is relevant and needed information [10]. The bug tracker should tell a user what is relevant information and how a user can find it [10], or there should be no open bug reporting at all, but users should use support and feature request forums and moderators should pick up all found bugs into the internal bug tracker [38].

Usability and overall user experience (UX) issues are different than other bugs and issues. Bugs and UX-issues should be segregated, but they both may be visible in the same bug tracker and linked together. [5]. HCI specialists should also be welcomed to a project [4] and they should have an acknowledged role on the project web portal [4] which should have its own tab for UX-issues [5]. 'Confusion reports' and 'surprise reports' are suggested to be used additionally to bug reports. These are reports by active, reflective users and contain descriptions on what the user tried to do and what caused their surprise, whether derived from confusion or not. [8]. Also groups such as 'reference', 'core' and 'bleeding edge' users are brought up. They can be used for initially testing the solution or new features. They can be offered a more advanced tool set for providing feedback, while all users can use some basic tools related to which some automatic summaries are generated to lessen the burden for the developers to analyze the results. In addition, Open Content projects organized around using certain FLOSS tools are brought up. During them users and developers of the tools collaborate on the task and developers gain feedback. [60].

As mentioned, one problem related to the communication means used in FLOSS projects is that they do not support communicating visually. Related to usability problems and user interface design, textual descriptions are not enough - it may not be sufficient to articulate these issues only textually or it may take too much of effort to do it [15, 48, 49, 61]. Natural language and visualizations should be preferred as well as cooperation during design work supported [33]. In time and location distributed development coders are happy with emails and a version control system, because they are working with a source code, which is text, but domain specialists and HCI experts need also other means of communication, including face to face communication [1, 15, 48, 49].

User interface design by blogs has been mentioned as a way to communicate design solutions in a distributed environment [49, 60]. Dedicated mailing lists for usability discussion have also been brought up [12, 48]. In addition, specific design areas supporting brainstorming and discussion of user related issues have been mentioned [60, 61], as well as a 'usability system' enabling easy to use usability bug reporting with the possibility to use multimedia [15, 49]. Moreover, (remote) usability evaluation and user data gathering tools have been recommended [47–49, 60].

By relying on the traditional HCI literature, the importance of paper prototyping during early design phases has also been highlighted [15]. There should also be user requirements and profiles produced, and users might be interviewed or questionnaires used to inquire them and their needs [15, 60]. It is, however, unclear what kind of communication means there should be supporting these activities. Another problem related to these suggestions is that they seem to assume that the development proceeds similarly like to traditional proprietary software development, with certain phases sequentially following each other, with HCI experts hired to do the job.

All in all, FLOSS has many characteristics, which harness usability work. Typically FLOSS programs, also development tools, are highly modular via plugins and customizable via many configurations. This increases complexity related to installing and using them. Documentation is fragmented over forums, personal web pages, and source code. [62]. FLOSS development is rapid and iterative, so rapid that it might look like one code-and-fix attempt from the outside [13], and thus user centric design is challenging [5]. But rapid prototyping can be also good for UX-testing [4].

The lack of coordination causes more frustration to HCI specialists than to coders [15]. Non-hierarchical decision making is not good for HCI specialists, because then it is hard to get their ideas accepted and implemented [15]. The core developers typically make all the decisions related to what to include in the code base, more peripheral developers and users having no decision-making power regarding this [69]. HCI experts may have difficulties in being able to affect design decisions in FLOSS projects [1, 9, 12, 15, 49, 61]. Coders work for merit, due to which also usability/UX tasks should be made visible and merited, so that these specialists could gain a more authoritative position [5]. Important for HCI specialists is particularly to make themselves known and visible in the project, educate the developers in HCI matters and offer usability feedback [60]. Building trust, providing opportunities to show merit and developing a new workflow with HCI specialists' and developers' work integrated are important for HCI work to become accepted and practiced in FLOSS projects. The new workflow should include a special phase for design, following users suggestion or feedback, preceding actual development. The design should be iterated until finding a satisfactory solution, users' feedback being utilized during the process. [5].

HCI specialists should be acknowledged and empowered by creating another layer of roles into the traditional onion model that is used to depict the decision making structure in FLOSS projects. While the technical stakeholders can be identified as users, contributors, committers and core team members, the human oriented layer introduces positions for non-technical users, usability evaluators, usability designers and a HCI core team, respectively (note that one person can act in several roles). Since the viewpoint to software and produced information differs from source code, also communication means should be different within that layer. The most important communication channel between technical and non-technical project members is shared decision making between the technical and HCI core teams, but it should not be limited to that. [33].

Some FLOSS projects have voting mechanism for bugs and new features that enable users and possible HCI experts to have a say. They differ from project to project. KDE's and some other Bugzilla based bug tracker have "Most hated bugs" and "Most wanted features" and each registered user can give votes to bugs or features (https://bugs.kde.org/). Then there are also brainstorm forums (e.g. http://brainstorm.ubuntu.com/), where users can make any suggestion relating to requirements or design and others can comment and vote for (or against) them. However, these solutions are derived from the FLOSS world, not from HCI oriented research, but they are mentioned here since they somewhat enable HCI specialists and end users to take part in the decision-making process in FLOSS projects.

4   HCI Research Guidelines

In this paper the focus is on supporting user-developer communication, which has been particularly addressed by the field of HCI. Generally one can say that in HCI methods and textbooks the development of interactive systems has been separated into three main phases: 1) requirements construction, 2) producing design solutions, 3) evaluating the design solutions [11, 18, 44, 55]. User-developer communication is needed during all these phases. During the requirements construction phase, developers need to be in contact with users for the purpose of understanding the users, their needs and problems, and their context of use. Typically, this is achieved through face-to-face contact: users are interviewed or observed in their context of use, even though some methods also mention e.g., surveys as one possibility to inquire users and their needs [11, 18, 44, 55].

During the design phase, different kinds of design solutions are produced for human-computer interaction. Developers should initially carefully redesign users tasks or work practices, before considering software or user interface design [11, 18, 44, 55]. Part of the literature highlights the importance of user contact also during this phase: users may be invited as design partners to produce the HCI solutions together with developers. Typically also this is assumed to take place in a co-located setting, utilizing representations such as scenarios or storyboards to capture the design ideas [11, 18, 44, 55].

One should start the evaluation as early as possible. Typically this entails the use of low-tech prototypes and such, which can be produced as well as modified very fast and easily. On the other hand, it is also important to evaluate the finished or almost finished solution and to check whether it is ready for release. The evaluations typically include users as test participants. [44, 55]. Usability testing is the most widely known and used method. Typically it is again carried out in a co-located setting, in a usability laboratory or in a field setting, but also remote usability testing has been brought up in the literature [32]. Studies in the actual use context after the release are also recommended. This should be done in order to improve the next version. Methods such as interviews, observation, surveys or focus groups can be used, implying again user contact but not necessarily a face-to-face one. [44].

HCI literature recommends developers to cooperate with users, and also emphasizes that HCI experts should take part in the development and 'represent the users' in the development [17, 35, 36]. Typically it is recommended that HCI experts take the responsibility of user contact during these different phases: they observe and interview users during the requirements construction, they invite users to take part in the design process and they organize the usability evaluations involving users as test participants. However, HCI experts and users may find it difficult to have any impact on the solution being developed [35, 37]. They both might only be in an informative role, acting as providers of information or in a consultative role commenting on predefined design solutions, when developers alone proceed to make all the decisions [35]. HCI literature, nevertheless, suggests a more authoritative role for users and for HCI experts representing them - the literature maintains that users and HCI experts should be allowed to have a participative role, taking part in the design process and having decision making power regarding the solution [35].

5   Recommended Communication Means for FLOSS Projects

In this chapter our FLOSS related findings are connected to the HCI literature presented. We acknowledge all the different phases associated with interactive systems development, but we also emphasize that FLOSS development does not follow the waterfall model. These phases can be found from FLOSS development, but they do not appear sequentially, but are intertwined and overlapping. Scacchi argues that requirements construction in FLOSS development does not consist of the traditional steps following each other, but instead the requirements are asserted or implied in a multitude of textual descriptions in the FLOSS environment, the descriptions being discussed, negotiated and made sense of in a continuous, evolving manner. During the process the requirements are condensed, hardened and concentrated, the requirements construction being comingled with design, implementation and testing. Sometimes the requirements may actually be only implicitly produced as a by-product of implementation. [56]. We acknowledge that requirements construction as well as HCI design and HCI evaluation all take place in FLOSS development one way or the other, but within the boundaries of constraints and characterizations mentioned above. We call these tasks rather than phases due to issues mentioned above.

During the requirements construction task developers may ask for and collect information about users and their needs, while users (or HCI experts representing them) may provide information about these to the development. During the HCI design task, all these parties may produce design solutions (ranging from ideas or rough mockups to finished software) and wish to communicate those to others. During the evaluation task, furthermore, developers may ask for and collect feedback, while users (or HCI experts) may provide it to the development. The actual use can not be separated from FLOSS development but instead both continuously take place after the initial release of the software. During it, developers may provide user support while users may ask for it, this being already labeled as a user-centric strength of FLOSS development [40, 69].

Table 1 summarizes the results of our literature analysis on 1) What kinds of means for communication exist in FLOSS projects for user-developer communication? and 2) What kinds of means should there be? The results were derived by contrasting the existing FLOSS literature on user-developer communication with prescriptive HCI literature on user-developer communication. In FLOSS projects these tasks are overlapping and not even explicitly acknowledged.

Therefore, communication during the evaluation task is somewhat supported in FLOSS projects, but design and requirements construction tasks are clearly in need for further support, even though ideas have already been presented. However, we maintain that all these tasks need to be supported future and plan to experiment with certain solutions in the further. FLOSS development, as a special case of distributed software development, is very tool-centric. Networked ways to communicate, capture and manipulate information are essential to everyday work. At the same time the tools also affect the ways tasks are performed.

Furthermore, we argue that like in other development contexts, also in FLOSS development users and HCI experts should be able to enter the participative role, i.e. the role in which they are considered as equal partners in the design process with decision-making power regarding the solution [23, 35]. By relying on the model introduced by Hedberg and Iivari [33], Table 2 lists the recommended

Table 1. Summary of Communication Means Provided and Required

Tasks	Support provided	Areas for Improvement
Requirements construction (requirements, needs, problems asked for/provided)	Currently weak support. Feature requests may be provided in bug trackers, mailing lists, IRC or forums, but requirements are typically not constructed but asserted or only implied in the actual source code [56].	Tools supporting distributed requirements construction (gathering data on users, their needs, their contexts of use etc). 'Confusion reports' and 'surprise reports' [8], usability and user related discussion forums [12, 48, 61] suggested
Design (ideas, designs, implementations asked for/provided)	Currently weak support. Ideas may be provided in bug trackers, mailing lists, IRC or forums, but lack support for visualization or for collaboration	Tools supporting visualization and collaboration during distributed design work [30, 33]. E.g. user interface design by blogs [49, 60], open content projects [60], specific design areas supporting brainstorming [49, 60], a separate tab for usability/UX issues [5] suggested
Evaluation (feedback asked for/provided)	Currently some support. Bug trackers, mailing lists, IRC, forums etc. available for users to provide feedback, but may be difficult to use properly.	Tools supporting user and usability feedback gathering. E.g. moderators handling the user reported bugs [38], bug tracker guiding users in bug reporting [10], a tool for usability bug reporting [15, 49], reference/bleeding edge/core users providing feedback [60], open content projects [60], and remote usability testing [47–49] suggested

roles for human layer work in FLOSS projects, and connects these roles with the recommended communication tools (and practices), listing also numerous problems or needs they still have in connection to communication.

Even though there are communication means suggested for all the human layer roles, also problems and needs related to each role can be identified. For example, the UX-Tab for supporting the work of HCI core is explained in the article [5], but never implemented. Altogether, communication, coordination and decision

Table 2. Summary of communication means and problems connected to human layer roles

Role/Task	Tools (and practices)	Needs
HCI core	UX-tab [5]	An implementation of the UX-tab, a means to communicate with the technical core, ensuring decision-making power, tools supporting coordination related to UX and technical development
HCI designers	blogs [49, 60], brainstorming areas [49, 60]	Support for synchronous, collaborative, visual design work [33], tools supporting coordination related to UX and technical development
HCI evaluators	remote usability testing tools [47–49]	Tools supporting coordination related to UX and technical development
Technical users (additionally to non-technical user)	confusion reports [8], surprise reports [8], brainstorming areas [49, 60]	
Non-technical users	usability-forum [12, 48, 61], usability-tracker [15, 49], remote usability testing tools [47–49], open content projects [60], moderators [38]	More user-friendly tools [10]

making between HCI core and technical core need tool support and thus more investigating. Also there should be better support for coordination between the work of HCI designers and evaluators and technical development. The technical users in FLOSS projects are able to use the tools used by developers, but the non-technical users need more user-friendly solutions. By providing new tools for FLOSS development, the user-developer discussion could be facilitated, or even forced if it is built in a tool. However, it is not an easy task, because the FLOSS way of doing - the FLOSS philosophy - must be taken into account.

6   Conclusions

This paper examined the following questions. "What kinds of means for communication exist in FLOSS projects for user-developer communication? What kinds of means should there be?" The results indicate that user-developer communication during the evaluation task is somewhat supported in FLOSS projects, but design and requirements construction tasks are badly in need for further support. Especially there is a need to support working of the HCI core team, as well as to support the coordination between HCI designers and evaluators and technical development.

The tools used in FLOSS development are not the best for managing user/UX related issues. Thus it seems that there is still room for tool development. However, we emphasize in line with Wilson that practices are in a bigger role than tools in software development [66]. Therefore, not only tool support is needed in FLOSS projects, but also new practices for user-developer communication. The tools can however enhance and stimulate to use those practices. Therefore, we have presented the status quo of user-developer communication tools and deductions on how to make the situation better by enhancing the tool set of FLOSS development.

The FLOSS philosophy must be taken into account as well. The developers should accept these new tools. Fogel presents two important points on the planning and taking in to use of new tools. First, it should not take too much effort. Secondly, users must think that the additional effort is worth it [27]. If this effort is too much, the new tool (or a feature of it) is omitted. It is necessary that new tools are not aimed to replace FLOSS practices, but to complement and make them better. Additionally they should be as transparent as the existing tools [21]. It seems that FLOSS developers value simple but versatile tools. Simple textual emails, wiki pages and bug tracker discussions do not dazzle with glorious visual effects, but they simply enable functions that are really needed.

The limitation of FLOSS research is that researchers tend to focus on large and well-known projects and communities. Small projects may not have same issues, but they should be studied also. Common trend is to make quantitative studies utilizing a data mass provided by FLOSS development support sites. For small projects qualitative case studies carried out might be better. Another limitation connected to this literature review is the varying terminology used in the studies addressing user-developer communication. It was very difficult to settle the appropriate keywords, since the studies have varyingly used terms such as user, usability, HCI, UX, usability engineering, user-centered design to describe their research focus. Even non-FLOSS projects are studied under FLOSS topic.

In addition to tool development, the paths for future work include looking for existing FLOSS projects. It would be helpful to extract the best practices, conventions and tools that have led to successful user-developer communication in FLOSS development from real cases. Even though there are some exploratory studies performed and enhancement ideas raised, it seems that the results are not used in practice in FLOSS development yet.

References

1. Andreasen, M., Nielsen, H., Schrøder, S., Stage, J.: Usability in open source software development: opinions and practice. Information Technology and Control 35(3A), 303–312 (2006)

2. Ankolekar, A., Herbsleb, J., Sycara, K.: Addressing Challenges to Open Source Collaboration With the Semantic Web. In: The 3rd Workshop on Open Source Software Engineering, the 25th International Conference on Software Engineering, ICSE, Portland OR, USA (2003)

3. Anvik, J., Hiew, L., Murphy, G.: Who Should Fix This Bug? In: Proceedings of the 28th International Conference on Software Engineering, ICSE 2006, pp. 361–370 (2006)

4. Bach, P.: Supporting the user experience in free/libre/open source software development, Ph.D. dissertation, Pennsylvania State University (2009)

5. Bach, P., DeLine, R., Carroll, J.: Designers Wanted: Participation and the User Experience in Open Source Software Development, Boston, MA, April 4-9, pp. 985–994. ACM, USA (2009)

6. Bach, P., Kirschner, B., Carroll, J.: Usability and Free/Libre/Open Source Software SIG: HCI Expertise and Design Rationale. ACM, New York (2007)

7. Bach, P., Twidale, M.: Lucky Seven: How Can the Crowd Help Design?, Penn State College of IST. University Park, PA (2007)

8. Bach, P., Twidale, M.: Involving reflective users in design. In: Proceedings of the 28th International Conference on Human Factors in Computing Systems, CHI 2010, April 10 - 15, ACM, New York (2040)

9. Benson, C., Müller-Prove, M., Mzourek, J.: Professional usability in open source projects: GNOME, OpenOffice.org, NetBeans. In: Extended Abstracts of CHI 2004, pp. 1083–1084. ACM, New York (2004)

10. Bettenburg, N., Just, S., Schrter, A., Weiss, C., Premraj, R., Zimmermann, T.: What makes a good bug report? In: Proceedings of the 16th ACM SIGSOFT international Symposium on Foundations of Software Engineering. ACM, New York (2008)

11. Beyer, H., Holtzblatt, K.: Contextual Design: Defining Customer-Centered Systems. Morgan Kaufmann Publishers, San Francisco (1998)

12. Bødker, M., Nielsen, L., Orngreen, R.N.: Enabling User Centered Design Processes in Open Source Communities. In: Aykin, N. (ed.) HCII 2007. LNCS, vol. 4559, pp. 10–18. Springer, Heidelberg (2007)

13. Bollinger, T., Nelson, R., Self, K., Turnbull, S.: Open Source Methods: Peering through the Clutter. IEEE Software 16(4), 8–11 (1999)

14. Carmel, E., Agarwal, R.: Tactical approaches for alleviating distance in global software development. IEEE Software 18(2), 22–29 (2001)

15. Cetin, G., Verzulli, D., Frings, S.: An analysis of involvement of HCI experts in distributed software development: Practical issues. In: Schuler, D. (ed.) HCII 2007 and OCSC 2007. LNCS, vol. 4564, pp. 32–40. Springer, Heidelberg (2007)

16. Cheung, G., Chilana, P., Kane, S., Pellett, B.: Designing for discovery: opening the hood for open-source end user tinkering. In: Proceedings of the 27th international Conference Extended Abstracts on Human Factors in Computing Systems, CHI 2009, pp. 4321–4326. ACM, New York (2009)

17. Cooper, C., Bowers, J.: Representing the users: notes on the disciplinary rhetoric of human-computer interaction. In: Thomas, P.J. (ed.) The Social and Interactional Dimension of Human-Computer Interfaces, pp. 48–66. Cambridge University Press, Cambridge (1995)

18. Cooper, A., Reimann, R.: About Face 2.0: The essentials of interaction design. In: Information Visualization, vol. 3, pp. 223–225. New Wiley Pub., Indianapolis (2004)

19. Crowston, K., Annabi, H., Howison, J., Masango, C.: Effective work practices for floss development: A model and propositions. In: Proceedings of the 38th Hawaii International Conference On System Sciences, HICSS 2005, IEEE Press, Piscataway (2005)
20. Crowston, K., Howison, J.: The social structure of free and open source software development (2005),
    http://firstmonday.org/htbin/cgiwrap/bin/ojs/index.php/fm/article/viewArticle/1207/1127 (retrieved on September 1, 2009)
21. Čubranić, D., Booth, K.S.: Coordinating Open-Source Software Development. In: Proceedings of the 8th IEEE International Workshops Enabling Technologies: Infrastructure for Collaborative Enterprises, WET ICE 1999, pp. 61–65. IEEE CS Press, Los Alamitos (1999)
22. Curtis, B., Krasner, H., Iscoe, N.: A Field Study of the Software Design Process for Large Systems. Communications of the ACM 31(11), 1268–1287 (1988)
23. Damodaran, L.: User involvement in the systems design process - a practical guide for users. Behaviour & Information Technology 15(16), 363–377 (1996)
24. Fischer, G.: Communities of Interest: Learning through the Interaction of Multiple Knowledge Systems. In: 24th Annual Information Systems Research Seminar In Scandinavia, IRIS 24, Ulvik, pp. 1–14 (2001)
25. Fitzgerald, B., Ågerfalk, P.: The Mysteries of Open Source Software: Black and White and Red All Over? In: Proceedings of the 38th Annual Hawaii International Conference on System Sciences, HICSS 2005. IEEE Press, Piscataway (2005)
26. Fogel, K.: Producing Open Source Software. O'Reilly, Sebastopol (2005),
    http://producingoss.com/ (retrieved on December 15, 2009)
27. Fogel, K.: Beautiful Teams: Inspiring and Cautionary Tales from Veteran Team Leaders. ch. 21. O'Reilly, Sebastopol (2009),
    http://www.red-bean.com/kfogel/beautiful-teams/bt-chapter-21.html (retrieved on September 1, 2009)
28. Free Software Foundation, Inc., The Free Software Definition (2008),
    http://www.gnu.org/philosophy/free-sw.html (retrieved on: September 1, 2009)
29. Ge, X., Dong, Y., Huang, K.: Shared Knowledge Construction in an Open-Source Software Development Community: An Investigation of the Gallery Community. In: Proceedings of the International Conference on Learning Sciences, Bloomington, IN, June27-July 1, pp. 189–195 (2006)
30. Geisler, C., Rogers, E.: Technological Mediation for Design Collaboration. In: Proceedings of the IEEE Professional Communication Society International Professional Communication Conference and Proceedings of the 18th Annual ACM International Conference on Computer Documentation: Technology & Teamwork, IPCC/SIGDOC 2000. IEEE Educational Activities Department, Piscataway (2000)
31. Halloran, T.J., Scherlis, W.L.: High Quality and Open Source Practices. Presented at the 2nd Workshop on Open Source Software Engineering, Orlando, FL (2002)
32. Hartson, H.R., Castillo, J.C., Kelso, J., Neale, W.C.: Remote evaluation: the network as an extension of the usability laboratory. In: Proceedings of the SIGCHI Conference on Human Factors in Computing Systems: Common Ground, CHI 1996, pp. 228–235. ACM, New York (1996)

33. Hedberg, H., Iivari, N.: Integrating HCI Specialists into Open Source Software Development Projects. In: Boldyreff, C., Crowston, K., Lundell, B., Wasserman, A.I. (eds.) OSS 2009. IFIP AICT, vol. 299, pp. 251–263. Springer, Heidelberg (2009)
34. Howison, J.: Studying Free Software with Free Software and Free methods. In: A paper for the Australian Open Source Development Conference, Melbourne, Australia (December 1-3, 2004)
35. Iivari, N.: Representing the User' in software development - a cultural analysis of usability work in the product development context. Interact. Comput. 18(4), 635–664 (2006)
36. Iivari, N.: Constructing the users' in open source software development: An interpretive case study of user participation. Information Technology & People 22(2), 132–156 (2009)
37. Iivari, N., Molin-Juustila, T.: Listening to the Voices of the Users, in Product Based Software Development. International Journal of Technology and Human Interaction 5(3), 54–77 (2009)
38. Ko, A., Chilana, P.: How Power Users Help and Hinder Open Bug Reporting, In. In: Proceeding of the 28th International Conference on Human Factors in Computing Systems, CHI 2010. ACM, New York (2010)
39. Kittur, A., Kraut, R.E.: Harnessing the wisdom of crowds in wikipedia: quality through coordination. In: Proceedings of the 2008 ACM Conference on Computer Supported Cooperative Work, CSCW 2008, ACM, New York (2008)
40. Lakhani, K., von Hippel, E.: How open source software works: 'free' user-to-user assistance. Research Policy 32, 923–943 (2003)
41. Lievrouw, L.: Oppositional and activist new media: remediation, reconfiguration, participation. In: Proceedings of the Participatory Design Conference, pp. 115–124. CPSR, Palo Alto (2006)
42. Ljungberg, J.: Open Source Movements as a Model of Organising. In: Proceedings of 8th European Conference on Information Systems, Vienna, pp. 208–216 (2000)
43. Massey, B.: Where Do Open Source Requirements Come From (And What Should We Do About It)? In: Proceedings of the 2nd Workshop on Open Source Software Engineering, ICSE (2001)
44. Mayhew, D.: The usability engineering lifecycle: a practitioner's handbook for user interface design. Morgan Kaufmann Publishers Inc., San Francisco (1999)
45. Mockus, A., Fielding, R., Herbsleb, J.: A Case Study of Open Source Software Development: The Apache Server. In: Proceedings of the 22nd International Conference on Software Engineering, ICSE 2000. ACM, New York (2000)
46. McConnell, S.: Open source methodology: ready for prime time? IEEE Software 16(4), 6–8 (1999)
47. Nichols, D., McKay, D., Twidale, M.: Participatory Usability: supporting proactive users. In: Proceedings of the 4th Annual Conference of the ACM Special Interest Group on Computer Human Interaction, pp. 63–68. ACM, Dunedin (2003)
48. Nichols, D., Twidale, M.: The Usability of Open Source Software. First Monday 8(1) (2003)
49. Nichols, D., Twidale, M.: Usability processes in open source projects. Software Process Improvement and Practice 11, 149–162 (2006)
50. Ogawa, M., Ma, K., Bird, C., Devanbu, P., Gourley, A.: Visualizing Social Interaction in Open Source Software Projects. In: Proceedings of Asia-Pacific Symposium on Visualization, APVIS, pp. 25–32 (February 2007)

51. Open Source Initiative, The Open Source Definition (2006), http://www.opensource.org/docs/osd (retrieved on September 1, 2009)
52. Parkinson, C.N.: Parkinson's law: or, the pursuit of progress / C. Northcote Parkinson J. Murray, London (1957)
53. Paul, C.: A Survey of Usability Practices in Free/Libre/Open Source Software. In: Boldyreff, C., Crowston, K., Lundell, B., Wasserman, A.I. (eds.) OSS 2009. IFIP AICT, vol. 299, pp. 264–273. Springer, Heidelberg (2009)
54. Raymond, E.: The Cathedral and the Bazaar: Musings on Linux and Open Source by an Accidental Revolutionary. OReilly, US (1999)
55. Rosson, M., Carrol, J.: Usability Engineering: Scenario-Based Development of Human-Computer Interaction. Morgan Kaufmann, New York (2002)
56. Scacchi, W.: Understanding the requirements for developing open source software systems. IEE Proceedings - Software 149(1), 24–39 (2002)
57. Schwartz, D., Gunn, A.: Integrating user experience into free/libre open source software: CHI 2009 special interest group. In: Proceedings of the 27th International Conference Extended Abstracts on Human Factors in Computing Systems, CHI 2009, pp. 2739–2742. ACM, New York (2009)
58. Singh, V., Twidale, M.B., Nichols, D.M.: Users of Open Source Software - How Do They Get Help? In: Proceedings of the 42nd Hawaii International Conference on System Sciences, HICSS 2009, Big Island, Hawaii, January 5-8 (2009)
59. Stamelos, I., Angelis, L., Oikonomou, A., Bleris, G.: Code quality analysis in open source software development. Information Systems Journal 12, 43–60 (2002)
60. Terry, M., Kay, M., Lafreniere, B.: Perceptions and Practices of Usability in the Free/Open Source Software (FOSS) Community. In: Proceedings of the Conference on Human Factors in Computing Systems, pp. 999–1008. ACM, New York (2010)
61. Twidale, M., Nichols, D.: Exploring usability discussions in open source development. In: Proceedings of the 38th Hawaii International Conference on System Sciences, HICSS 2005. IEEE Press, Piscataway (2005)
62. Viorres, N., Xenofon, P., Stavrakis, M., Vlachogiannis, E., Koutsabasis, P., Darzentas, J.: Major HCI challenges for open source software adoption and development. In: Schuler, D. (ed.) HCII 2007 and OCSC 2007. LNCS, vol. 4564, pp. 455–464. Springer, Heidelberg (2007)
63. von Hippel, E., von Krogh, G.: Open source software and the "private-collective" innovation model, Issues for organization science. Organ. Sci. 14(2), 209–223 (2003)
64. Wiggins, A., Howison, J., Crowston, K.: Social dynamics of FLOSS team communication across channels. In: Proceedings of the IFIP 2.13 Working Conference on Open Source Software (OSS), Milan, Italy, pp. 131–142 (2008)
65. Williams, S.: Free as in Freedom: Richard Stallman and the Free: Richard Stallmans Crusade for Free Software. O'Reilly Media, Sebastopol (2002); ISBN: 0596002874
66. Wilson, G.: Is the Open-Source Community Setting a Bad Example? IEEE Software 16(1), 23–25 (1999)
67. Yamauchi, Y., Yokozawa, M., Shinohara, T., Ishida, T.: Collaboration with Lean Media: how open-source software succeeds. In: Proceedings of the 2000 ACM Conference on Computer Supported Cooperative Work, CSCW 2000, pp. 329–338. ACM, New York (2000)
68. Yatani, K., Chung, E., Jensen, C., Truong, K.N.: Understanding how and why open source contributors use diagrams in the development of Ubuntu. In: Proceedings of the 27th International Conference on Human Factors in Computing Systems, CHI 2009, pp. 995–1004. ACM, New York (2009)

69. Ye, Y., Kishida, K.: Toward an understanding of the motivation of open source software developers. In: Proceedings of the 25th International Conference on Software Engineering (ICSE), pp. 419–422. IEEE Press, Piscataway (2003)
70. Zhao, L., Deek, F.: Improving open source software usability. In: Proceedings of the 11th Americas Conference on Information Systems, AMCIS, Omaha, NE, August 11-14, pp. 923–928 (2005)

Knowledge Homogeneity and Specialization in the Apache HTTP Server Project

Alexander C. MacLean, Landon J. Pratt,
Charles D. Knutson, and Eric K. Ringger

Computer Science Department, Brigham Young University, Provo, Utah
{amaclean,landonjpratt}@byu.edu, {knutson,ringger}@cs.byu.edu

Abstract. We present an analysis of developer communication in the Apache HTTP Server project. Using topic modeling techniques we expose latent conceptual sub-communities arising from developer specialization within the greater developer population. However, we found that among the major contributors to the project, very little specialization exists. We present theories to explain this phenomenon, and suggest further research.

1 Introduction

Private information is "information possessed by a relatively small segment of the population" [8]. For example, in development organizations certain individuals specialize and become *de facto* leaders within conceptual domains. Latent roles develop and define structure and vulnerability within an organization. In practice, these individuals become centers of private information sub-communities and thereby control the shape and flow of information within that sub-community. The identities of these individuals are latent in that there is often no obvious correlation between the overt organizational structure and the centers of knowledge within that structure.

We should be clear that private information is neither inherently good nor bad. Instead, its influence is context specific. Private information refers to information known to an individual or small set of individuals, not necessarily information that is hoarded or deliberately withheld. Rather, private information exists as a natural byproduct of organizational learning. A deeper discussion is presented by Krein, et al [8].

1.1 Specialization

Developer specialization is a form of private information in which developers within an organization become expert in a particular concept domain. Example domains might include UI developers, "the database guy," or any small group of developers who are essential to interacting with a particular piece of the product.

Specialization is often found in large organizations where the scope of the project is such that no single person has the time or capacity to master all

S.A. Hissam et al. (Eds.): OSS 2011, IFIP AICT 365, pp. 106–122, 2011.

aspects of a product. Organizations benefit from specialization by minimizing the overlap of skill acquisition since it can be time consuming and expensive. Individuals benefit from specialization through job security and decreased initial training requirements.

Although beneficial when viewed from a schedule and budget perspective, specialization introduces risk into an organization. For example, in an organization with two core developers, is it appropriate to put both of them on the same plane for a business trip? Or, what contingency should the organization put into place in the event that these two developers decide to retire? While development organizations could certainly recover from the untimely departure of specialized developers, and other developers exist who are capable enough to fill the roles, could the overall organization recover from the delays imposed by the resulting loss in productivity and stay competitive in an aggressive marketplace?

Developing a contingency plan for unexpected change is relatively straightforward when the specialized roles are overtly expressed in the organizational structure or are generally understood among the developers. However, danger arises when the roles are latent and therefore difficult to identify.

1.2 The Apache HTTP Server Project

The Apache HTTP Server (httpd) is an open-source HTTP server implementation and the leading HTTP server with 59% market share as of January 2011 [3]. The project is maintained by a group of volunteers from around the world who collaborate through the use of online tools such as email and chat [2]. Most importantly for this analysis, project tenets stipulate that all development and managerial communication must happen in the publicly available mailing lists.

We are interested in this project because of its open-development philosophy (anyone who proves their worth within the meritocracy can contribute) as well as its success. We would like to use open source projects such as those maintained by the Apache Foundation as an analogue for all software development organizations. Whereas it is difficult and rife with legal issues to obtain source code and developer communications from closed source enterprises, it is comparatively trivial to obtain the same information from the Apache Foundation. If we can show that open source projects behave like their closed source counterparts we can circumvent this roadblock.

1.3 Reference Organizations

The authors are personally familiar with three large software development organizations that each exhibit evidence of private information. Each organization is unique with regards to the degree of geographic distribution and size, yet all contain easily identifiable private information. These organizations are presented as references against which we may compare the Apache HTTP Server project. For consistency we refer to the organizations by number, even though we also identify Organization 2 by name.

For each of these organizations we note the degree of geographic distribution in an effort to illustrate that private information exists whether or not developers are collocated. Since the httpd project is massively distributed, it is important to show that private information is not necessarily caused by lack of face-to-face communication.

Organization 1. Organization 1 builds and maintains a large suite of software products comprised of both legacy and new components. The legacy engine is highly complex and is central to the success of the suite. Although a large organization develops the products and builds new features, only two developers may change the legacy engine. Both of the developers have been with the company for over twenty years. Most products in this organization target either consumer grade computers or cloud-based computing systems.

This organization is somewhat distributed (and is growing more distributed over time). However, a large portion of the developers are still collocated.

Organization 2. Organization 2, IBM, is large (426,751 employees, with gross revenues of $99 billion in 2010 [1]) and builds many disparate products. Despite well-established corporate practices, Krein, et al, still found that "loss of specialized information arises from both reorganization and loss of employees" and that "employee loss. . . creates information gaps" [8]. Their study specifically targeted extended stakeholders[1], and therefore is not a direct analogue to a development organization. However, it illustrates some of the organizational fragility that arises from private information.

IBM is largely distributed, and has employees in more than 40 countries [1]. Many IBM employees telecommute, further increasing the distributed nature of the organization.

Organization 3. Organization 3 is large but still smaller than both organizations 1 and 2. Developers often specialize and become centers of conceptual communities surrounding hardware interfaces, user interfaces, network communication, fault tolerance, and other product specific topics. Unlike Organization 1, this organization primarily develops software that is shipped in embedded devices.

This organization is lightly distributed across half a dozen different locations. However, members of individual product groups are all collocated.

1.4   Private Information

Krein, et al, presented *the problem of private information*, a framework for understanding communication dynamics within a distributed organization. They studied extended stakeholders in an effort to understand how specialized information flows between stakeholders [8]. While inspired by Hayek's work in

---

[1] Non-developers.

economics [6], Krein's work represents the first application of the notion of private information to software organizations. Since Krein, et al, is an introductory study, few methods exist to aid us in practice. Still, the concept of private information is a powerful guiding metaphor and provides a framework within which to study knowledge homogeneity and specialization.

1.5   Goal

In order to discover pockets of private information within the Apache HTTP Server community, and thereby identify developer specialization, we analyzed committer email records and commit history. The goal was to identify two phenomenon: 1) committers who write exclusively about a particular topic, or 2) topics that are only discussed by a small set of committers. The first case indicates committers who specialize in a particular topic. The second case indicates topics that are dominated by particular committers.

2   Data

We gathered data for this study from the Apache Foundation during February of 2011. Our data set consists of the commit history and email archives for the Apache HTTP Server Project, spanning sixteen years (2/27/1995 - 1/31/2011).

2.1   Mailing Lists

The mailing list archives for the Apache Software Foundation are freely available. They consist of files stored in the mbox format that contain all of the communication on a given channel for a particular month. For this paper we only analyze the "dev" channel of the mailing lists for the httpd project and refer to it as "the mailing list." This channel should represent communication regarding project development. It consists of 124,938 messages and 166 developers[2] (see Section 3.1). The mailing lists were imported into PostgreSQL.

Developer contribution to the mailing list is not uniform. Instead, a small subset of developers generate most of the traffic (see Figures 1 and 2). Figure 1 shows the number of messages sent per developer for the entire time period. Note the power law distribution of developer email traffic. Figure 2 shows the same activity in 2009, illustrating that the activity patterns are consistent regardless of time window.

2.2   Subversion Repository

The commit history for the project consists of 46,336 revisions by 134 developers. However, as with mailing list activity, a small group of developers committed a majority of the changes to the project (see Figure 3). This behavior has been demonstrated in previous studies [7].

---

[2] Not all registered committers actually made changes to the project.

Fig. 1. Messages per developer. Each bar represents a single developer, sorted by number of messages.

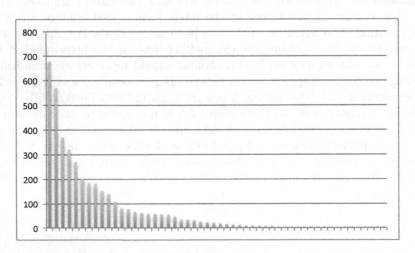

Fig. 2. Messages per developer in 2009. Each bar represents a single developer, sorted by number of messages.

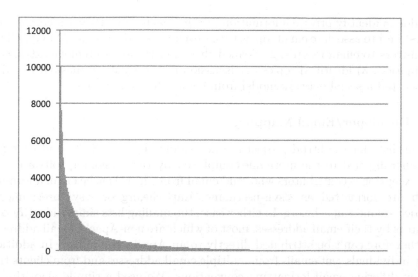

Fig. 3. Developer commits. The x-axis is individual developers, sorted by commit volume.

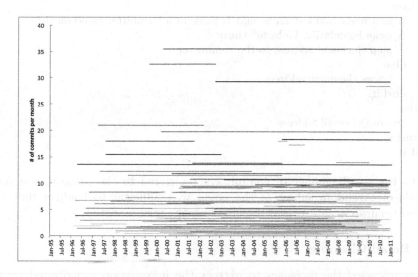

Fig. 4. Developer tenure. Each horizontal line represents a single developer. The length of the line represents the span of a developer's commits, from the first to the last.

3   Methods

In order to identify private information in the Apache HTTP Server community, we first had to associate email topics with developers. To do so, we 1) mapped the committers to email records, 2) cleaned the email records to remove extraneous information, 3) identified topics of discussion in the resulting messages, and 4) constructed a social network model from committers and topics.

3.1   Developer/Email Mapping

Mailing lists for the httpd project are not associated with SVN, and therefore are generally tied to the usernames employed by the versioning software. The only exception is contributors who send email from their Apache email addresses, which are formatted as <svn-username>@apache.org or <svn-username>@ <project-name>.apache.org. In either case, the mailing lists simply identify contributors by their email addresses, most of which are non-Apache email addresses and therefore can't be attributed directly to an Apache committer. In addition, many individuals sent emails from multiple email addresses and from clients that employ differing email formatting conventions. We used a simple algorithm to map email addresses to committers:

> **if** email address ends with *apache.org* **then**
>   use all text before the "@" symbol as the committer username
> **else**
>   **if** from header contains a name (e.g., "John Lawrence Smith" <jls@nowhere.com>)
>   **then**
>   **if** first name and last name match those of a committer listed on the
>   Apache Foundation Website[3] **then**
>   map the email address to the committer
>   **else**
>   ignore the email address
>   **end if**
>   **else**
>   ignore the email address
>   **end if**
> **end if**

Due to the small number of committers registered to the project, a manual analysis of a random subset of the email was sufficient to confirm that this method accurately categorized the email on the mailing list.

3.2   Email Cleaning

We preprocessed the messages to extract the information contributed by the sender and remove information contributed by others, thereby improving the

---

[3] http://people.apache.org/committer-index.html

accuracy of the identified topics. The email messages are in MIME format. Preprocessing consisted of seven steps:

1. Remove all sections of the multipart emails that were not "text/plain."
2. Remove all header lines.
3. Remove lines that were most likely quoted text from a previous email. We identified four patterns that denoted quoted text (see Table 1).
4. Remove signatures by deleting all contiguous lines following --\n until encountering an empty line.
5. Remove all instances of the author's name (see Section 3.3).
6. Remove a list of stopwords[4].
7. Tokenize the messages such that words consist of groups of letters and underscores.

Table 1. Quoted text identification

Identification	Action
Lines that started with ">"	Line removed

Sections that started with
 	
`-------- forwarded message`
 	
or
 	
`---------- forwarded message`
	Following text removed

Lines like `<so-and-so> (wrote|writes):`	Line removed
Lines like `On <date> <so-and-so> (wrote|writes):`	Line removed

Wang and McCallum [10] used a similar cleaning scheme but indicated that doing so may remove inline responses. However, we found that in this dataset inline responses were generally made to lines of text that started with ">," and therefore only the quoted text was removed. Of a random sample of 100 emails, this process did not remove any lines of text that were not quotes, and only failed to remove all of the quoted text in three instances.

3.3   Topic Identification

In order to identify latent roles and groups in the httpd community that may suggest private information, we extracted topics from messages sent by developers on the mailing list. These topics were identified in an unsupervised/unbiased

---

[4] Removing stopwords improves the accuracy of LDA by removing commonly occuring terms that are not topic specific.

> **regex match pattern regexp string cmd regex_t expression preg regular register pcre posix compiled regexec regcomp library pmatch rm_so ap_pregcomp**

Fig. 5. The top twenty terms in the regular expression topic. Term importance decreases from left to right, where "regex" is the most important term.

fashion using Latent Dirichlet Allocation (LDA), a Bayesian probabilistic topic model that clusters words from the corpus into topics [4]. "Topics" are probability distributions over the vocabulary words (terms) in the corpus, based upon word collocation within messages. In the process of topic identification, words in the emails are attributed to topics and a mixture of topics is deduced for each message in a manner that depends upon collocation of words within messages. We utilized the MALLET [9] implementation of LDA to identify 200 topics from the email corpus and assign message topic probabilities. Figure 5 shows the top twenty terms in the regular expression topic.

Our initial tests using LDA to identify topics from the email messages resulted in topics that contained author names. We removed the author's name from each email to avoid identifying these name-heavy topics. Although grouping author names with terms provided interesting insight into developer prominence, including the names diluted the topics and decreased the likelihood that less prolific developers would be associated with a given topic.

3.4   Social Network Analysis

From the topics assigned to email messages by LDA we created a two-mode social network consisting of topic nodes and committer nodes. Connections only exist between topic nodes and committer nodes: no inter-topic or inter-committer edges exist. Edge weight between a topic committer and a topic is determined by the weighted proportion of the messages attributed to the committer that refer to the topic, where weighted proportion is

$$\sum_{i=1}^{M_c} p_t(m_i) \tag{1}$$

M_c: All messages attributed to committer c,
i: Message index,
m_i: Message i from M_c,
$p_t(x)$: Proportion of message x attributed to topic t.

This calculation allows committers whose communications on the mailing list are voluminous to dwarf those who contribute less frequently. This seems appropriate in light of the intent of the metric to identify those who have or control

Fig. 6. Proportion of communication for which each topic is the dominant topic in a message

knowledge about individual topics. However, it is worth noting that the metric does not capture subtleties, such as committers who specialize in a particular subject, but whose contributions to the project are relatively minor.

4   Results

If specialization exists within the httpd community, we should see distinct communities develop around topics. In addition, unique groups of developers should congregate around specialized subtopics. We examined the data from both angles: topical affinity and topic communities.

4.1   LDA Results

We used LDA to identify 200 topics that summarize the communication on the mailing list. The most prevalent topic described 2% of the text contained in the messages. Figure 6 shows a stacked bar chart of topic proportions over the entire corpus, where each segment of the stacked bar represents a single topic. The median topic described 0.3% of the text contained in the messages. Notably no topics dominated the discussion.

Despite stripping developer names from the emails, a few topics exist that are centered around prominent committers to the project. These topics likely result from other members of the community referring to these more prominent members. We ignored developer-centric topics in our analysis. Although they may indicate private information about a developer, they do not indicate private information about the project.

4.2   Topical Affinity

One measure of developer communication is topical affinity—how much a developer discusses a particular topic. In Figure 7 we see a network that contains the top 22 developers by topical affinity. Note how clusters of topics gather around certain developers. Note also that a small number of developers discuss these topics in such volume that most of the developers are filtered out of this network. This threshold was discovered visually. There is a clear threshold at which the network stabilizes into a small subset of prolific authors.

Regardless of the edge weight threshold, the network looks the same, albeit more or less connected. This trend directly contradicts experience in the three reference organizations (see Section 1.3). If private information exists we would expect to see distinct groups appear within the network as the lower threshold for edge inclusion increased. Instead, we see a core group of developers to whom all topics are connected more heavily than to any specialized group.

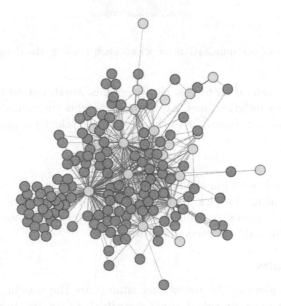

Fig. 7. The top 22 developers on the mailing list by topical affinity. Dark nodes are topics, light nodes are developers. Note that there are strong clusterings of topics around certain developers.

4.3   Topic Communities

We examined the communities that form around major topics, such as voting, SSL, licenses, security, patching, module development, CGI, configuration, regular expressions, and error handling. In each case, the resulting community was comprised almost exclusively of the core developers. Figure 8 is indicative of the groups that form around topics. In the network, all of the top ten developers are attached to the topic. This pattern repeats itself for all non-developer specific topics. In several cases concepts were split among multiple topics. However, the communities that congregated around the combined topics were also composed of the core developers.

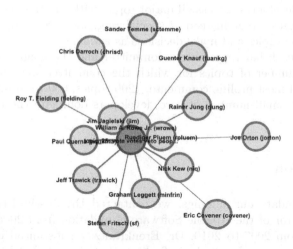

Fig. 8. Developers who discuss the voting topic. Proximity indicates volume of communication.

4.4   Specialization

Because there are no prominent groups in our network of topics and developers, we must conclude that there is little specialization among the core group of developers in the Apache HTTP Server project. This result is especially surprising in light of our reference organizations which exhibit clear specialization. Instead,

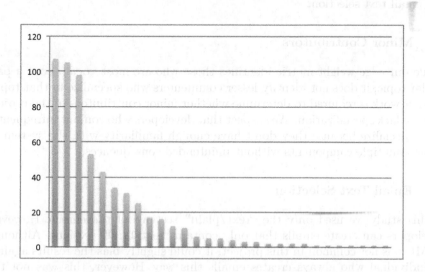

Fig. 9. Each bar represents the number of times an individual was the first, second, or third most prolific communicator in a topic in 2010. There is one bar per developer.

all of the core developers discuss all major topics with varying degrees of prominence. This suggests that instead of specializing, the core group of developers discuss all of the topics and make decisions as a whole.

In Figure 9 each bar represents a committer and the magnitude of the bar indicates the number of topics for which the committer was either the first, second, or third most prolific commenter (200 topics, 600 total first, second, or third ranks). A small number of core developers dominate the discussions on each topic.

5   Validation

In order to validate our findings, we contacted Dr. Justin Erenkrantz who served as Director of the Apache Software Foundation from 2005 to 2010 and as President from 2007 to 2010. Dr. Erenkrantz, a substantial contributor to the httpd project, corroborated our findings that suggest that little specialization exists among the core developers. This validated our interpretation of the distribution of topic usage in the project. In addition, his corroboration was in stark contrast to the results obtained in the study by Krein, et al [8], where the stakeholders were well aware of specialization and private information within the organization.

6   Threats to Validity

Here we mention two threats to the validity of our study: minor contributors and email text selection.

6.1   Minor Contributors

While our edge weight metric identifies those who are most vocal about a particular topic, it does not identify lesser commenters who specialize in that topic. Further work is required to determine whether minor contributors to the project likewise lack specialization. We suspect that developers who commit infrequently must specialize because they don't have enough familiarity with the system to change multiple components without unintended consequences.

6.2   Email Text Selection

In this study we used only the "text/plain" sections of the emails. However, developers can create emails that only contain "text/html" sections. Although HTML was not common in this project, it could slightly bias the results against an individual who always creates emails this way. However, this was not the case with the most prolific contributors, and therefore should not bias the study significantly.

7 Future Work

This analysis has unearthed a multitude of additional questions, some of which we list here.

7.1 Hidden Private Information

In this study we used topics gleaned from email messages to determine that there was little specialization in the community and corroborated our results through member checking. However, our results were based upon communication and members' impressions. To more fully explain specialization in the httpd project we must look at the commit patterns of developers. If the commit patterns indicate that the core developers work on everything, then our results are confirmed. If they indicate the opposite, then there is an assumption by the developers of a lack of specialization when in fact they do specialize. This result would have fascinating social implications.

7.2 Developer Tenure

During the dot com boom, conventional wisdom maintained that the average developer tenure in a job was 18 months. A study by Fallick, Fleischman, and Rebitzer [5] seems to indicate that average tenure in Silicon Valley has grown to around 4.2 years. However, their work revolved around developer mobility, not tenure, and extrapolating 4.2 years is statistically unsound[5]. Nevertheless, the numbers used in their study still cast heavy doubt on the previous figure of 18 months.

In contrast, the median tenure of developers on the Apache HTTP Server project is 3.7 years. Is this tenure significantly different from the median tenure of developers in traditional, closed source, closed development organizations? Also, when a developer is employed by a company, the company can expect the developer to spend a certain amount of time each week on assigned projects. Is the effort exerted by volunteer[6] developers analogous to that of an employee?

7.3 Topic Analysis in a Mid-Sized Organization

Only 134 developers have ever committed to the Apache HTTP Server project. Of those 134 developers, the top 30 are responsible for 75% of the commits. Comparing the httpd project to more traditional, closed source organizations of comparable size would provide insight into whether or not the behavior in this FOSS community differs significantly. Ideally this study would involve both collocated development organizations and distributed organizations in order to identify any effect that geographic distribution may have.

---

[5] Fallick, Fleischman, and Rebitzer did not make this extrapolation in their paper.

[6] We realize that most of the prominent developers in an open source projects are employed by external organizations and paid to contribute to the project. However, the question is still legitimate.

7.4   Cross Project Comparison in Apache

If knowledge homogeneity is a symptom of open development, we would expect to see the same kind of knowledge distribution in other projects in the Apache Foundation. All projects within the foundation are governed by a similar set of rules, and all are maintained by volunteer developers. Future work should explore the degree to which specialization appears within other projects.

7.5   Knowledge and Email Communication

In this paper we have been primarily concerned with topic specialization in email messages in the Apache HTTP Server project. However, further work is required to determine the degree to which analysis of topics in email messages identifies knowledge distribution in an open source organization.

8   Conclusions

Lack of specialization within this community may result from a myriad of reasons. Here we explore several factors that may perpetuate this homogeneous knowledge base.

8.1   Organizational Resiliency

Development of the Apache HTTP Server is performed by volunteers, most of whom are paid by external organizations to develop the product. Incentives for the external organizations vary, but one common element holds: developers are potentially transient. At any point a sponsoring entity may decide that a developer is required for an internal project either temporarily or permanently. Although the developer may continue to contribute on personal time, it is unlikely that he or she will be able to contribute anywhere near previous levels. If that developer holds unique specialized knowledge, the knowledge is essentially lost.

By spreading the knowledge across the core members of the organization, this organizational threat is mitigated. We doubt that the organization consciously promulgated a policy of knowledge homogeneity, but rather reacted to organizational fluidity.

8.2   Small Group of Core Developers

A second, related factor may be the relatively small size of the core group of developers. The question becomes: on what are the developers working? If the core group spends much of their time working on bug fixes, then the management of bug priority could homogenize the developer communication. If the developers work together on new features, feature timeline could lead to a homogeneous workgroup—all developers work on the same feature to get it out the door for the next release.

8.3   Small Project Size

A third, trivial, reason, may be that the project is simply too small to require developer specialization. It may be the case that the conceptual domain of the project is small enough that a single developer can, and does, comprehend that entire space. If so, then there is no need for specialization in a project of this size.

However, this reason seems highly unlikely. The conceptual domain contains complicated subdomains such as SSL, user authentication, OS specific tailoring, and transfer protocols. We don't believe that the required depth of knowledge in these subdomains is sufficiently shallow that developers can operate with a broad overview of knowledge.

8.4   Voting

Lastly, organizational culture within the httpd project could *require* homogeneity. Releases and major commits to the project must be approved by a vote of the Project Management Committe (PMC), a group of the more seasoned developers. Informed votes require that members of the PMC are familiar with all aspects of the project and ensure that they are kept abreast of developments.

8.5   Software Engineering Taxonomy

Regardless of the reasons behind the observed knowledge homogeneity, the implications are fascinating and motivate further exploration. By better understanding the similarities and differences between various types of development organizations—be they open or closed source, open or closed development, geographically distributed or collocated, large or small, embedded, desktop, or cloud—we continue to develop a working taxonomy of the software engineering landscape. This taxonomy informs our analysis of future projects and enables better comparison between domains.

Acknowledgements. The authors would like to thank Dr. Justin Erenkrantz for his willingness to provide validation and insight regarding the Apache HTTP Server community.

References

1. 2010 form 10-k, international business machines corporation. United States Securities and Exchange Commission
2. Apache http server project (April 2011)
3. January 2011 web server survey (January 2011)
4. Blei, D.M., Ng, A.Y., Jordan, M.I.: Latent dirichlet allocation. The Journal of Machine Learning Research 3, 993–1022 (2003)
5. Fallick, B., Fleischman, C.A., Rebitzer, J.B.: Job-hopping in Silicon Valley: Some evidence concerning the microfoundations of a high-technology cluster. The Review of Economics and Statistics 88(3), 472–481 (2006)

6. Hayek, F.A.: The use of knowledge in society. The American Economic Review 35(4), 519–530 (1945)
7. Krein, J.L., MacLean, A.C., Delorey, D.P., Knutson, C.D., Eggett, D.L.: Impact of programming language fragmentation on developer productivity: a sourceforge empricial study. International Journal of Open Source Software and Processes (IJOSSP) 2, 41–61 (2010)
8. Krein, J.L., Wagstrom, P., Sutton Jr., S.M., Williams, C., Knutson, C.D.: The problem of private information in large software organizations. In: International Conference on Software and Systems Process. ACM Press, New York (2011)
9. McCallum, A.K.: Mallet: A machine learning for language toolkit (2002), http://mallet.cs.umass.edu
10. Wang, X., McCallum, A.: Topics over time: a non-Markov continuous-time model of topical trends. In: Proceedings of the 12th ACM SIGKDD International Conference on Knowledge Discovery and Data Mining, pp. 424–433. ACM, New York (2006)

Building Knowledge in Open Source Software Research in Six Years of Conferences

Fabio Mulazzani, Bruno Rossi, Barbara Russo, and Maximilian Steff

Center for Applied Software Engineering (CASE),
Free University of Bozen-Bolzano,
Piazza Domenicani, 3, 39100 Bolzano, Italy
{fmulazzani,brrossi,brusso,maximilian.steff}@unibz.it

Abstract. Since its origins, the diffusion of the OSS phenomenon and the information about it has been entrusted to the Internet and its virtual communities of developers. This public mass of data has attracted the interest of researchers and practitioners aiming at formalizing it into a body of knowledge. To this aim, in 2005, a new series of conferences on OSS started to collect and convey OSS knowledge to the research and industrial community. Our work mines articles of the OSS conference series to understand the process of knowledge grounding and the community surrounding it. As such, we propose a semi-automated approach for a systematic mapping study on these articles. We automatically build a map of cross-citations among all the papers of the conferences and then we manually inspect the resulting clusters to identify knowledge building blocks and their mutual relationships. We found that industry-related, quality assurance, and empirical studies often originate or maintain new streams of research.

Keywords: Systematic Mapping Study, Cross-citations.

1 Introduction

Since its origins, the diffusion of the OSS phenomenon and the information about it has been entrusted to the Internet and its virtual communities of developers. As such, information on OSS has grown exponentially resulting in a vast quantity of data readily available. This data has attracted the interest of researchers and practitioners aiming at formalizing this information into a body of knowledge (e.g. [12], [13]). For this reason, in 2005, a new series of conferences on OSS (OSS conference series[1]) and, in 2009, a new journal[2] have been established. These initiatives have substantially contributed to initiate a long process to ground knowledge in OSS that masters data from different sources and consolidates them into well-accepted concepts and their mutual relations. As such, they represent a valuable source for understanding how OSS knowledge has been created and how it will evolve in the future. Our work tackles this issue proposing a systematic mapping study of the papers of the OSS conference series.

---

[1] International Symposium on Open Source Software initiated in 2005, in Genoa, Italy.
[2] International Journal of Open Source Software and Processes, http://www.igi-global.com

S.A. Hissam et al. (Eds.): OSS 2011, IFIP AICT 365, pp. 123–141, 2011.

Systematic Mapping Studies (SMS) and Systematic Literature Reviews (SRL) are techniques of knowledge synthesis. These techniques are typically based on manual inspections of articles ([23], [24], [25], and [37]). Manual inspection requires significant effort in mining large sets of articles. On the other hand, a complete automated inspection can produce inaccurate results. In our work, we propose a semi-automated approach to mine articles' repositories for systematic mapping studies. We automatically inspect articles to build a map of cross-citations and then we manually inspect the resulting clusters to identify the building blocks of knowledge in OSS and the social network of the community maintaining it. We selected the complete database of articles of the OSS conference series since its origin in 2005[3]. Our choice is driven by three criteria: 1) papers are peer reviewed - this excludes for example the MIT repository[4], Apache conferences[5], OpenOffice.org conferences[6], etc...; 2) the series' mission is to disseminate knowledge in OSS - this excludes traditional journals in software engineering; and 3) papers report more than a group discussion - this excludes workshops or one day events co-located with larger non OSS events. The result of this study aims at answering two major questions:

RQ1. Is there any social network underlying the research production at the OSS conference series?

RQ2. What are the major streams of research proposed at the OSS conference series?

Our answer to RQ1 will identify the cornerstone papers and the links among them across the years. Links will express the relation among the authors by means of the connection of their research production. The analysis will reveal unexpected and undeclared connections among authors as well as lack of connections among conceptually related papers. This will also illustrate the self-sustainability of the OSS conference series and the value that it provides to the OSS community. In addition, using the results in [37], our research will also discuss how empirical studies fit the network. An answer to RQ2 will help build the baseline for future investigations in OSS research or to extend existing ones.

In the following section, we introduce related work and motivate our work. Section 3 presents our method of SMS and Section 4 explains our analysis methodology. In Section 5, we explore the results of our analysis and describe the clusters of papers we identified, followed by a summary of our findings in Section 6. We close with the conclusions and limitations.

2   Background and Motivation

The software engineering community has been increasingly adopting Evidence-Based Software Engineering (EBSE) approaches to build discipline-specific bodies of knowledge such as Inspection, Testing, and Requirements Engineering ([7], [39]). Apart from the traditional ways of doing literature review, also called ad-hoc reviews,

---

[3] https://pro.unibz.it/staff/brusso/PapersUsed.html
[4] http://opensource.mit.edu/
[5] http://na11.apachecon.com/
[6] http://www.ooocon.org

the EBSE practice uses Systematic Literature Reviews (SLRs) ([6], [24], [25], and [26]) and Systematic Mapping Studies (SMSs) ([7], [23], and [29]) as robust methodologies of searching, selecting, analyzing, and synthesizing literature and aggregate evidence on a specific topic. As such, SLRs and SMSs are called secondary studies as they aggregate research of other, so-called primary studies. SLR is "a means of evaluating and interpreting all available research relevant to a particular research question, topic area or phenomenon of interest" ([11], [26]). Research in SE has provided guidelines and lessons learned for performing SLR ([5], [6], [10], [25], and [34]). SMS is used to draw a landscape of reported research on a particular topic ([14], [23], [25], and [29]). Being less specific, SMS requires significantly less effort than SLR; however, it provides only a course-grained overview of the published literature. An SMS can also serve as a preparation activity before doing an SLR. SLR and SMS in OSS has been typically used to provide evidence of practices and methods for a more general SE research purpose Only recently, secondary studies have been published to investigate specific areas of OSS. In 2010, Hauge et al. performed an SLR of research on OSS adoption [16]. In 2009, Stol et al. ([36], [37]) presented an SLR on empirical papers published at the OSS conference series. To our knowledge, the first secondary study that investigated OSS as a holistic phenomenon is the work in [2]. In their work, the authors have published taxonomy of OSS mining 623 journal papers - excluding conference papers, though. In this context, our work provides an SMS on OSS as a holistic phenomenon. Our work is complementary to the work in [2] in terms of papers investigated, method of analysis, and research goal.

3 Research Method

Our method follows the concepts of a systematic mapping study [29]. In the introduction, we illustrate our search criteria for inclusion and exclusion. Following them, we select all the papers of the OSS conference series. In this section, we describe how we identify and apply classification criteria on the selected papers. Classification categories are taken from the Calls for Papers of the OSS conference series and are used to label cross-citation clusters. In particular, we automate papers classification to reduce effort of articles' inspection and enable future replications. Following [3], to increase the transparency of our method, we also detail the tools we used and the process we follow.

Worth noticing that this approach differs from the one proposed by Kitchenham in 2010 ([23]). Kitchenham uses citations to identify most and least cited papers. We propose here to extend this approach using citations to determine streams of research.

3.1 Creating the Directed Graph of Cross-Citations

We collected the PDF files of papers from the Springer repository. We developed an application (PDF Analyzer[7]) that (a) converts the papers from the PDF format to a textual representation, and then (b) injects this representation into an XML file with nodes corresponding to the paper's sections. The application allows user intervention during the conversion process. We have also sampled part of the XML files to verify

---

[7] Apache PDFBOX library to convert PDF to TXT and DOM4J library to create XML the file.

and validate the tool output. When problems occurred, we manually corrected them with the aid of the original PDF file.

We create a Python script that 1) extracts all papers' titles and conference years from the XML files, 2) parses all the references for all the paper titles, 3) for every hit, extracts conference names, and 4) goes over the references again to identify possibly missing titles using the different variations of conferences' names.

We noticed that conferences' titles significantly vary in that we identified 17 different variations. At the end, we also manually further checked for missed citations. The final output of the Python tool classifies papers by year of conference and by citations and passes them to GraphViz[8] to display the final graph (Fig. 11).

3.2 Descriptive Analysis of Cross-Citations

Before performing any inspection of the clusters, we have analyzed the citations of the papers we found. Table 1 shows the distribution of citations over the years. Articles refer to full and short papers if any.

Table 1. Number of articles cited by or citing another article of the OSS conference series

	2005	2006	2007	2008	2009	2010
Cited	25	15	16	15	8	-
Citing	-	12	13	17	11	13
Total	83	41	51	42	28	40
Isolated	58	23	33	21	16	27
% citing/cited	31%	44%	35%	50%	43%	32%
# articles cited >2	11	3	4	3	0	0

In particular, Table 1 shows that there are a good number of *isolated articles* in that they do not cite other papers.

3.3 Inspecting the Graph

Fig. 11 illustrates the complete directed graph of articles that cite or are cited by other articles. Each article is a node and citations are edges. Articles-nodes mapping is provided at https://pro.unibz.it/staff/brusso/PapersUsed.html. We define *fan in* as the number of citations to a paper and *fan out* as the number of citations from a paper. We define the *distributor of knowledge* as a node in the graph that has at least fan-in equals two and an *attractor of knowledge* a as node with at least fan-out equals two. A pure distributor has zero *fan out* and a pure attractor has zero *fan in*. A node can also simultaneously be a distributor and an attractor. A node that is a distributor with fan out one is a *router* as it branches the knowledge of the single citation into different articles. For example, node #107 is a router that distributes the knowledge of paper #82 to five other articles (Fig. 2). Fig. 1 displays the types of nodes. A path is a set of nodes connected by edges following the direction of the graph.

---

[8] http://www.graphviz.org/

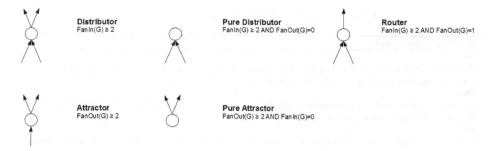

Fig. 1. Types of nodes

To determine a research area in OSS, we make three assumptions:

i)   Pure distributors determine research areas;

ii) A path originating from a pure distributor and leading to a pure attractor or a dead end (a paper that has only one fan out) determines an area of research in OSS;

iii) Paths starting from a pure distributor determine a cluster.

Thus, we start from a pure distributor, for example #82, and then follow one of its links. We follow links in opposite direction downwards in the graph until we reach a pure attractor or a dead end. Then we aggregate all the paths from a pure distributor to define a cluster. Finally, we add all the dead ends cited by an attractor of the cluster. For example, the edge linking #82 to #107 determines an area of research originated from paper #82 including the dead ends node #41 and #81. The five paths originated by paper #82 determine a cluster.

To label paths, we have used the taxonomy of the Call for Papers (CfP) of the OSS conference series[9]. Typically, a CfP includes the major topics of the conference. Consequently, accepted papers concern topics listed in the CfP. We have also considered the classification proposed in [37]. Unfortunately, the classification was too high-level for our analysis.

We have identified twenty - one clusters in the graph (Fig. 11). Four are bipoles - clusters of two articles - and two are isolated clusters. Eleven in 2005, two in 2006, and two in 2007 originate the fifteen clusters. Thirty-four are empirical papers according to [37]. Note that the majority of the pure distributors that originate the largest clusters are empirical. Table 2 lists pure distributors and attractors that define the largest clusters - papers with more than three fan-out or fan-in. It also classifies them as empirical according to [37].

To identify the reason of the citation, we have manually looked up each citation in the text. We have used the wording of the authors and the position of the citation within the article structure to understand the reason for each citation. If, for example, a citation is located in the "Background" section only and it is given to justify the work, then we label the corresponding link as motivation of work.

---

[9] http://ossconf.org/

Table 2. Major Distributors, Attractors, and Routers

Distributors and routers
2005
#5 pure distributor and empirical paper ([1]), #8 pure distributor and empirical paper ([30]), #17 pure distributor and empirical paper ([27]), #44 pure distributor ([31]), #82 pure distributor ([22])
2006
#84 pure distributor and empirical paper ([39]), #107 router and empirical paper ([21]), #119 router ([8]), #121 pure distributor and empirical paper ([32])
2007
#127 pure distributor ([38]), #128 pure distributor and empirical paper ([14]), #138 router and empirical paper ([20])
2008
#180 distributor/attractor and empirical paper ([19])
2009
#234 router ([9])
Attractors
2008
#175 Platform for research ([15])
2009
#230 Framework ([33]), #233 Extensive background ([18])
2010
#305 Includes SLR ([17]), #312 Framework ([35])

4   Classification of the Articles

We have read all the articles following the patterns defined by the clusters. The reading confirms that each path originated from a pure distributor determines a well-defined perspective of research that takes its motivation from the distributor. In many cases, we are also able to identify authors that contributed the most to a given research area and determine the semantic of the cross-citations besides the motivation of work. In the following, we report of this analysis per cluster. The number of the pure distributor names clusters.

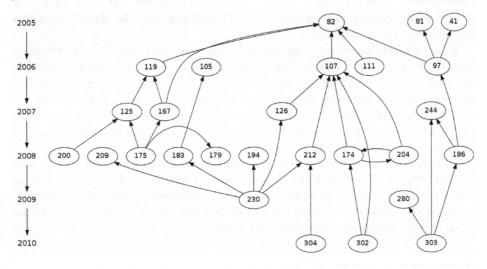

Fig. 2. Citations cluster originated from paper #82

Cluster #82. The largest cluster originates from node #82. Paper #82 introduces the OSSmole project (later called FLOSSmole). OSSmole is a repository of data, scripts, and analysis of data collected from OSS projects. The cluster is then branched into five links (Fig. 2). Three links identify three major sub-clusters, defined by router #119, router #107, and node #97. The sub-cluster defined by router #107 is generally concerned with developer communities and social network analysis of these communities. Links are rather strong. Paper #174 cites paper #107 to motivate its research and the use of a metric (outdegree centrality), and cites paper #204 as example of method of analysis. Two of the articles in this cluster focus on the same project (Apache).

Howison and Crowston have been the major contributors. The branch ends in 2010 with the work of Conaldi and Rullani that proposes a global perspective of F/OSS network structure mining the SourceForge repository. Paper #119 on the future of OSS data mining starts a new branch that focuses on analyses and improvements of project mining tools. In this sub-cluster, we found citations motivated by the use of the same repository or the same research goals. The branch ends in 2008 with recommendations for the design of research infrastructure in OSS by Gasser and Scacchi (paper #175). Paper #97 originates a research branch on the analysis of code artefacts for modelling maintenance processes and specializes over the years in bug fixing processes. Although article #97 cites #82, it does not use OSSmole directly mining CVS log files. The major contributors in this branch are Dalle and den Besten and the majority of the articles focus on the Firefox community.

Cluster #44. Paper #44 introduces to practices for quality assurance in OSS projects (Fig. 3). Paper #44 generates three sub-clusters on OSS deployment, project's activity, and community building and participation. Article #156 connects the last two topics by means of increase of the community size and their activity growth. In cluster #44, we were not able to identify major contributors as different authors contributed to the research streams and all the citations were to motivate the work.

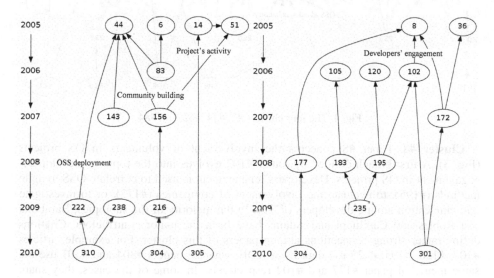

Fig. 1. Citations cluster originated from papers #44 and #8

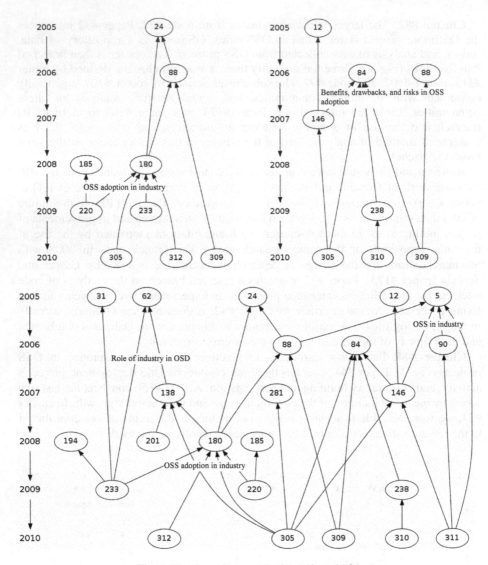

Fig. 4. The four clusters #5, #24, #62, and #84

Cluster #8. Paper #8 concerns the involvement of volunteers in OS projects (Fig. 3). A first branch defined by router #102 evolves into the topic on developers' engagement in OS projects. Developers' engagement is used to correlate OSS to agile methods (#195), to evaluate the involvement of companies (#172), or to investigate communication among developers (#235). In the majority of the subtopics, Barahona and Robles and Capiluppi and Adams have been the major contributors. Citations define rather strong connections among papers of this cluster. For example, articles #102, #120, #195, #235 use data from KDE, whereas papers #304 and #301 use the same metric of paper #177 and #102 respectively. In some of the cases, they share some of the authors, too.

Cluster #5. Cluster #5 is a big cluster connected to three other sub-clusters (#24, #84, and #62 in Fig. 4). These clusters focus on the relation between OSS and industry. The empirical pure distributor #5 summarizes the works of an international workshop on pros and cons of the use of OSS in industry. This paper spins off into two contributions of people that participated in the workshop (#90) and of some of its authors (#146 and #88). Looking at the affiliations of the authors, we call cluster #24 the industrial "Scandinavian case": it starts with a case of interest in OSS in the Finnish industry (#24), then includes the Swedish one in 2006 (#88) and closes with the Norwegian one in 2008 (#180). From paper #180, Hauge, Sorensen and Conradi initiate a new research theme on OSS adoption in industry (#233, #220, #309, #312, and #305). Paper #62 uses the Capability Maturity Model to assess the migration from closed to open software development in industry. This paper initiates a research investigation on the role of industry in Open Source Development (OSD) (router #138) and in particular, the migration of an existing business model to an open source one (#233). It further evolves in OSS adoption in industry (#180). Cluster #84 originates from the paper of Ven and Verelst (#84) and gives impulse to the specific perspective of OSS adoption in industry that relates to benefits, drawbacks, and risks. The authors that contribute the most to cluster #84 are Hauge and Conradi. The majority of the citations of the four clusters concerns motivation of work, but there are links that connect papers by the same method of analysis (e.g. #90 and #311). Worth noticing is that all the three papers of 2006 (#84, #88, #90) report of a case study in industry, but only the first two are as empirical according to [37].

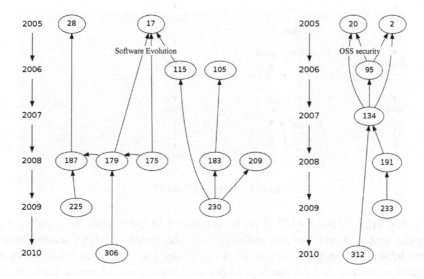

Fig. 5. Clusters #17 and #20_?

Cluster #17. Cluster #17 is originated by the work on software evolution by Koch (Fig. 5). The research has developed into software evolution as total growth of software (Riehle et al., #179, #187, and #225) and evolution of OS communities as measure of their governance (#230). An independent sub-cluster (originated by paper

#28) is connected with the software evolution theme as it deals with the analysis of commits in distributed development. The link is provided by a chain of citations in 2008 that connects the work of Gasser and Scacchi (#175) on a research framework for multi-disciplinary studies in OSS (including the "laws of software evolution") with the study of continuous integration in OS development as an example of mining multiple data sources (#187). In this cluster, we identify a citation from the paper of Sirkkala et al. (#230) that have used part of the approach in Weiss et al. (#115) and a citation from Deshpande and Riehle (#179) that uses the results of another work of them (#187) to validate their conclusions on total growth of software.

Cluster #20 and #2. Two pure distributors (#2 and #20) originate this cluster (Fig. 5). We call it the "Italian case" as the major contributors are Ardagna, Damiani, Frati and other Italians. In their three works, they deal with security of OSS. In the last two years, the area has evolved to the larger problem of selecting and integrating OSS for industry (#233 and #312) where security is especially relevant (like telecom applications).

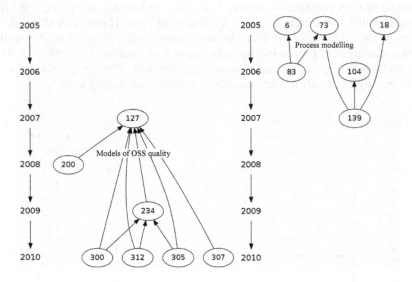

Fig. 6. Clusters #127 and #73

Cluster #127. Cluster #127 (Fig. 6) originated in 2007 with the results of the European project QualiPSo and enforced with the results on OSS trustworthiness (router #234). The majority of the works proposes and compares frameworks and comparative models of OSS quality. Major contributors are Morasca, Lavazza, and Taibi, members of the above project. In this cluster, the majority of the papers and in particular, all those in 2010 cite #127 and #234 to use the model defined there.

Cluster #73 and Cluster #121. These clusters (Fig. 6 and 7) concern process modelling and they are connected through the paper of Jensen and Scacchi in 2007 on guiding the discovery of OSS processes with a reference model (#139). Two branches derive from the pure distributor #73: on modelling communication and information exchange in processes and on modelling the process as a whole. Paper #139 motivates its research citing the problem of managing information in distributed development (#73) and issues in creating theoretical models of OS processes (#104, #18, and #121). The research goal is organizing knowledge in OSD and providing guidance in allocating resources, selecting tools, and performing activities of OSD. Following Cluster #121, the research culminates in 2009 into an investigation of the selection of OSS products in industry as indication of reuse (#220).

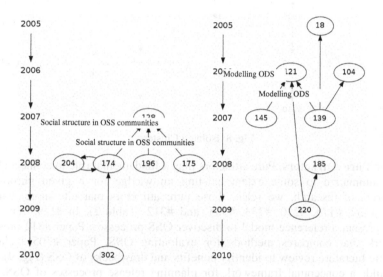

Fig. 7. Clusters #128, #121

Cluster #128. A study of network analysis on SourgeForge originates this cluster (Fig. 7). The cluster concerns social structures of OSS communities, as membership networks or communication networks. In particular, Gasser and Scacchi propose an infrastructure for research in social networks of OSS communities whereas Balieiro et al. introduce a study on the meso-level structure of OSS development as collaboration environment. Papers are authored by various members of the OSS research community and are connected by a net of citations that concerns mainly the motivation of work.

Isolated Clusters. There are six isolated clusters on the right of Fig. 11 (Fig. 8). The four bipoles concern innovation (#164 - #193), adoption (#149 - #181), services (#87 - #137) and requirements (#211 - #241). In all but the bipole on adoption, the two papers share part or all of the authors. Cluster #7 concerns measures of success of OS projects. Over the years, the concept of success evolves from a static meaning – e.g. number of hits of web searches - to an evolutionary perspective of development efficiency as the code produced over time. All the citations motivate the work and no specific author can be uniquely associated to the cluster. Cluster #68 concerns teaching OSS at university level. All the citations motivate the work.

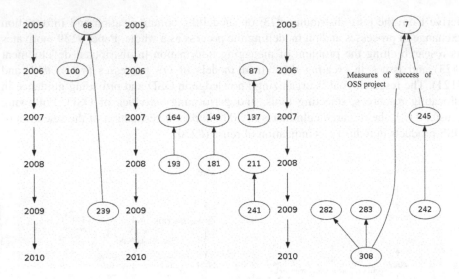

Fig. 8. Isolated Clusters

Major Pure Attractors. Pure attractors play a fundamental role to connect clusters and to summarize to some extent existing knowledge for a given purpose. As synthesizers of research, we select those pure attractors that cite more than four articles: #139, #175, #230, #233, #305, and #312 (Table 2). In #139, Jensen and Scacchi present a reference model to discover OSS processes. Paper #312 presents a framework that compares methods for evaluating OSS. Paper #305 includes a systematic literature review to identify benefits and drawbacks of OSS (Fig. 9). Paper #230 builds a conceptual framework for planning release processes of OSS. Paper

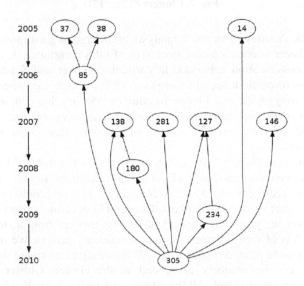

Fig. 2. Major Pure Attractor #305

#175 defines the design of a platform for investigating the OSS phenomenon through multidisciplinary research. This work also provides recommendations and critical issues in OSS research. Paper #233 includes an extensive background section that motivates the work.

4.1 Inter-cluster Connections

There are several interlinks among clusters (Figure 10). We mark a link between two clusters if there is a node that directly connects them, i.e. a pure attractor citing papers of two clusters. Connections among clusters describe mutual influence among topics and have the potential to converge to a single cluster[10]. The major findings in Figure 10 show that:

1. "Tools for data mining" is a cross cutting topic connecting various research areas, from the analysis of social structures and networks of OSS communities and their developers' engagement, software evolution, OSS security, and models of OSS quality.

2. OSS in industry specifies into "adoption in industry," "role of industry," "case studies in industry" and a stream in "benefits, risks, and drawbacks of OSS."

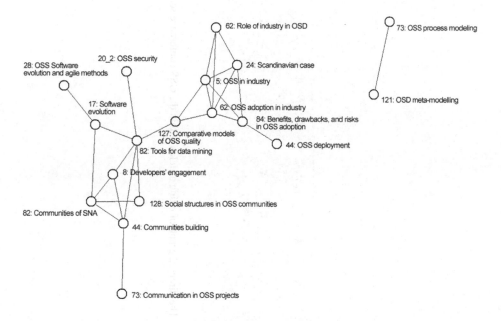

Fig. 10. Links between research streams

---

[10] Figure 10 does not show papers that connect clusters. This information is accessible at https://pro.unibz.it/staff/brusso/LinkingPapers.html

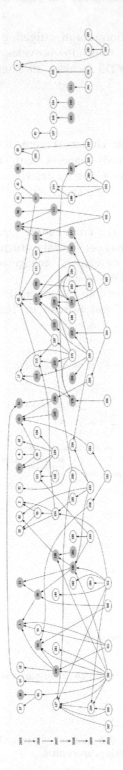

Fig. 11. The directed graph of citing articles within the OSS conference series. Grey nodes are empirical papers according to [37].

5 Discussion

Our analysis reports of well-scoped clusters of citations that highlight major areas of research in OSS and their evolution across the six years of OSS conferences. We report of a lively social network of citations (RQ1), and several streams of research established across years (RQ2).

In particular, we have found that the creation of a big repository for data mining (FLOSSmole) has originated research in social network analysis, tools for data mining, and analysis of code artefacts to understand maintenance processes, specifically bug fixing. The topic "tools for data mining" also appears to be a cross-cutting theme that connects several other research areas.

Quality assurance is another hot topic. It motivates studies on community building (as a means for QA), OSS deployment (as a factor to consider in QA), project's activity (as a measure of QA) and on comparative models (for QA).

A specific concern at the OSS conference series is the perception of OSS in industry. Research has focused on the role of industry in OSS development and after 2008 on OSS adoption. A report of a large workshop in OSS that joined researchers and practitioners of the industry in 2005 initiated this research.

A good number of papers at the OSS conference series have been dedicated to developers' engagement. Developers' engagement is used to correlate OSS methods to agile methods, to evaluate the involvement of companies, or to investigate communication among developers. This topic is also cited in works on software evolution and community building and participation.

There are themes still little cited that might have some future potential: OSS process (meta-) modelling, OSS security, Agile and OSS development methods, and teaching OSS in universities.

We have found the majority of large clusters to have a group of authors that contribute to the research stream over the years. Clusters also reveal authors' interrelations that define informal research groups and outline social network structures, like in the "Scandinavian case."

Large clusters are initiated by empirical papers with the only exception being the paper on the FLOSSmole repository.

Papers with a large number of citations are synthesizers of research often presenting a framework or a platform to guide research in OSS.

As a final remark, we want to stress that we are aware that the sample we selected does not include all the crucial papers that may have contributed to ground knowledge in OSS. Some of the relevant contributions have been published in journals not dedicated to OSS or have simply been reported on the Internet. History of knowledge synthesis is full of such examples as illustrated by the FODA paper [28], a technical report cited more than 1500 times although not peer reviewed. Our future work will extend this analysis to OSS article hubs like for example http://flosshub.org or http://pascal.case.unibz.it or the MIT repository.

We also acknowledge that fact that cross-citations are indicators of research connections, but are not unique and not necessarily the best ones. A textual similarity analysis could give finer results. Namely, we have already applied some known algorithms (e.g. cosinus, Jaro Winkler) for text similarity on the XML sections of the papers. Despite some preliminary good results on titles and authors, for longer sections the algorithms proved to be inaccurate. We plan to use more sophisticated techniques of string similarity and a better data cleansing to get finer results.

6 Conclusions

In our research, we aim at understanding the major research topics outlined in six years of conferences dedicated to OSS. We have clustered articles by their cross-citations and inspected each single cluster to search for major research themes, evolution of research topics, and major initiators or synthesizers of research in OSS.

We have also introduced a social network underlying the research production. This network assumes that authors are connected by their common research interest revealed by citations as a community of practice. This network is different from a social network defined by co-authorship. In this case, links may be transversal to research topics and are generated by existing collaborations. Differently, our network may reveal hidden and potential collaborations among researchers.

Finally, we have used a semi-automated approach for systematic mapping studies that automatically creates clusters from cross-citations and manually inspect the clustered papers. Our research has identified cornerstone papers and links among them revealing unexpected and undeclared connections among authors as well as lack of connection among potentially connected topics. Large clusters are mostly initiated and sustained by empirical papers. Since 2008, synthesizers of research have introduced frameworks and platforms to perform OSS research paving the way for future work. The analysis of non-cited papers indicates that significant research has not been exploited, yet. Therefore, we recommend the OSS community to exploit further the potential provided by the OSS conference series while maintaining the interest in its major research streams.

Acknowledgements. We would like to thank Muhammad Ali Babar and Klaas Jan Stol and the reviewers for their valuable comments.

References

1. Ågerfalk, P.J., Deverell, A., Fitzgerald, B., Morgan, L.: Assessing the Role of Open Source Software in the European Secondary Software Sector: A Voice from Industry. In: Proceedings of the 1st International Conference on Open Source Systems (OSS 2005), Genoa, Italy, pp. 82–87 (2005)
2. Aksulu, A., Wade, M.R.: A Comprehensive Review and Synthesis of Open Source Research. Special Issue, Journal of Association for Information Systems 11(11), 576–656 (2010)
3. Anel, J.A.: The Importance of Reviewing the Code. Communication of the ACM, 40–41 (May 2011)
4. Ayala, C., Hauge, Ø., Conradi, R., Franch, X., Li, J., Velle, K.S.: Challenges of the Open Source Component Marketplace in the Industry. In: Boldyreff, C., Crowston, K., Lundell, B., Wasserman, A.I. (eds.) OSS 2009. IFIP AICT, vol. 299, pp. 213–224. Springer, Heidelberg (2009)

5. Biolchini, J., Mian, P. G., Natali, A. C. C., Travassos, G. H.: Systematic Review in Software Engineering, University of Rio de Janeiro:TR-ES 679/05 (2005)
6. Brereton, P., Kitchenham, B., Budgen, D., Turner, M., Khalil, M.: Lessons from applying the systematic literature review process within the software engineering domain. Journal of Systems and Software 80, 571–583 (2007)
7. Budgen, D., Charters, S., Turner, M., Brereton, P., Kitchenham, B., Linkman, S.: Investigating the applicability of the evidence-based paradigm to software engineering. In: Proceedings of the 2006 International Workshop on Workshop on Interdisciplinary Software Engineering Research, Shanghai, China, May 20 (2006)
8. Conklin, M.: Beyond Low-Hanging Fruit: Seeking the Next Generation in FLOSS Data Mining. In: Proceedings of the 2nd International Conference on Open Source Systems (OSS 2006), Como, Italy, pp. 47–56 (2006)
9. Del Bianco, V., Lavazza, L., Morasca, S., Taibi, D.: Quality of Open Source Software: The QualiPSo Trustworthiness Model. In: Boldyreff, C., Crowston, K., Lundell, B., Wasserman, A.I. (eds.) OSS 2009. IFIP AICT, vol. 299, pp. 199–212. Springer, Heidelberg (2009)
10. Dybå, T., Dingsøyr, T.: Applying Systematic Reviews to Diverse Study Types: An Experience Report. In: Proceedings of the Proceedings of the First International Symposium on Empirical Software Engineering and Measurement (2007)
11. Dybå, T., Kitchenham, B., Jorgensen, M.: Evidence-Based Software Engineering for Practitioners. IEEE Software 22, 58–65 (2005)
12. Feller, J., Fitzgerald, B.: Understanding Open Source Development. Addison-Wesley, Reading (2001)
13. Feller, J., Fitzgerald, B., Hissam, A.S., Lakhani, K.R.: Perspectives on Free and open Source Software. MIT Press, Cambridge (2007)
14. Gao, Y., Madey, R.G.: Network Analysis of the SourceForge.net Community. In: Proceedings of the 3rd International Conference on Open Source Systems (OSS 2007), Limerick, Ireland, pp. 187–200 (2007)
15. Gasser, L., Scacchi, W.: Towards a Global Research Infrastructure for Multidisciplinary Study of Free/Open Source Software Development. In: Proceedings of the 4th International Conference on Open Source Systems (OSS 2008), Milano, Italy, pp. 143–158 (2008)
16. Hauge, Ø., Ayala, C.P., Conradi, R.: Adoption of open source software in software-intensive organizations - A systematic literature review. Information & Software Technology 52(11), 1133–1154 (2010)
17. Hauge, Ø., Cruzes, D., Conradi, R., Sandanger Velle, K., Skarpenes, T.A.: Risks and Risk Mitigation in Open Source Software Adoption: Bridging the Gap between Literature and Practice. In: Ågerfalk, P., Boldyreff, C., González-Barahona, J.M., Madey, G.R., Noll, J. (eds.) OSS 2010. IFIP AICT, vol. 319, pp. 105–118. Springer, Heidelberg (2010)
18. Hauge, Ø., Ziemer, S.: Providing Commercial Open Source Software: Lessons Learned. In: Boldyreff, C., Crowston, K., Lundell, B., Wasserman, A.I. (eds.) OSS 2009. IFIP AICT, vol. 299, pp. 70–82. Springer, Heidelberg (2009)
19. Hauge, Ø., Sørensen, C., Conradi, R.: Adoption of Open Source in the Software Industry. In: Proceedings of the 4th International Conference on Open Source Systems (OSS 2008), Milano, Italy, pp. 211–221 (2008)
20. Hauge, Ø., Sørensen, C., Røsdal, A.: Surveying Industrial Roles in Open Source Software Development. In: Proceedings of the 3rd International Conference on Open Source Systems (OSS 2007), Limerick, Ireland, pp. 259–264 (2007)

21. Howison, J., Inoue, K., Crowston, K.: Social Dynamics of Free and Open Source Team Communication. In: Proceedings of the 2nd International Conference on Open Source Systems (OSS 2006), Como, Italy, pp. 319–330 (2006)
22. Howison, J., Conklin, M., Crowston, K.: OSSmole: A collaborative repository for FLOSS research data and analyses. In: Proceedings of the 1st International Conference on Open Source Systems (OSS 2005), Genoa, Italy, pp. 54–60 (2005)
23. Kitchenham, B.: What's up with software metrics? - A preliminary mapping study. Journal of Systems and Software 83(1), 37–51 (2010)
24. Kitchenham, B., Brereton, O.P., Budgen, D., Turner, M., Bailey, J., Linkman, S.: Systematic literature reviews in software engineering – A systematic literature review. Information and Software Technology 51, 7–15 (2009)
25. Kitchenham, B., Charters, S.: Guidelines for Performing Systematic Literature Reviews in Software Engineering, Keele University, UK EBSE-2007-1 (2007)
26. Kitchenham, B.: Procedures for Performing Systematic Reviews, Keele University Technical Report TR/SE-0401 (2004)
27. Koch, S.: Evolution of Open Source Software Systems – A Large-Scale Investigation. In: Proceedings of the 1st International Conference on Open Source Systems (OSS 2005), Genoa, Italy, pp. 148–153 (2005)
28. Kyo, C.K.: FODA: Twenty years of Perspectives on feature Models. In: Keynote at 13th International Product Line Conference, SPLC (2009)
29. Petersen, K., Feldt, R., Mujtaba, S., Mattsson, M.: Systematic mapping studies in software engineering. In: Proceedings of the 12th International Conference on Evaluation and Assessment in Software Engineering, pp. 71–80 (2008)
30. Robles, G., Gonzales Barahona, J.M., Michlmayr, M.: Evolution of Volunteer Participation in Libre Software Projects: Evidence from Debian. In: Proceedings of the 1st International Conference on Open Source Systems (OSS 2005), Genoa, Italy, pp. 100–107 (2005)
31. Rossi, B., Scotto, M., Sillitti, A., Succi, G.: Criteria for the non invasive transition to OpenOffice. In: Proceedings of the 1st International Conference on Open Source Systems (OSS 2005), Genoa, Italy, pp. 250–253 (2005)
32. Simmons, G.L., Dillon, T.S.: Towards an Ontology for Open Source Software Development. In: Proceedings of the 2nd International Conference on Open Source Systems (OSS 2006), Como, Italy, pp. 65–75 (2006)
33. Sirkkala, P., Aaltonen, T., Hammouda, I.: Opening Industrial Software: Planting an Onion. In: Boldyreff, C., Crowston, K., Lundell, B., Wasserman, A.I. (eds.) OSS 2009. IFIP AICT, vol. 299, pp. 57–69. Springer, Heidelberg (2009)
34. Staples, M., Niazi, M.: Experiences using systematic review guidelines. Journal of Systems and Software 80, 1425–1437 (2007)
35. Stol, K., Ali Babar, M.: A Comparison Framework for Open Source Software Evaluation Methods. In: Ågerfalk, P., Boldyreff, C., González-Barahona, J.M., Madey, G.R., Noll, J. (eds.) OSS 2010. IFIP AICT, vol. 319, pp. 389–394. Springer, Heidelberg (2010)
36. Stol, K.-J., Ali Babar, M.: Reporting empirical research in open source software: The state of practice. In: Boldyreff, C., Crowston, K., Lundell, B., Wasserman, A.I. (eds.) OSS 2009. IFIP AICT, vol. 299, pp. 156–169. Springer, Heidelberg (2009)
37. Stol, K.J., Ali Babar, M., Russo, B., Fitzgerald, B.: The use of empirical methods in Open Source Software research: Facts, trends and future directions. In: ICSE Workshop on Emerging Trends in Free/Libre/Open Source Software Research and Development, pp. 19–24 (2009)

38. Taibi, D., Lavazza, L., Morasca, S.: OpenBQR: a framework for the assessment of OSS. In: Proceedings of the 3rd International Conference on Open Source Systems (OSS 2007), Limerick, Ireland, pp. 173–186 (2007)

39. Ven, K., Verelst, J.: The Organizational Adoption of Open Source Server Software by Belgian Organizations. In: Proceedings of the 2nd International Conference on Open Source Systems (OSS 2006), Como, Italy, pp. 111–122 (2006)

40. Zennier, C., Melnik, G., Maurer, F.: On the success of empirical studies in the international conference on software engineering. In: Proceedings of the 28th International Conference on Software Engineering (ICSE 2006), Shanghai, China, pp. 341–350 (2006)

The Importance of Architectural Knowledge in Integrating Open Source Software

Klaas-Jan Stol[1], Muhammad Ali Babar[2], and Paris Avgeriou[3]

[1] Lero—The Irish Software Engineering Research Centre
University of Limerick, Ireland
[2] IT University of Copenhagen, Denmark
[3] University of Groningen, the Netherlands
klaas-jan.stol@lero.ie, malibaba@itu.dk, paris@cs.rug.nl

Abstract. Open Source Software (OSS) is increasingly used in Component-Based Software Development (CBSD) of large software systems. An important issue in CBSD is selection of suitable components. Various OSS selection methods have been proposed, but most of them do not consider the software architecture aspects of OSS products. The Software Architecture (SA) research community refers to a product's architectural information, such as design decisions and underlying rationale, and used architecture patterns, as Architecture Knowledge (AK). In order to investigate the importance of AK of OSS components in integration, we conducted an exploratory empirical study. Based on in-depth interviews with 12 IT professionals, this paper presents insights into the following questions: 1) what AK of OSS is needed? 2) Why is AK of OSS needed? 3) Is AK of OSS generally available? And 4) what is the relative importance of AK? Based on these new insights, we provide a research agenda to further the research field of software architecture in OSS.

Keywords: Open Source Software integration, component-based development, architectural knowledge, software architecture, OSS Integrator, survey.

1 Introduction

Software development organizations can adopt Open Source Software (OSS) in various ways [1]. One way is to integrate OSS components in Component-Based Software Development (CBSD), which has become a standard way of building large-scale systems. In such a setting, a software development organization has the role of *OSS integrator* [2]. Building systems from externally and independently developed components is not without risks. Garlan et al. reported their experiences of using only four components and experienced significant integration problems. They identified the root cause of their problems to be *architectural mismatch* [3]. The increasing use of OSS components as an alternative to Commercial Off-The-Shelf (COTS) components in CBSD [4] can result in more architectural mismatch challenges for OSS integrators. Despite the common concerns about OSS licenses and a lack of

S.A. Hissam et al. (Eds.): OSS 2011, IFIP AICT 365, pp. 142–158, 2011.

support and documentation, the use of OSS offers various benefits such as access to the source code and low costs [5], which means that industry is likely to continue to use OSS products.

Various authors have reported studies of CBSD with OSS products [6-9], or more generally with Off-The-Shelf (OTS) components (both OSS and COTS) [10, 11]. Most of these studies focus on evaluation and selection of components, which is a critical aspect in CBSD. While software architecture (SA) has long been recognized as an important success factor in CBSD [12], little attention has been paid so far to architectural aspects of OSS components, such as a component's architectural styles or patterns, and architectural design decisions and their rationale. This type of information is called Architectural Knowledge (AK) [13]. We assert that AK of OSS products can provide valuable insights to integrators, not only during the evaluation and selection phase, but also in later phases of the software lifecycle. However, there is no empirical evidence about the AK needs of OSS integrators. This motivated us to carry out an empirical study to investigate this topic. The contributions of this paper are an empirical understanding of:

- The architectural knowledge needs of OSS integrators;
- Why architectural knowledge is needed and how it can help OSS integrators;
- The availability of architectural knowledge to OSS integrators;
- The relative importance of architectural knowledge.

Based on these results, we also identify a number of opportunities for further research to bridge the gap between the OSS and SA research communities.

This paper proceeds as follows. Section 2 presents background and motivation of this study. Section 3 presents the design of this study. Section 4 presents results. Section 5 presents discussion and conclusion of this paper.

2 Background and Motivation

In this section, we present relevant related work and motivate our study. Section 2.1 provides an overview of CBSD with OSS. Section 2.2 introduces the field of SA and AK. Section 2.3 provides an overview of the research on SA in the context of OSS.

2.1 Component-Based Development with Open Source Software

OSS products have become more commercially viable as an alternative to COTS products [14]. In the last decade or so, OSS has been increasingly used by software-development organizations in CBSD [1]. While OSS components are often used as if they were closed-source components [11], there are some differences from an integrator's point of view. The availability of the source code eases component integration and allows white-box testing [15]. Di Giacomo reviewed the activities of CBSD and concludes that these activities need to be extended to accommodate the different nature of OSS [16]. Furthermore, the relationship with the supplier (COTS vendor versus OSS community) is different; one reason is the fact that the OSS integrator does not pay an OSS community [17].

Li et al. list three main phases in the CBSD lifecycle: *selection*, *integration*, and *maintenance* [11]. The remainder of this subsection briefly reviews the literature on OSS-based CBSD according to these lifecycle phases.

Component Selection. A critical success factor in CBSD is the selection of appropriate OSS products, and both researchers and industry have proposed a variety of OSS evaluation and selection methods [18]. However, studies have also shown that practitioners typically do not use these "normative" selection methods [19, 20]. Hauge et al. observed a *first fit* rather than a *best fit* strategy [19].

Component Integration. The source code's availability allows direct modification by the OSS integrator [16], or "glue code" can be written to make integration easier. However, integrators typically do not make changes to the source code, for various reasons [11]. One obstacle is that it often requires an in-depth knowledge of the component to be able to make useful changes [11].

Component Maintenance. Once OSS components are integrated, they must be maintained as part of an application. If custom changes were made during the integration phase, these may have to be separately maintained if the changes are not incorporated back into the source base that is maintained by the OSS community [21]. An OSS integrator can choose to actively participate in the development or roadmap planning of the OSS product [17, 21].

2.2 Software Architecture and Architectural Knowledge

The field of SA has emerged as an important CBSD sub-discipline within the research area of software engineering [22]. SA as a product has been recognized as an important design artifact in software development activities, including analysis, design and evaluation activities as well as implementation and evolution [23]. A system's SA typically constitutes the design decisions about a system [24], such as the use of certain architectural styles or patterns [22], e.g., layers or model-view-controller (MVC). Architecture patterns are common solutions to recurring system design problems [25], and affect the system-wide quality attributes (QAs, sometimes referred to as a system's *"ilities"*) such as performance and reliability [26]. For instance, a "layered" architecture is likely to improve maintainability, as it facilitates a clear separation of concerns. However, passing large numbers of "messages" (e.g., function calls) up and down a layer "stack" may negatively impact performance. Various patterns have been documented in detail, for instance, in [25].

Information such as used patterns, design decisions, etc. is referred to as AK. We note that there is no commonly accepted definition of AK. In recent years, the SA research community has recognized the value of AK and has started to focus on recording, managing and sharing AK[1] [13].

2.3 Software Architecture in Open Source Software Research

Software architecture has received little attention in OSS research. The remainder of this section presents a brief overview of research on SA in the context of OSS.

---

[1] We are aware of the semantic difference between "knowledge" and "information". Even though "information" may be the appropriate term, we use the term "knowledge" as is custom in the software architecture community.

Tran et al. [27] presented an experience report on "architectural drift", when concrete architecture (implementation) differs from the conceptual architecture (design). They argue that a system can be more easily understood by developers if its architecture is repaired and demonstrate their approach to architectural repair for two OSS products. Nakagawa et al. presented a case study of an OSS web-based system, and also found that the architecture had drifted after approximately two years of development [28]. They highlight that the architecture affects the product quality, and propose architecture refactoring to repair the architecture.

Capiluppi and Knowles report on architecture recovery of three OSS products in the same domain (instant messaging) [29] and found that a common architecture for these products had emerged. They argue that architecture recovery may facilitate OSS developers to understand the design as well as a sharing of tacit knowledge of other OSS developers.

Merilinna and Matinlassi investigated the state of the art and practice of OSS integration [20]; while they did not specifically focus on SA, they found that architectural mismatch was either preempted by selection of OSS components from a list of "fluently integrating" components, or that practitioners were not aware of the concept of architectural mismatch.

Matinlassi performed a survey to investigate the role of SA as perceived by 15 OSS communities, and found that "architecture" was mostly considered to be a high-level, coarse grained abstraction of the system's structure [30]. Modularity was considered to be the most important QA of an SA, whereas other QAs such as performance and ease of integration were not considered important by most respondents.

Ali Babar et al. [31] argued that organizations may be able to more confidently use OSS components in a software product line if OSS communities pay more attention to the architectural aspects during development and evolution of those components. Previously, researchers and practitioners have organized SA-related workshops [32] to draw attention to this topic in OSS, and proposed research roadmaps [33, 34]. Recently, AK Management (AKM) in OSS projects is discussed from the perspective of OSS developers [35].

So far, there have only been a few empirical studies on the role and importance of SA in OSS. These mostly focused on architectural repair and recovery. However, there has been no study that has focused on OSS integration from the OSS integrator's point of view, and in particular with respect to a component's AK.

2.4 Research Objectives

Hauge et al. identified a lack of empirical research in the OSS research area and highlighted the need for studies related to integration of OSS components [1]. The objective of our study was to empirically explore the observations and experiences of OSS integrators for gaining and disseminating insights into the architectural aspects of OSS integration.

Much of the literature on OSS selection highlights the importance of evaluation criteria such as the OSS license, support, functionality, and so on. However, there are few studies from the perspective of OSS integrators. Since software architecture plays an important role in component integration, it is important to investigate what architectural knowledge OSS integrators need. Therefore, research question one is:

RQ1: What AK of OSS Products Do Integrators Need?

Besides an understanding of *what* AK is needed, it is also useful to understand how AK can help OSS integrators. Hence, our second research question is:

RQ2: Why Do Integrators Need AK of OSS Products?

Furthermore, it is also important to find out whether the required AK about OSS component is usually available. Hence, our third research question is:

RQ3: Is AK of OSS Products Generally Available to Integrators?

In this paper, we assert that AK about an OSS product is important. However, it is equally important to find out OSS integrators' perceived importance of AK in integration. In order to get an empirically-based understanding of the importance of the role of architecture in OSS integration, we defined the fourth research question:

RQ4: What Is the Relative Importance of AK of OSS Components?

3 Research Design

3.1 Research Method and Data Collection

Since the role and importance of software architecture in OSS is still a largely unexplored research area in its nascent phase, a qualitative research approach is appropriate [36]. We decided to conduct an in-depth exploratory, qualitative survey using semi-structured interviews to collect data from practitioners. The use of interviews gives a researcher the flexibility to go deeper into unforeseen types of information that may emerge during an interview [37].

We invited IT practitioners through our professional network to participate in our study; our sample was therefore a *convenience sample* [38]. Table 1 lists background information of the participants. All but one participant worked at organizations located in different countries in Europe; P5 was located in the USA. We do not report on specific details for confidentiality reasons. Two participants (P3 and P4) worked both at organization C, but at different branches or departments. We indicated this by a number suffix (i.e., C1, C2). The participants at organization E were contacted through our professional contact at this organization, who was also one of the participants. The participants had various positions, and all had extensive experience in the field. In total, we drew data from 12 participants who worked at five different organizations. Though participants P6-P12 all worked at organization E, they worked in different teams or departments (E1 to E6).

The participants worked at organizations in different domains that integrate OSS products in various degrees. Organization A is a large consultancy organization that develops business process support systems. Organization B develops mostly embedded software. Organization C is a research and innovation institution that develops both proof-of-concept prototypes and final products for customers. Participant P5 (organization D) is an independent consultant, and works on business

Table 1. Participants of our study

ID	Position	Experience	Org.	Domain	Interview
P1	Software architect	13 years	A	Business	Phone
P2	Software developer	5 years	B	Embedded	Face-to-face
P3	R&D developer	13 years	C1	Telecom	Phone
P4	R&D developer	13 years	C2		Face-to-face
P5	Independent consultant	10 years	D	Business	Instant messenger
P6	Software architect	20 years	E1		Face-to-face
P7	Project leader	26 years	E2		Face-to-face
P8	Sr software designer	10 years	E3	Safety	Face-to-face
P9	Team leader*	25 years	E4	critical	Face-to-face
P10	Software architect	12 years	E5	systems	Face-to-face
P11	Technology manager	25 years	E2		Face-to-face
P12	Team leader*	15 years	E6		Face-to-face

* Participants have also experience as a software architect.

process support systems. Organization E develops hardware and software for safety critical systems.

Prior to conducting the interviews, we designed an interview guide [39]. All face-to-face interviews were conducted at the premises of the participants' organizations. All but one interview lasted approximately one hour; the interview with P5 (through instant messenger (IM)) lasted approximately two hours. We digitally recorded all face-to-face and phone interviews with the participants' consent. All interviews were transcribed verbatim by the interviewer (the transcript of the interview with P5 was recorded by the IM client), resulting in more than 120 pages text (A4 size).

3.2 Data Analysis

Due to the qualitative nature of the collected data, we chose to analyze the data using qualitative data analysis methods [37]. We thoroughly read all interview transcripts, during which we extracted phrases of interest that were relevant to one of our research questions. The extracted data was stored in a spreadsheet, along with the page number of the source transcript that allowed us to trace back phrases to their original context. We annotated each entry with a code that reflected the contents of the entry. After this, the data were sorted and grouped based on the codes. Per group, we used the constant comparison technique [37] to identify common themes for answering our research questions.

4 Results

4.1 RQ1: Architectural Knowledge Needs of OSS Integrators

We identified four categories of information that integrators would like to have. Table 2 presents an overview of the findings using the categories that emerged from the coding of the participants' answers. It is interesting to note that the first type of AK (N1) affects the other three types of AK (N2-N4); for instance, a component's

Table 2. Types of architectural knowledge needed by OSS integrators

ID	Description	Reported by
N1	**Component structure.** Patterns, partitioning, structure	P1, P2, P3, P4, P5, P6, P7, P10, P12
N2	**Quality attributes and behavior.** Information about performance, e.g., bottlenecks, processor usage, disk usage, and other resources; reliability, stability, robustness, predictability, use of tactics.	P2, P4, P5, P6, P7, P8, P10, P11
N3	**Architectural fit.** Ease of integration, interface, API compliance, dependencies	P2, P3, P4, P5, P6, P7, P8, P10, P11
N4	**Component usage.** Insight in how the component can be used and how it performs the task	P2, P4, P10

structure directly affects its quality attributes and behavior. We address this in more detail in Section 5.

4.1.1 Component Structure

Most participants were interested in the internal structure of components. When speaking of a component's internal structure, participants often spoke of 'pattern'. In particular, some participants mentioned the presence of patterns such as layers and model-view-controller to be interesting for their purpose. One participant stated: *"Well, the first thing that I would do is to look at the documentation, in which they explain the structure there, preferably in pictures. And when those are not available, then it's a matter of extracting a zip file [jar file] and see how the directory structure would look like."* (P4).

4.1.2 Quality Attributes and Behavior

Several participants explicitly indicated specific runtime QAs, such as performance, reliability and stability. Like the literature on QAs, the participants also agreed that these QAs are directly affected by a component's behavior related to the use of resources (e.g., memory, processor, interrupts, database connections). Integrators consider it to be important to understand how a component behaves regarding these system resources; one participant reported to have used a "sandbox" to measure performance, processor usage, disk usage, and similar parameters.

One factor that affects a component's behavior is the use of architectural tactics [26, 40]. A tactic is a common technique to achieve a certain quality attribute. For instance, in order to improve the performance of software that connects to a database system, a developer may apply the "connection pooling" tactic. Such information is valuable to integrators, as one participant described: *"if it is a component that uses relatively expensive or limited system resources, then you'd like to know the strategy of those components to deal with resources. If I'm integrating it and it's using 10 out of 11 available database connections [...] that's certainly handy to know."* (P10).

We did not quantitatively analyze the most important quality attributes such as in [41], as our sample size was limited, and QAs are likely to be dependent on the domain in which software operates.

4.1.3   Architectural Fit

The most common concern of participants was the architectural fit of a component; in other words, does a component fit in the existing architecture? One participant described it clearly: *"[I look at] whether it can be used in the architecture that I had envisioned. To what extent do I need to adapt my own software in order to be able to use that product? [And] if I have to make certain changes in an OSS product, in order to be able to use it, how much."* (P3).

Participants used the term "architectural fit", but also referred to the interface and API of a component. It was interesting to find that a large majority of the participants distinguished a component's architectural fit from its internal structure or used patterns (see subsection 4.1.1), and generally considered the former to be more important than the latter. This confirms common knowledge that patterns have a direct influence on a component's architectural fit [42].

4.1.4   How to Use a Component

A few participants mentioned the need to understand how they can use a component. Besides insight in whether the component fits (see subsection 4.1.3), practitioners need to understand how it can be used within the system they build, and how to access the functionality provided. The participants prefer to have examples of how to use the software, or even the availability of a test environment that demonstrates how a component can be used, as one participant suggested: *"If the component came together with a sort of test environment, a sort of test application, that could show how the components can be used, and how it performs the tasks that it is supposed to do. That would help a lot."* (P2).

4.2   RQ2: Why Is Architectural Knowledge Needed?

The previous section addressed the question *what* AK integrators need; in this section we addresses the question *why* integrators need AK. We identified a number of different reasons how such knowledge can help integrators. Table 3 presents an overview of the findings to answer this question.

Table 3. Reasons why architecture knowledge is important to integrators

ID	Description	Reported by
W1	**Quality assessment.** Architecture, partitioning and patterns indicate certain maturity and qualities and help to analyze a component from a performance and functional perspective.	P2, P5, P7, P8, P10
W2	**Assess architectural fit.** Internal structure and patterns indicate what architecture is used and whether it fits in the existing architecture	P1, P2, P3, P6, P7, P10
W3	**Improve maintainability.** Patterns improve maintainability and replacing of components	P1, P2, P7, P8, P10, P12
W4	**Help to use component.** Architectural information and patterns help you to use the component in a useful way	P1, P10

4.2.1 Quality Assessment

An important use of AK is that it can be used in the assessment of a component with respect to its quality attributes. Participants particularly referred to the structure and patterns used in this context. Patterns are common solutions with predictable effects on a component's QAs. This could refer to a product's runtime attributes, such as performance and reliability, but also with respect to its build-time properties, such as the ease of integration. One participant explained: *"[Information of internal structure is] certainly very useful information, because such a pattern indicates what kind of architecture was used, and such a pattern indicates a certain quality. That could be configurability, but also the ease of integrating, those patterns secure a certain stability."* (P7).

4.2.2 Architectural Fit Assessment

Several participants indicated that architectural information provides useful insights into whether or not an OSS product is compatible at the architectural level with the main system in which it may be integrated. As noted by one participant: *"Well at least it indicates what style is being used, and what are the fundamental concepts, and your own product, or platform in our case, also has certain styles, and then you can see, whether it fits together. Does it make sense to try to put it together, or will we need a whole lot of glue code to be able to use it. And if so, is it worth our while, or is it easier to develop ourselves. At a high level it indicates whether something is a good match with your product, with the existing software."* (P10).

4.2.3 Maintenance and Evolution

A number of participants indicated that knowledge of the patterns used in a component could help in the maintenance phase of a product, including defect fixing. Patterns increase the understandability of how a component is constructed, which makes it easier to change the component. One participant described this as follows: *"If you have some OSS that's constructed using good patterns, then of course it'll help you to make changes. If you get OSS that's constructed in an ad-hoc manner, organically grown, and you decide to make changes to make it fit better, then that's completely useless. The more it is structured, thought about it in terms of layers and standard structures, the easier it will be to adopt it and change it."* (P12).

While the actual presence of patterns improves a product's maintainability, it is important to understand *which* patterns have been used, in order to make changes that do not degrade the pattern's integrity and consistency.

4.2.4 Using the Component

A good knowledge of architectural aspects of a component can help an integrator to understand the appropriate use of that component. During our study, a few participants also indicated that AK such as the patterns used provides valuable insights into whether or not a component will be suitable in the context of a given system and how to use that component. One of the participants explained: *"if you wouldn't know those [architectural concepts and patterns], it's difficult to use such a framework in a useful way. So in that case it would be difficult to find everything in the code, so at least you'd like to see some high level descriptions of the concepts and design patterns."* (P10).

4.3 RQ3: The Availability of Architectural Knowledge

In this section, we address the question whether or not AK is usually available to OSS integrators. We identified two themes, namely (1) the general availability of AK and (2) how integrators deal with the lack of AK. We decided to present this analysis in a descriptive way, rather than a tabular representation as used in 4.1 and 4.2.

4.3.1 Availability Differs Per Product

The results indicate that whether or not AK is available, depends on the product. One participant highlighted that: *"[for] open source it's quite easy to figure out how it fits into my architecture. If I look at Spring, how does this fit into my architecture, is it MVC or part of MVC... I also think that is indirectly a big plus for OSS, because they actually need to be compatible, they need to be able to integrate with other parts at the architecture level, otherwise they wouldn't survive."* (P1). This suggests that AK is sufficiently available. Another participant confirmed that for well-known frameworks such as the Spring framework[2], this information is indeed quite easily available, however, this may not be the case with other components: *"in case of Spring yes, but in other components not so much, and you just have to look in the documentation and in some cases even in the implementation."* (P3).

Participants P5 and P7 explicitly claimed that there is typically not much AK of OSS products available. One participant emphasized that a good design will result in good quality code, and makes the component author's intents clear: *"I don't usually have much architectural info when investigating an OSS component, unless it's in the docs, on the website etc. Historically, there hasn't been a focus on architecture patterns by OSS component authors. It's not a best practice to include architectural patterns information."* (P5)

4.3.2 How Integrators Deal with a Lack of Architectural Knowledge

The participants of our study mentioned different approaches to deal with a lack of AK of an OSS product. We present the main findings under the following categories:

Look at Others. One way to assess the quality of a product's QAs, is to look at other customers of the component, as explained by one participant: *"Our line of thought with respect to the 'ilities' was, as long as a rather large group of people uses it, you may assume that most of the trouble with the 'ilities' are solved."* (P6).

Assume the Worst. Another approach to deal with a component's characteristics, such as security, is to assume the worst, and make sure that the rest of the system compensates for its shortcomings: *"I don't think I would try to understand whether a certain OSS product, whether it's secure enough. I think I would assume that it's not secure enough and then I would make sure that the environment in which the software is run takes care of the security."* (P3).

Try to Recover. If architectural information in the documentation is missing, then the only solution to this is to look into the implementation or to ask the community for

---

[2] http://www.springsource.org/

more information. However, in such a case, the time to get a reply was considered to be very important due to development schedules and deadlines. Furthermore, while studying the implementation was mentioned as a solution, it should be considered as a last resort: *"When you have to go into the code to know how it works, then I do think you have a problem. [...] I think open source has much potential but the investment in knowledge is quite an issue."* (P9).

4.4 RQ4: The Relative Importance of Architectural Knowledge

Our last research question investigates the participants' perceived importance of AK. This puts the need for AK in perspective compared to other selection criteria, which are known to be important, such as availability of support and license. One participant considered AK as just one type of information that is needed, but stated that: *"the more you know, the better"* (P10). As for RQ3, we identified a few themes, which we discuss next.

4.4.1 Missing AK Hinders Usage

In subsection 4.3.1, participant P1 indicated that the availability of AK of an OSS product is important for the project's survival. Three participants explicitly indicated that the lack of an OSS component's AK means that it drops on the list of candidates, and hinders its usage. This means that a lack of AK negatively affects the selection decision of whether to use the OS product. One participant explained: *"Well, then you're making a big investment if you do that. So I can think it can hinder you in using it. When you have to go into the code to know how it works, then I do think you have a problem."* (P9).

4.4.2 Need for AK Depends on Product Type and Size

A recurring factor that influenced the participants' interest in having AK of an OSS product is the *type* and *size* of the component. The information needs heavily depend on whether it is a library that provides functionality or non-critical parts, such as graphical widgets, or that it is a central component that makes up an important part of a system. One participant described it as follows: *"If I'm looking for an MVC framework, then yes [I'll be interested in architectural information]. But if I'm looking for a foo-munging module, and there are 12, then the last 3 release dates, smoke reports, browsing source, etc., are likely more important."* (P5).

A few participants explicitly highlighted the architectural impact that an OSS component may have on the existing system architecture. One participant described: *"But for instance the reporting engine, if we choose product A or B, that was more of a feature level, and not typically on the architecture level, even if we would pick product A or B, it wouldn't ruin our existing architecture because this is more on the side."* (P1).

Related to this issue is whether practitioners value a component's *functionality* or *architecture*. On the one hand, some participants highlight the focus on functional compliance; one participant explained this concisely: *"My software MUST meet the functional requirements, and MAY have a good architecture."* (P5, emphasis by participant in IM transcript).

Furthermore, practitioners' interest in a component's functionality or architecture also depends on the type of the component. One participant explained: *"Well, frankly it's always been the functionality for us than that we're looking at the architecture. That may have to do with the type of components that we integrate. They're typically not complete subsystems, but rather limited libraries. Just those things that don't really count."* (P10).

On the other hand, some participants preferred a well-designed component with a good architecture to the functionality provided by the component. Extensibility and size of the component are decisive factors, as one participant explained: *"I would tend to choose for good quality and architecture, but perhaps a lack of functionality. But there should be a good possibility to extend that functionality, and when that's not there, then I may decide to take a chance and select the thing with all functionality but less quality, and to fix that thing myself. That also depends on the size by the way, when it's a huge project, then I won't start on that."* (P4).

4.4.3 The Relative Importance of Architectural Knowledge

So far, the answer to the question whether or not AK is an important factor is: "it depends". Factors are the type and size of a component (e.g., library versus framework). In general, the participants of our study seem to be quite interested in a component's AK, which can provide valuable insights that may affect the decision to use the component. One participant summarized this as follows: *"It depends on what it is, but I think if it is something that you're interested in anyway, in how it's constructed, which can be part of the evaluation of the piece of open source, then it's certainly interesting. It can give you a good feeling that people have thought about it. It fits or doesn't fit with what we have. So yes, it's certainly valuable."* (P10).

The participants generally agreed that the quality of a component can have many aspects, but that the critical selection factors depend on the context. One participant highlighted the importance of the implementation language in these words: *"The use of patterns in OSS components played some part when considering their use, but the flexibility of Perl allows integration of widely-varying programming models, so for Perl components, the test coverage, documentation and community support were more important."* (P5).

5 Discussion and Conclusion

In Section 4, we have identified the types of AK of OSS needed by OSS integrators and the main reasons for why the AK is important. It is interesting to note that the categories identified in Section 4.1 (*what* AK is needed) and 4.2 (*why* is AK needed) are quite similar. Participants indicated the need of understanding a component's structure, its quality attributes and behavior regarding e.g., system resources, whether or not components are an architectural "fit" with the main system's design, and how to use a component. Following are the main reasons *why* OSS integrators consider architectural knowledge of a component to be valuable:

- Assess the quality of the component;
- Assess the architectural fit of the component;
- Affect the maintainability of the component;
- Understand how the component can be used.

Knowledge of a component's structure (including its patterns) seems to be the most important aspect for integrators. While the participants indicated a desire to know a component's QAs, its architectural fit (compatibility), and how to use it, it seems that the architectural structure of a component is valuable input to satisfy most AK needs. A component's architecture structure, including its patterns, directly affects its QAs, architectural compatibility, and can provide insight on how to use the component.

The four categories of why an integrator would like to have AK of a component can be mapped to the three main phases in CBSD briefly outlined in Section 2.1. Quality assessment and assessment for architectural fit are both activities that are performed in the *Evaluation* phase, when components are evaluated and selected. Understanding of how to use a component is important in the *Integration* phase, when components are integrated into a product. Finally, having AK to improve a component's maintainability supports the *Maintenance* phase of OTS integration, after the main system has been deployed. Therefore, we can conclude that AK can support the OSS integrator in all three phases of CBSD.

However, in Section 4.3 we found that availability of AK of OSS products may vary, and just how much AK is needed also varies. When AK is not available, OSS integrators typically do not try to recover it, which means they have to take a different strategy to deal with such lack of information. From Section 4.4 it has become clear that practitioners do consider AK to be valuable, but that depends on the type and size of the product. A lack of AK, however, was shown to be a potential obstacle for using a component. These results highlight the importance to investigate how OSS integrators can be supported. We suggest a research agenda along the following lines:

- We found that the type and size of a product affect whether or not an OSS integrator needs to acquire AK of the product. It would be valuable to gain a deeper insight into how these characteristics affect the need for AK, what other factors are at play, and to develop a systematic method to assist OSS integrators in identifying what AK is needed and how AK can be identified in an effective way.

- Some OSS projects are more successful in publishing AK of the product than others, and one of our participants suggested that it is vital for a project's survival. What factors affect whether a community makes such AK available, and how can other OSS communities be supported in this activity?

- Our results suggest that if AK is not available, OSS integrators do not try hard to recover it. Furthermore, looking into the source code seemed to be the only solution. However, it would be valuable to, based on a larger scale survey, get better insights into whether OSS integrators recover AK, how they use this AK, whether they store it, and whether this AK is contributed back to the OSS community.

- Related to using OSS components is the integration of so-called "Inner Source Software" (ISS); ISS is closed-source software developed within an organization that has adopted OSS development methods, a phenomenon known as "Inner Source" [43]. In Inner Source, departments can be consumers and producers in an internal software market ("bazaar"). It would be very informative to understand what kind of AK integrators need and have available in Inner Source, and what lessons can be drawn from Inner Source to use in OSS integration (and vice versa).

5.1 Limitations of This Study

We are aware of a few limitations of this study. Firstly, we based this paper on data gathered through 12 interviews, which is insufficient to draw general conclusions. However, since the role of SA has not been studied extensively in the context of OSS integration, we decided to perform an *exploratory* study. Once this field has matured and specific hypotheses have been defined based on initial findings, we argue that other types of studies with larger numbers of participants will be more appropriate, such as questionnaire-surveys.

Secondly, our sample of participants was a convenience sample, which means there is a selection bias. Participants were contacted through our professional network. Furthermore, seven participants worked at one organization that is active in a safety critical domain; this may have biased the participants' views towards certain concerns. However, we did not find significant differences with respect to the participants' opinions and needs of AK. We argue that, since this is an exploratory study, these results can be used as input to identify hypotheses to design studies based on larger samples of participants from a variety of product domains.

5.2 Conclusion

This paper presents the results of an exploratory interview-based survey of software architects and other IT professionals to investigate the importance of architectural knowledge (AK) in the integration of OSS products CBSD. In particular, this paper presents a classification of different types of AK that is considered to be useful, why AK is needed, whether AK is available, and the relative importance of AK in the integration of OSS components. Knowledge of a component's partitioning and its patterns used seems to be particularly important to satisfy the various reasons of why AK is needed (e.g., assessment of quality, architectural fit, etc.).

Despite an increasing attention for SA in software engineering, SA has received little attention in the OSS research community. This paper provides empirical findings to demonstrate that integrators have a need for AK of OSS products. Based on our findings, we suggested a number of open research questions in order to bridge the gap between the OSS and SA research communities.

Acknowledgements. This work is partially funded by IRCSET under grant no. RS/2008/134 and by Science Foundation Ireland grant 03/CE2/I303_1 to Lero (www.lero.ie). We are grateful to the participants of our study for their time and enthusiasm.

References

[1] Hauge, Ø., Ayala, C., Conradi, R.: Adoption of Open Source Software In Software-Intensive Organizations A Systematic Literature Review. Information and Software Technology 52(11), 1133–1154 (2010)

[2] Hauge, Ø., Sørensen, C.-F., Røsdal, A.: Surveying Industrial Roles in Open Source Software Development. In: Feller, J., Fitzgerald, B., Scacchi, W., Sillitti, A. (eds.) Open Source Development, Adoption and Innovation, pp. 259–264. Springer, Heidelberg (2007)

[3] Garlan, D., Allen, R., Ockerbloom, J.: Architectural mismatch: why reuse is so hard. IEEE software 12(6), 17–26 (1995)

[4] Hissam, S.A., Weinstock, C.B.: Open Source Software: The Other Commercial Software. In: Feller, J., Fitzgerald, B., van der Hoek, A. (eds.) 1st Workshop on Open Source Software Engineering, ICSE (2001)

[5] Morgan, L., Finnegan, P.: Benefits and Drawbacks of Open Source Software: An Exploratory Study of Secondary Software Firms. In: Feller, J., Fitzgerald, B., Scacchi, W., Sillitti, A. (eds.) Open Source Development, Adoption and Innovation, pp. 307–312. Springer, Heidelberg (2007)

[6] Ayala, C., Hauge, Ø., Conradi, R., Franch, X., Li, J., Velle, K.S.: Challenges of the Open Source Component Marketplace in the Industry. In: Boldyreff, C., Crowston, K., Lundell, B., Wasserman, A.I. (eds.) OSS 2009. IFIP AICT, vol. 299, pp. 213–224. Springer, Heidelberg (2009)

[7] Chen, W., Li, J., Ma, J., Conradi, R., Ji, J., Liu, C.: An empirical study on software development with open source components in the chinese software industry. Software Process: Improvement and Practice 13(1), 89–100 (2008)

[8] Jaaksi, A.: Experiences on Product Development with Open Source Software. In: Feller, J., Fitzgerald, B., Scacchi, W., Sillitti, A. (eds.) Open Source Development, Adoption and Innovation, pp. 85–96. Springer, Heidelberg (2007)

[9] Madanmohan, T.R., De, R.: Open source reuse in commercial firms. IEEE Software 21(6), 62–69 (2004)

[10] Ayala, C., Hauge, Ø., Conradi, R., Franch, X., Li, J.: Selection of third party software in Off-The-Shelf-based software development: An interview study with industrial practitioners. The Journal of Systems and Software 84(4), 620–637 (2011)

[11] Li, J., Conradi, R., Slyngstad, O.P.N., Bunse, C., Torchiano, M., Morisio, M.: Development with Off-the-Shelf Components: 10 Facts. IEEE software 26(2), 80–87 (2009)

[12] Bosch, J., Stafford, J.A.: Architecting Component-Based Systems. In: Crnkovic, I., Larsson, M. (eds.) Building Reliable Component-Based Software Systems. Artech House Publishers, Norwood (2002)

[13] Ali Babar, M., Dingsøyr, T., Lago, P., van Vliet, H.: Software Architecture Knowledge Management: Theory and Practice. Springer, Heidelberg (2009)

[14] Fitzgerald, B.: The transformation of open source software. MIS Quarterly 30(3), 587–598 (2006)

[15] Mäki-Asiala, P., Matinlassi, M.: Quality Assurance of Open Source Components: Integrator Point of View. In: Proceedings of the 30th Annual International Computer Software and Applications Conference (COMPSAC), pp. 189–194 (2006)

[16] Di Giacomo, P.: COTS and open source software components: Are they really different on the battlefield? In: Franch, X., Port, D. (eds.) ICCBSS 2005. LNCS, vol. 3412, pp. 301–310. Springer, Heidelberg (2005)

[17] Norris, J.S.: Mission-critical development with open source software: Lessons learned. IEEE Software (2004)

[18] Stol, K., Ali Babar, M.: A Comparison Framework for Open Source Software Evaluation Methods. In: Ågerfalk, P., Boldyreff, C., González-Barahona, J.M., Madey, G.R., Noll, J. (eds.) OSS 2010. IFIP AICT, vol. 319, pp. 389–394. Springer, Heidelberg (2010)

[19] Hauge, Ø., Østerlie, T., Sørensen, C.-F., Gerea, M.: An Empirical Study on Selection of Open Source Software - Preliminary Results. In: Capiluppi, A., Robles, G. (eds.) 2nd workshop on Emerging Trends in FLOSS Research and Development (ICSE), Vancouver, Canada (2009)

[20] Merilinna, J., Matinlassi, M.: State of the Art and Practice of Open Source Component Integration. In: Proceedings of the 32nd Euromicro Conference on Software Engineering and Advanced Applications (SEAA), pp. 170–177 (2006)

[21] Ven, K., Mannaert, H.: Challenges and strategies in the use of Open Source Software by Independent Software Vendors. Information and Software Technology 50(9-10), 991–1002 (2008)

[22] Shaw, M., Garlan, D.: Software Architecture: Perspectives on an Emerging Discpline. Prentice-Hall Inc., New Jersey (1996)

[23] Tang, A., Avgeriou, P., Jansen, A., Capilla, R., Ali Babar, M.: A comparative study of architecture knowledge management tools. Journal of Systems and Software 83, 352–370 (2010)

[24] Jansen, A., Bosch, J.: Software Architecture as a Set of Architectural Design Decisions. In: 5th Working IEEE/IFIP Conference on Software Architecture (WICSA), Pittsburgh, PA, USA, pp. 109–120 (2005)

[25] Buschmann, F., Meunier, R., Rohnert, H., Sommerlad, P., Stal, M.: Pattern-oriented Software Architecture - A System of Patterns. J. Wiley and Sons Ltd., Chichester (1996)

[26] Bass, L., Clements, P., Kazman, R.: Software Architecture in Practice, 2nd edn. Addison-Wesley, Boston (2003)

[27] Tran, J.B., Godfrey, M.W., Lee, E.H.S., Holt, R.C.: Architectural repair of open source software. In: Proceedings of the 8th International Workshop on Program Comprehension, IWPC (2000)

[28] Nakagawa, E., de Sousa, E., de Brito Murata, K., de Faria Andery, G., Morelli, L., Maldonado, J.: Software Architecture Relevance in Open Source Software Evolution: A Case Study. In: Proceedings of the 32nd International Computer Software and Applications Conference (COMPSAC), pp. 1234–1239 (2008)

[29] Capiluppi, A., Knowles, T.: Software engineering in practice: Design and architectures of FLOSS systems. In: Boldyreff, C., Crowston, K., Lundell, B., Wasserman, A.I. (eds.) OSS 2009. IFIP AICT, vol. 299, pp. 34–46. Springer, Heidelberg (2009)

[30] Matinlassi, M.: Role of Software Architecture in Open Source Communities. In: Proceedings of the Sixth Working IEEE/IFIP Conference on Software Architecture (WICSA), Mumbai, India (2007)

[31] Ali Babar, M., Fitzgerald, B., Ågerfalk, P.J., Lundell, B.: On the Importance of Sound Architectural Practices in the Use of OSS in Software Product Lines. In: Second International Workshop on Open Source Software and Product Lines, collocated with the 11th International Software Product Line Conference (2007)

[32] Ali Babar, M., Lundell, B., van der Linden, F.: A Joint Workshop of QACOS and OSSPL. In: Boldyreff, C., Crowston, K., Lundell, B., Wasserman, A.I. (eds.) OSS 2009. IFIP AICT, vol. 299, pp. 357–358. Springer, Heidelberg (2009)

[33] Lennerholt, C., Lings, B., Lundell, B.: Architectural issues in Opening up the advantages of Open Source in product development companies. In: Proceedings of the 32nd Annual IEEE International Computer Software and Applications Conference, pp. 1226–1227. IEEE Computer Society, Washington, DC, USA (2008)

[34] Arief, B., Gacek, C., Lawrie, T.: Software architectures and open source software-where can research leverage the most? In: Feller, J., Fitzgerald, B., van der Hoek, A. (eds.) 1st Workshop on Open Source Software Engineering, Collocated with the 23rd International Conference on Software Engineering, ICSE (2001)

[35] Stamelos, I., Kakarontzas, G.: AKM in Open Source Communities. In: Ali Babar, M., Dingsøyr, T., Lago, P., van Vliet, H. (eds.) Software Architecture Knowledge Management: Theory and Practice, pp. 199–215. Springer, Heidelberg (2010)

[36] Edmondson, A.C., McManus, S.E.: Methodological Fit in Management Field Research. Academy of Management Review 32(4), 1155–1179 (2007)

[37] Seaman, C.B.: Qualitative methods in empirical studies of software engineering. IEEE Transactions on Software Engineering 25(4), 557–572 (1999)

[38] Robson, C.: Real World Research: A Resource for Social Scientists and Practitioner-Researchers, 2nd edn. Blackwell Publishing, Malden (2002)

[39] Taylor, S.J., Bogdan, R.: Introduction to Qualitative Research. John Wiley & Sons, New York (1984)

[40] Harrison, N.B., Avgeriou, P.: How do architecture patterns and tactics interact? A model and annotation. The Journal of Systems & Software 83(10), 1735–1758 (2010)

[41] Ameller, D., Franch, X.: How Do Software Architects Consider Non-Functional Requirements: A Survey. In: Wieringa, R., Persson, A. (eds.) REFSQ 2010. LNCS, vol. 6182, pp. 276–277. Springer, Heidelberg (2010)

[42] Shaw, M.: Architectural issues in software reuse: it's not just the functionality, it's the packaging. SIGSOFT Softw. Eng. Notes, 20 (SI), 3-6 (1995)

[43] Wesselius, J.: The Bazaar inside the Cathedral: Business Models for Internal Markets. IEEE Software 25(3), 60–66 (2008)

Successful Reuse of Software Components: A Report from the Open Source Perspective

Andrea Capiluppi[1], Cornelia Boldyreff[1], and Klaas-Jan Stol[2]

[1] University of East London, United Kingdom
[2] Lero—the Irish Software Engineering Research Centre
University of Limerick, Ireland
{a.capiluppi,c.boldyreff}@uel.ac.uk, klaas-jan.stol@lero.ie

Abstract. A promising way of software reuse is Component-Based Software Development (CBSD). There is an increasing number of OSS products available that can be freely used in product development. However, OSS communities themselves have not yet taken full advantage of the "reuse mechanism". Many OSS projects duplicate effort and code, even when sharing the same application domain and topic. One successful counter-example is the FFMpeg multimedia project, since several of its components are widely and consistently reused into other OSS projects. This paper documents the history of the libavcodec library of components from the FFMpeg project, which at present is reused in more than 140 OSS projects. Most of the recipients use it as a black-box component, although a number of OSS projects keep a copy of it in their repositories, and modify it as such. In both cases, we argue that libavcodec is a successful example of reusable OSS library of components.

Keywords: Software reuse, OSS components, component-based software development.

1 Introduction

Reuse of software components is one of the most promising assets of software engineering [5]. Enhanced productivity (as less code needs to be written), increased quality (since assets proven in one project can be carried through to the next) and improved business performance (lower costs, shorter time-to-market) are often pinpointed as the main benefits of developing software from a stock of reusable components [35,31].

Although much research has focused on the reuse of Off-The-Shelf (OTS) components, both Commercial OTS (COTS) and OSS, in corporate software production [25,36], the reusability "of" OSS projects "in" other OSS projects has only started to draw the attention of researchers and developers in OSS communities [22,28,8]. A vast amount of code is created daily, modified and stored in OSS repositories, yet, software reuse is rarely perceived by OSS developers as a critical success factor in their projects or processes. For different and composite reasons [34], using other OSS projects as components is typically not

S.A. Hissam et al. (Eds.): OSS 2011, IFIP AICT 365, pp. 159–176, 2011.
© IFIP International Federation for Information Processing 2011

considered as a way to build new OSS products. As an example, a search for the *"FTP client"* topic in the SourceForge repository[1] results in more than 350 different projects, each implementing similar features in the same domain. As a result, much functionality is duplicated in similar products, with little sharing of existing components.

The interest of practitioners and researchers in the topic of software reuse has focused on two predominant questions: (1) how to select an OSS component to be reused in another (potentially commercial) software system, and (2) how to provide potential re-users with a level of objective "trust" in available OSS components. This interest is based on a sound reasoning; given the increasing amount of source code and documentation created and modified daily, it starts to be a (commercially) viable solution to browse for components in existing code and select existing, working resources to reuse as building blocks of new software systems, rather than building them from scratch.

Among the reported cases of successful reuse within OSS systems, components with clearly defined requirements, and hardly affecting the overall design (i.e., the "S" and "P" types of systems following the original S-P-E classification by Lehman [24]) have often proven the typical reused resources by OSS projects. Reported examples include the "internationalization" component (which produces different output text depending on the language of the system), or the "install" module for Perl subsystems (involved in compiling the code, test and install it in the appropriate locations) [28]. Little is known about successful cases of OSS reuse, and an understanding of internal characteristics of what makes a component reusable in the OSS context is lacking.

The main focus of this paper is to report on the successful reuse of the components of the `FFMpeg` project. This project is a cornerstone component in the multimedia domain; several dozens of OSS projects reuse parts of `FFMpeg`, and this wide-spread of reuse is mostly based upon the `libavcodec` library of components. In the domain of OSS multimedia applications, this library is now established as the most widely adopted and reused audio/video codec (**co**ding and **dec**oding) resource. Its reuse by other OSS projects is so widespread since it represents a cross-cutting resource for a wide range of systems, from single-user video and audio players to converters and multimedia frameworks.

This paper makes two contributions: first, it establishes that the `libavcodec` component (contained in `FFMpeg`) is an "evolving and reusable" component (an "E" type of system [24]), and as such it poses several interesting challenges when other projects integrate it. Second, it presents two scenarios that have emerged in the reuse of this resource: on the one hand, the majority of the cases the `libavcodec` component is reused as a "black-box", as such incurring into the synchronization issues due to the co-evolution "project+component". On the other hand, a subset of OSS projects apply a "white-box" reuse strategy, by maintaining a private copy of `libavcodec`. The latter scenario is empirically analyzed in order to obtain a better understanding of how the component not only is reused, but also integrated into the main system. The two scenarios are

---

[1] http://sourceforge.net/

summarized in Figure 1: as an example, the MPlayer project keeps a "copy" of the library in its repository (white-box reuse), while the VLC project, at compilation time, requires the user to provide the location of an up-to-date version of the FFMpeg project (black-box reuse).

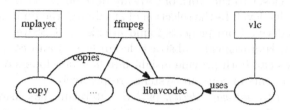

Fig. 1. Black-box and white-box reuse

This paper proceeds as follows. Section 2 provides an overview of the related work on software components and OSS systems. Section 3 provides the definitions and the empirical approach used throughout the paper. Section 4 presents the results of the empirical study of the OSS projects showing a white-box reuse strategy of the libavcodec component. Section 5 discusses the threats to validity of this study. Section 6 concludes.

2   Background and Related Work

The OSS approach to software development has gained much attention in the empirical Software Engineering research community, mostly due to the availability of software and non-software artifacts (*e.g.*, bug tracking systems and mailing lists). Although the majority of published works have a non OSS-related rationale, some researchers have started to collect evidence specifically related to OSS systems. Among these late emerging areas, the topics of OSS components and architectures have been investigated both within research works [27,18,7,25], and through specifically funded EU projects (QualiPSo[2] – Quality Platform for Open Source Software and QUALOSS[3] – QUALity in Open Source Software). This research directly responds to the need of identifying and extracting existing OSS components [2], or of providing options for choosing the best OSS component for inclusion in a software system [18].

This work is also related to the study of *software architectures*, in the forms of hierarchical and coupling views. Previous works ([19,21,38]) have defined and used different views of architecture of a software system. For example, Kruchten [21] refers to a "4+1" view model to describe a system involving logical, process, physical, development views, and use-cases. This model defines different perspectives for different stakeholders; the present work uses the concepts of logical

---

[2] http://www.qualipso.org/
[3] http://www.qualoss.org/

("hierarchical") and process ("coupling") views to establish a comparison between these two views. Similarly, [19] defines four architectural views of software systems, which in turn focus on coarser degrees of granularity (conceptual, or the abstract design level; module, or the concrete design level; code, or components level; and execution level). As stated above, the present research focuses on the views which are closer to the work of software developers, such as, for instance, the folder or the file level. In the selection of attributes, the limit is on those that it is possible to derive from projects found in existing OSS repositories with a reasonable effort. Hierarchical ("abstract design level") and coupling ("component level") views can both provide insight into how developers deal with macro and micro-components of software systems, respectively.

With reference to *software decay*, past SE literature has firmly established that software architectures and the associated code degrade over time [13], and that the pressure on software systems to evolve in order not to become obsolete plays a major role in their evolution [23]. As a result, software systems have the progressive tendency to loose their original structure, which makes it difficult to understand and further maintain them [33]. Among the most common discrepancies between the original and the degraded structures, the phenomenon of highly coupled, and lowly cohesive, modules has already been known since 1972 [30] and is an established topic of research. *Architectural recovery* is one of the recognized counter-measures to this decay [12]. Several earlier works have been focused on the architectural recovery of proprietary [12], closed academic [1], COTS [4] and FLOSS [6,17,37] systems; in all of these studies, systems were selected in a specific state of evolution, and their internal structures analysed for discrepancies between the *folder-structure* and *concrete* architectures [37]. Repair actions have been formulated as frameworks [32], methodologies [20] or guidelines and concrete advice to developers [37].

3   Empirical Approach

The approach of building by decomposition into, and the composition of, several components is a common scenario when considering OSS systems. Perhaps the best-known example are Linux distributions, which are collections of projects, libraries and components, which request or provide services to components via connections. This has been reported in various studies [15,25], especially relating to the issues of OSS licenses [16]. Apart from systems composed of subsystems which are already OSS projects, it is essential that empirical knowledge on reuse and domain engineering is based on finer-grained components, smaller than entire systems (as in the LAMP – Linux, Apache, MySQL, Python/Perl/PHP – stack of reuse).

The FFMpeg project has been chosen as an example of software reuse for several reasons:

1. It has a long history of evolution as a multimedia player, that has grown and refined several build-level components throughout its life-cycle. Some of

these components appear like "E" type systems, instead of traditional "S" or "P" types, with lower propensity for software evolution.

2. Several of its core developers have been collaborating also in the MPlayer[4] project, one of the most commonly used multimedia players across OSS communities. Eventually, the libavcodec component has been incorporated (among others from FFMpeg) into the main development trunk of MPlayer, increasing FFMpeg's visibility and wide-spread usage.

3. Its components are currently reused on different platforms and architectures, both in static- and in dynamic-linking. Static linking involves the inclusion of source code at compile-time, while dynamic linking involves the inclusion of a binary library at runtime.

4. Finally, the static-linking reuse of the FFMpeg components presents two opposite scenarios: either a black-box reuse strategy, with "update propagation" issues reported when the latest version of a project has to be compiled against a particular version of the FFMpeg components [29]; or the white-box reuse strategy, with copies of the components being deployed in the repositories of other projects which are managed independently from the their original development branch.

3.1   Definitions and Operationalization

This paper is built on top of two basic architectural principles: the concept of *build-level components* [11] and the principle of *architectural decay* along the evolution of software systems [13]. The build-level components are *"directory hierarchies containing ingredients of an application's build process, such as source files, build and configuration files, libraries, and so on. Components are then formed by directories and serve as unit of composition"* [11], and these compose the "folder-structure" or "tree-structure" of a software system [9,10].

In this paper we use terminology and definitions provided in related and well-known past studies. The definition of *common coupling* (intended for both object-oriented [3,26] and procedural [14] languages). The following operational definitions have been used:

- **Coupling:** this is the union of all the *includes, dependencies* and *functions calls* (i.e., the common coupling) of all source files as extracted by the Doxygen tool[5]. Since the empirical study is based on the definition of build-level components, two further conversions have been made:

  1. The *file-to-file* and the *functions-to-functions* couplings have been converted into *folder-to-folder* couplings, considering the folder that each of the above elements belongs to. A stronger coupling link between folder A and B will be found when many elements within A call elements of folder B.

---

[4] MPlayer, http://www.mplayerhq.hu
[5] http://www.stack.nl/~dimitri/doxygen/

   2. Since the behavior of "build-level components" is studied here, the couplings to subfolders of a component have also been redirected to the component alone; hence a coupling $A \rightarrow B/C$ (with C being a subfolder of B) is reduced to $A \rightarrow B$.

– **Connection:** distilling the couplings as defined above, one could say, in a Boolean manner, whether two folders are linked by a *connection* or not, disregarding the strength of the link itself. The overall number of these connections for the FFMpeg project is recorded monthly in Figure 2; the connections of a folder to itself are not counted (for the encapsulation principle), while the two-way connection $A \rightarrow B$ and $B \rightarrow A$ is counted just once (since we are only interested in which folders are involved in a connection).

– **Cohesion:** for each component, the sum of all couplings, in percentage, between its own elements (files and functions);

– **Outbound coupling** (fan-out): for each component, the percentage of couplings directed from any of its elements to elements of other components, as in requests of services. A component with a large fan-out, or "controlling" many components provides an indication of poor design, since the component is probably performing more than one function.

– **Inbound coupling** (fan-in): for each component, the percentage of couplings directed to it from all the other components, as in "provision of services". A component with high fan-in is likely to perform often-needed tasks, invoked by many components, which is regarded as an acceptable design behavior.

The source code repository (CVS) of FFMpeg was parsed monthly, resulting in some 100 temporal points, after which the tree structures were extracted for each of these points. On the one hand, the number of source folders of the corresponding tree is recorded in Figure 2. On the other hand, in order to produce an accurate description of the tree structure as suggested by [37], each month's data has been further parsed using Doxygen, with the aim of extracting the common coupling among the elements (i.e., source files and headers, and source functions) of the systems. The analysis of size growth has been performed using the sloccount tool[6].

3.2   Description of the FFMPeg System

As mentioned above, the FFMpeg system has successfully become a highly visible OSS project partly due to its components, libavcodec in particular, which have been integrated into a large number of OSS projects in the multimedia domain.

   In terms of a global system's design, the FFMpeg project does not yet provide a clear description of either its internal design, or how the architecture is decoupled into components and connectors. Nonetheless, by visualizing its source tree composition [10], the folders containing the source code files appear to be semantically rich, in line with the definitions of *build-level components* [11], and *source tree composition* [9,10]. The first column of Table 1 summarizes which folders currently contain source code and subfolders within FFMpeg.

---

[6] http://www.dwheeler.com/sloccount/

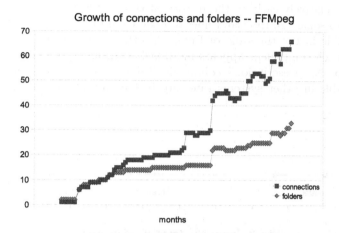

Fig. 2. Growth of folders and connections

As shown, some components act as containers for other subfolders, apart from source files, as shown in columns 2 and 3, respectively. Typically these subfolders have the role of specifying/restricting the functionalities of the main folder in particular areas (e.g., the libavutil folder which is further divided into the various supported architectures – x86, ARM, PPC, etc.). The fourth column also describes the main functionalities of the component. It can be observed that each directory provides the build and configuration files for itself and the subfolders contained, following the definition of build-level components. The fifth column of Table 1 lists the month when the component was first detected in the repository. Apart from the miscellaneous tools component, each of these are currently reused as OSS components in other multimedia projects as development libraries, for example, the libavutil component is currently redistributed as the libavutil-dev package).

Table 1 shows that the main components of this system have originated at different dates, and that the older ones (i.e., libavcodec) are typically more articulated into several directories and multiple files. The libavcodec component

Table 1. FFMpeg (build-level) components

Name	Folders	Files	Description	Date
libavcodec	12	625	Extensive audio/video codec library	08/2001
libpostproc	1	5	Library containing video postprocessing routines	10/2001
libavformat	1	205	Audio/video container mux and demux library	12/2002
libavutil	8	70	Shared routines and helper library	08/2005
libswscale	6	20	Video scaling library	08/2006
tools	1	4	Miscellaneous utilities	07/2007
libavdevice	1	16	Device handling library	12/2007
libavfilter	1	11	Video filtering library	02/2008

was created relatively early in the history of this system (08/2001), and it has now grown to some 220 thousand lines of code (KSLOC) alone.

As is visible in the time-line of Figure 3, other components have coalesced since then; each component appears modularized around a specific "function", according to the "Description" column in Table 1, and as such have become better reusable in other systems (and are in fact repackaged as distinct OSS projects).

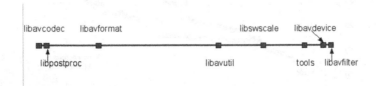

Fig. 3. Inception dates of components

4   Results and Discussion

This section provides the results of the empirical investigation into both the growth in size, and the evolution of connections between the components of FFMpeg. For each build-level component summarized in Table 1, a study of its relative change in terms of the contained SLOC (source lines of code) along its life-cycle has been undertaken. In addition, a study of the architectural connections has been performed, by analyzing temporally:

1. How many couplings were actually involved with elements of the same component (as per the definition of *cohesion* given above), and
2. How many couplings consisted of links to or from other components (as per the definition of *inbound* and *outbound* couplings).

4.1   Size Growth of FFMpeg Components

As a general result, two main behaviors can be observed, which have been clustered in the two graphs of Figure 4; on the top graph, three components (libavcodec, libavutil and libavformat) show a linear growth as a general trend (relative to the maximum size achieved by each). In the following, these components will be referred to as "E-type". On the other hand, the rest of FFMpeg components show a traditional library behavior, and will be referred as either "S-type" or "P-type" systems.

Size Growth in E-Type Components. Considering Figure 4 (top), the libavcodec component started out as a medium-sized component (18 KSLOCs), but currently its size has reached over 220 KSLOCs, an increase of 1, 100%. Also, the libavformat component has moved through a comparable pattern of growth (250% increase), but with a smaller size overall (from 14 to 50 KSLOC). Although reusable resources are often regarded as "S-type" and "P-type" systems, since their evolutionary patterns manifest a reluctance to growth (as in the typical behavior of software libraries), these two components achieve an "E-type" evolutionary pattern even when heavily reused by several other projects. The studied cases appear to be driven mostly by adaptive maintenance, since new audio and video formats are constantly added and refined among the functions of these components.

Expressing these observations in biological terms, these software components appear and grow as "fruits" from the main "plant" ("trunk" in the version control system). Furthermore, these components behave as "climacteric" fruits, meaning that they ripen off the parent plant (and in some cases, they must be picked in order to ripen). These FFMpeg components have achieved an evolution even when separated from the project they belonged to, similarly to climacteric fruits.

Size Growth in S- and P-Type Components. On the other hand, the remaining components show a more traditional, library-style type of evolution: the bottom part of Figure 4 details the relative growth of these components. Libpostproc and libswscale appear hardly changing at all, even if they have been formed for several years in the main project. Libavdevice, when created, was already at 80% of its current state; libavfilter, instead, although achieving a larger growth, does so since it was created at a very small stage (600 SLOC), which has now doubled (1,4 KSLOCs). These resources are effectively library-type of systems, and their reuse is simplified by the relative stability of their characteristics, meaning the type of problem they solve. Using the same analogy as above, the components ("fruits") following this behavior are unlikely ripen any further once they have been picked. Outside of the main trunk of development, these components remain unchanged, even when incorporated into other OSS projects.

4.2 Architectural Growth of FFMpeg Components

The observations related to the growth in size have been used to cluster the components based on their coupling patterns. As mentioned above, each of the 100 monthly check-outs of the FFMpeg system have been analyzed in order to extract the common couplings of each element (functions or files), and these common couplings have been later converted into connections between components.

As observed also with the growth in size, the E-type components present a steadily increasing growth of couplings compared to the S- and P-type components. The former also display a more modularized growth pattern, resulting in a more stable and defined behavior.

Coupling Patterns in E-type Components. Figure 5 proposes the visualization of the three E-type components as identified above. For each component, 4 trends are displayed:

1. The overall amount of its common couplings;
2. The amount of couplings directed towards its elements (*cohesion*, labeled "self");

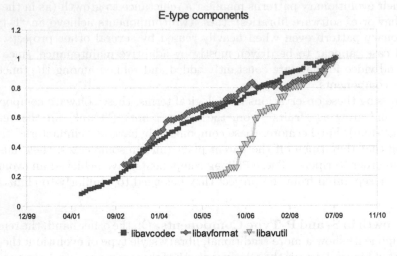

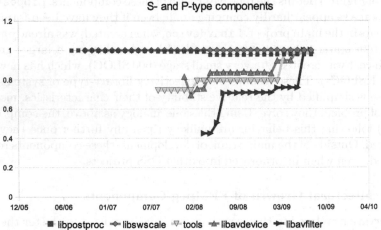

Fig. 4. E-type (top) and S- and P-type of components (bottom) – growth in size

3. The amount of its outbound couplings (*fan-out*, labeled "out");
4. The amount of its inbound couplings (*fan-in*, labeled "in");

Each component has a continuous growth trend regarding the number of couplings affecting it. The `libavutil` component has one sudden discontinuity in this growth, which will be later explained. As a common trend, it is also visible that both the `libavcodec` and `libavformat` components have a strong cohesion factor, which maintains over the 75% threshold throughout their evolution. This means that, in these two components, more than 75% of the total number of couplings are consistently between internal elements. The cohesion of `libavutil`, on the other hand, degrades until it becomes very low, revealing a very high fan-in; after the restructuring at around 1/5 of its lifecyle, this component becomes a pure server, fully providing services to other components (more than 90% of all its couplings – around 3,500 – come from external components).

When observing the three components as part of a common, larger system, the changes in one component become relevant to the other components as well. As an example, the general trend of libavcodec is intertwined to the other two components in the following ways:

1. The overall number of couplings towards its own elements decreased during a time interval when no further couplings were added, therefore its *cohesion* has degraded;
2. At the same time, its *fan-out* suddenly increased, topping some 17% at the latest studied point: observing carefully, the larger amount of requests of service were more and more directed towards libavutil, which around the same period experienced a sudden increase of its fan-in;
3. Also, the fan-in of libavcodec decreased: originally, the major cause of this was due to numerous requests from the libavformat component. Throughout the evolution, these links were converted into connections to libavutil instead.

Performing a similar analysis for libavformat, it becomes clear that its fan-out degrades, becoming gradually larger, the reason being an increasingly stronger link to the elements of both libavcodec and libavutil. This form of inter-component dependencies is a form of architectural decay: at the latest available data point (08/20009), this has been reproduced in Figure 6.

This graph shows the typical trade-offs between encapsulation and decomposition: several of the common files accessed by both libavformat and libavcodec have been lately moved to a third location (libavutil), that acts as a server to both. This in turns has a negative effect on reusability: when trying to use the functionalities of libavcodec, it will be necessary to import also the contents of libavutil. Even worse, when trying to reuse the attributes of libavformat, the connections to both libavutil and libavcodec have to be restored.

Coupling Patterns in S- and P-type Components. The characteristics of the E-type components as described above can be summarized as follows: large cohesion, fan-out under a certain threshold, and clear, defined behavior as a component (e.g., pure "server" as achieved by the libavutil component).

The second cluster of components identified above (the "S-" and "P-type") revealed several discrepancies from the results observed in subsection 4 2. A list of key results is summarized here.

1. As also observed for the growth of components, the number of couplings affecting this second cluster of components reveals a difference of one (libswscale, libavdevice and libavfilter) and even two (libpostproc) orders of magnitude with respect to the E-type components.

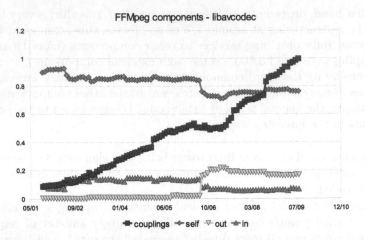

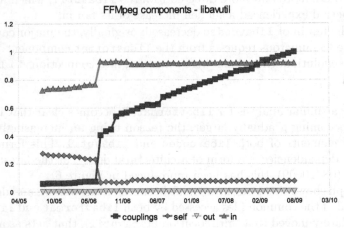

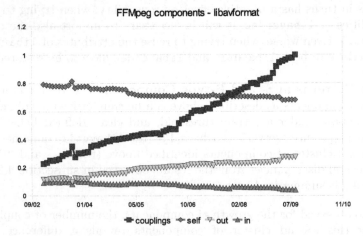

Fig. 5. E-type components – coupling patterns

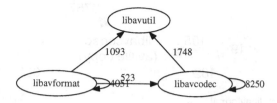

Fig. 6. Effects of excessive fan-out

2. Slowly growing trends in the number of couplings were observed in libavdevice and libavfilter, but their cohesion remains stable. On the other hand, a high fan-out was consistently observed in both, with values of 0.7 and 0.5, respectively. Observing more closely, these dependencies are directed towards the three E-type components defined above. This suggests that these components are not yet properly designed, also due to their relatively young age: their potential reuse is subsumed to the inclusion of other FFMpeg libraries as well.

As a summary, this second type of components can be classified as slowly growing, less cohesive and more connected with other components in the same system. They can be acceptable reusable candidates, but only in conjunction with the whole, hosting project (i.e., FFMpeg).

4.3   Deployment of Libavcodec in Other OSS Projects

The three components libavcodec, libavformat and libavutil have been characterized above as highly reusable, based on coupling patterns and size growth attributes. In order to observe how these components are actually reused and deployed in new hosting systems, this Section summarizes the study of the deployment of the libavcodec component in 4 OSS projects: avifile[7], avidemux[8], MPlayer and xine[9].

The selection of these projects for the deployment study is based on their current reuse of these components. Each project hosts a copy of the libavcodec component in their code repositories, therefore implementing a white-box reuse strategy of this resource. The issue to investigate is whether these hosting projects maintain the internal characteristics of the original libavcodec, hosted in the FFMpeg project. In order to do so, the coupling attributes of this folder have been extracted from each OSS project, and the number of connected folders has been counted, together with the total number of couplings.

Each graph in the Figure 7 represents a hosting project: the libavcodec copy presents some degree of cohesion (the re-entrant arrow), and its specific fan-in and fan-out (inwards and outwards arrows, respectively). The number of

---

[7] http://avifile.sourceforge.net/
[8] http://fixounet.free.fr/avidemux/
[9] http://www.xine-project.org/home

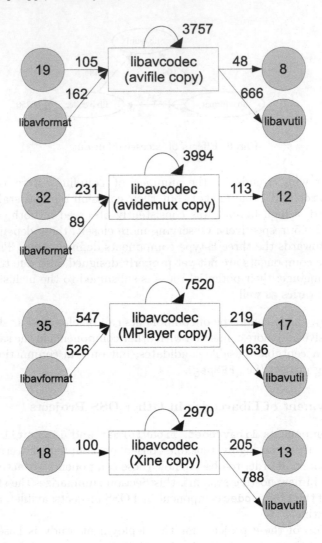

Fig. 7. Deployment and reuse of libavcodec

connections (i.e., distinct source folders) responsible for the fan-in and fan-out are displayed by the number in the circle. The following observations can be made:

- The total amount of couplings in each copy is always lower than the original FFMpeg copy: this means that not the whole FFMpeg project is reused, but only some specific resources;
- In each copy, the ratio $fan-in/fan-out$ is approximately 2:1. In the xine copy, this is reversed: this is due to the fact that apparently xine does not host a copy of the *libavformat* component;

- For each graph, the connections between `libavcodec` and `libavutil`, and between `libavcodec` and `libavformat` have been specifically detailed: the fan-in from `libavformat` alone has typically the same order of magnitude than all the remaining fan-in;
- The fan-out towards `libavutil` typically accounts for a much larger ratio. This is a confirmation of the presence of a consistent dependency between `libavcodec` and `libavutil`, which therefore must be reused together. The `avidemux` project moved the necessary dependencies to `libavutil` within the `libavcodec` component; therefore no build-level component for `libavutil` is detectable.

5   Threats to Validity

We are aware of a few limitations of this study, which we discuss next. Since we do not claim any causal relationships, we do not discuss threats to internal validity.

Construct Validity. We used common coupling to represent inter-software component connections. Furthermore, the build-level components presented in Table 1 are automatically assigned (though probably accurate), but could be only subcomponents of a larger component (e.g., composed of both `libavutil` and `libavcodec`).

External Validity. External validity is concerned with the extent to which the results of our study can be generalized. In our study, we have focused on one case study (`FFMPeg`), which is written mostly in C. Performing a similar study on a system written in, for instance, an object-oriented language, the results could be quite different. However, it is not our goal to present generalizations based on our results. Rather, the aim of this paper is to document a successful case of OSS reuse by other OSS projects.

6   Conclusions

Empirical studies of reusability of OSS resources should proceed in two strands: first, they should provide mechanisms to select the best candidate component to act as a building block in a new system; second, they should document successful cases of reuse, where an OSS component(s) has been deployed in other OSS projects. This paper attempts to give a contribution to the second strand by empirically analysing the `FFMpeg` project, whose components are currently widely reused in several multimedia OSS applications. The empirical study was performed on a monthly basis during the last 8 years of its development: the characteristics of its size, the evolutionary growth and its coupling patterns were extracted, in order to identify and understand the attributes that made its components a successful case of OSS reusable resources. After having studied these characteristics, 4 OSS projects were selected among the ones implementing a white-box reuse of the `FFMpeg` components: the deployment and the reuse of these components was studied from the perspective of their interaction with their hosting systems.

In the FFMpeg study, a number of findings were obtained: first, it was found that several of its build-level components make for a good start in the selection of reusable components. They coalesce, grow and become available at various points in the life cycle of this project, and all of them are currently available as building blocks for other OSS projects to use. Second, it was possible to classify at least two types of components: one set presents the characteristics of evolutionary (E-type) systems, with a sustained growth throughout. The other set, albeit with a more recent formation, is mostly unchanged, therefore manifesting the typical attributes of reusable libraries.

The two clusters were compared again in the study of the connections between components: the first set showed components with either a clearly defined behavior, or an excellent cohesion of its elements. It was also found that each of these three components becomes more connected to the others, as forming one single super-component. The second set appeared less stable, with accounts of large fan-out, which called for a poor design of the components.

One of the reusable resources found within FFMpeg (i.e., libavcodec) were analysed when deployed into 4 OSS systems performing a white-box reuse: its cohesion pattern appeared similar to the original copy of libavcodec, while it emerged with more clarity that at present its reuse is facilitated when the libavformat and libavutil components are reused too.

Acknowledgements. The authors would like to thank Daniel German for the clarification on the potential conflicts of licenses in the FFMpeg project, Thomas Knowles for the insightful discussions, and Nicola Sabbi for the insider knowledge of the MPlayer system. The work of Klaas-Jan Stol is partially funded by IRCSET under grant no. RS/2008/134 and by Science Foundation Ireland grant 03/CE2/I303_1 to Lero (www.lero.ie).

References

1. Abi-Antoun, M., Aldrich, J., Coelho, W.: A case study in re-engineering to enforce architectural control flow and data sharing. Journal of Systems and Software 80(2), 240–264 (2007)
2. Arief, B., Gacek, C., Lawrie, T.: Software architectures and open source software – Where can research leverage the most? In: Proceedings of Making Sense of the Bazaar: 1st Workshop on Open Source Software Engineering, Toronto, Canada (May 2001)
3. Arisholm, E., Briand, L.C., Foyen, A.: Dynamic coupling measurement for object-oriented software. IEEE Transactions on Software Engineering 30(8), 491–506 (2004)
4. Avgeriou, P., Guelfi, N.: Resolving architectural mismatches of COTS through architectural reconciliation. In: Franch, X., Port, D. (eds.) ICCBSS 2005. LNCS, vol. 3412, pp. 248–257. Springer, Heidelberg (2005)
5. Basili, V., Rombach, H.D.: Support for comprehensive reuse. IEEE Software Engineering Journal 6(5), 303–316 (1991)

6. Bowman, I.T., Holt, R.C., Brewster, N.V.: Linux as a case study: its extracted software architecture. In: Proceedings of the 21st International Conference on Software engineering (ICSE), pp. 555–563. IEEE Computer Society Press, Los Alamitos (1999)
7. Capiluppi, A., Boldyreff, C.: Identifying and improving reusability based on coupling patterns. In: Mei, H. (ed.) ICSR 2008. LNCS, vol. 5030, pp. 282–293. Springer, Heidelberg (2008)
8. Capiluppi, A., Knowles, T.: Software engineering in practice: Design and architectures of FLOSS systems. In: Boldyreff, C., Crowston, K., Lundell, B., Wasserman, A.I. (eds.) OSS 2009. IFIP AICT, vol. 299, pp. 34–46. Springer, Heidelberg (2009)
9. Capiluppi, A., Morisio, M., Ramil, J.F.: The evolution of source folder structure in actively evolved open source systems. In: Proceedings of the 10th International Symposium on Software Metrics (METRICS), pp. 2–13. IEEE Computer Society, Washington, DC, USA (2004)
10. de Jonge, M.: Source tree composition. In: Gacek, C. (ed.) ICSR 2002. LNCS, vol. 2319, pp. 17–32. Springer, Heidelberg (2002)
11. de Jonge, M.: Build-level components. IEEE Transactions on Software Engineering 31(7), 588–600 (2005)
12. Dueñas, J.C., de Oliveira, W.L., de la Puente, J.A.: Architecture recovery for software evolution. In: Proceedings of the 2nd Euromicro Conference On Software Maintenance And Reengineering (CSMR), pp. 113–120 (1998)
13. Eick, S.G., Graves, T.L., Karr, A.F., Marron, J.S., Mockus, A.: Does code decay? assessing the evidence from change management data. IEEE Transactions on Software Engineering 27, 1–12 (2001)
14. Fenton, N.E., Pfleeger, S.L.: Software metrics: a practical and rigorous approach. Thomson (1996)
15. German, D.M., Gonzalez-Barahona, J.M., Robles, G.: A model to understand the building and running inter-dependencies of software. In: Proceedings of the 14th Working Conference on Reverse Engineering (WCRE), pp. 140–149. IEEE Computer Society, Washington, DC, USA (2007)
16. German, D.M., Hassan, A.E.: License integration patterns: Addressing license mismatches in component-based development. In: Proceedings of the 2009 IEEE 31st International Conference on Software Engineering (ICSE), pp. 188–198. IEEE Computer Society, Washington, DC, USA (2009)
17. Godfrey, M., Eric, H.: Secrets from the monster: Extracting mozilla's software architecture. In: Proceedings of the 2nd Symposium on Constructing Software Engineering Tools, CoSET (2000)
18. Hauge, Ø., Østerlie, T., Sørensen, C.-F., Gerea, M.: An Empirical Study on Selection of Open Source Software - Preliminary Results. In: Capiluppi, A., Robles, G. (eds.) Proceedings of the 2009 ICSE Workshop on Emerging Trends in Free/Libre/Open Source Software Research and Development (FLOSS), Vancouver, Canada, May 18, pp. 42–47. IEEE Computer Society Press, Los Alamitos (2009)
19. Hofmeister, C., Nord, R., Soni, D.: Applied Software Architecture. Addison-Wesley, Reading (2000)
20. Krikhaar, R., Postma, A., Sellink, A., Stroucken, M., Verhoef, C.: A two-phase process for software architecture improvement. In: Proceedings of the IEEE International Conference on Software Maintenance (ICSM), p. 371. IEEE Computer Society, Washington, DC, USA (1999)
21. Kruchten, P.: The 4+1 view model of architecture. IEEE Software 12(5), 88–93 (1995)

22. Lang, B., Abramatic, J.-F., González-Barahona, J.M., Gómez, P., Pedersen, M.K.: Free and Proprietary Software in COTS-Based Software Development. In: Franch, X., Port, D. (eds.) ICCBSS 2005. LNCS, vol. 3412, p. 2. Springer, Heidelberg (2005)
23. Lehman, M.M.: Programs, cities, students, limits to growth? Programming Methodology, 42–62 (1978); Inaugural Lecture
24. Lehman, M.M.: Programs, life cycles, and laws of software evolution. Proc. IEEE 68(9), 1060–1076 (1980)
25. Li, J., Conradi, R., Bunse, C., Torchiano, M., Slyngstad, O.P.N., Morisio, M.: Development with off-the-shelf components: 10 facts. IEEE Software 26(2), 80–87 (2009)
26. Li, W., Henry, S.: Object-oriented metrics that predict maintainability. J. Syst. Softw. 23(2), 111–122 (1993)
27. Majchrowski, A., Deprez, J.-C.: An operational approach for selecting open source components in a software development project. In: O'Connor, R., Baddoo, N., Smolander, K., Messnarz, R. (eds.) EuroSPI. CCIS, vol. 16, pp. 176–188. Springer, Heidelberg (2008)
28. Mockus, A.: Large-scale code reuse in open source software. In: Proceedings of the First International Workshop on Emerging Trends in FLOSS Research and Development (FLOSS), p. 7. IEEE Computer Society, Washington, DC, USA (2007)
29. Orsila, H., Geldenhuys, J., Ruokonen, A., Hammouda, I.: Update propagation practices in highly reusable open source components. In: Russo, B., Damiani, E., Hissam, S.A., Lundell, B., Succi, G. (eds.) Open Source Development, Communities and Quality. IFIP, vol. 275, pp. 159–170. Springer, Boston (2008)
30. Parnas, D.L.: On the criteria to be used in decomposing systems into modules. Communications of the ACM 15(12), 1053–1058 (1972)
31. Sametinger, J.: Software engineering with reusable components. Springer, New York (1997)
32. Sartipi, K., Kontogiannis, K., Mavaddat, F.: A pattern matching framework for software architecture recovery and restructuring. In: Proceedings of the 8th International Workshop on Program Comprehension (IWPC), pp. 37–47 (2000)
33. Schmerl, B., Aldrich, J., Garlan, D., Kazman, R., Yan, H.: Discovering architectures from running systems. IEEE Transactions on Software Engineering 32(7), 454–466 (2006)
34. Senyard, A., Michlmayr, M.: How to have a successful free software project. In: Proceedings of the 11th Asia-Pacific Software Engineering Conference, pp. 84–91. IEEE Computer Society, Busan (2004)
35. Sommerville, I.: Software Engineering, 7th edn. International Computer Science Series. Addison Wesley, Reading (2004)
36. Torchiano, M., Morisio, M.: Overlooked aspects of cots-based development. IEEE Software 21(2), 88–93 (2004)
37. Tran, J.B., Godfrey, M.W., Lee, E.H.S., Holt, R.C.: Architectural repair of open source software. In: Proceedings of the 8th International Workshop on Program Comprehension (IWPC), pp. 48–59. IEEE Computer Society, Washington, DC, USA (2000)
38. Tu, Q., Godfrey, W.M.: The build-time software architecture view. In: Proceedings of 2001 International Conference on Software Maintenance, pp. 65–74. IEEE, Florence (2001)

License Update and Migration Processes in Open Source Software Projects

Chris Jensen and Walt Scacchi

Institute for Software Research,
University of California, Irvine Irvine, CA USA 92697
{cjensen,wscacchi}@uci.edu

Abstract. Open source software (OSS) has increasingly been the subject of research efforts. Central to this focus is the nature under which the software can be distributed, used, and modified and the causes and consequent effects on software development, usage, and distribution. At present, we have little understanding of, what happens when these licenses change, what motivates such changes, and how new licenses are created, updated, and deployed. Similarly, little attention has been paid to the agreements under which contributions are made to OSS projects and the impacts of changes to these agreements. We might also ask these same questions regarding the licenses governing how individuals and groups contribute to OSS projects. This paper focuses on addressing these questions with case studies of processes by which the Apache Software Foundation's creation and migration to Version 2.0 of the Apache Software License and the NetBeans project's migration to the Joint Licensing Agreement.

Keywords: Open source, license evolution, process, Apache, NetBeans.

1 Introduction

Software process research has investigated many aspects of open source software (OSS) development in the last several years, including release processes, communication and collaboration, community joining, and project governance. The central point of Lawrence Lessig's book "Code" is that the hardware and software that make up cyberspace also regulate cyberspace. He argues that code both enables and protects certain freedoms, but also serves as to control cyberspace. Software licenses codify these freedoms and regulations by setting forth the terms and conditions for software use, modification, and distribution of a system and any changes made to it. For that reason, others have suggested that licenses serve as contracts for collaboration. In the case of non-OSS licenses, that contract may indicate no collaboration, but rather strict separation between users and developers. OSS licenses, by contrast range widely in permissiveness, some granting more rights to the original authors and some granting more rights to consumers of OSS software. While research has examined OSS licenses to great detail, we are only beginning to understand license evolution. Just as OSS code is not static, neither are the licenses under which it is distributed. Research into license evolution is just beginning. However, when licenses change, so too the

S.A. Hissam et al. (Eds.): OSS 2011, IFIP AICT 365, pp. 177–195, 2011.

contracts for collaboration change. This paper seeks to provide an incremental step to understanding how changes in software licensing impact software development processes.

Why does understanding license update and migration matter? Companies using OSS software need to know how changes affect their use, modification, and distribution of a software system. License compatibility in OSS has long been a topic of debate. Research is only beginning to provide tools for assistance in resolving software license compatibility [1]. OSS project participants need to understand why changes are being made, whether the changes align with their values and business models (e.g., enabling new avenues of license compatibility offering strategic benefit or opening up new channels of competition). As a project sponsor or host, you may be concerned about how to best protect both the software system and your user community, but also your business model. You typically want a license that will attract a large number of developers to your project [2] while at the same time allowing you to make a profit and stay in business.

While licenses such as the GNU General Public License (GPL), the Berkeley Software Distribution (BSD) license, and the Apache License are well known, we rarely consider another type of license agreement critical to understanding collaboration in OSS projects: individual contributor license agreements (CLAs) and organizational contributor license agreements (OCLAs), for contributos from organized entities. In non-OSS software development, the contract for collaboration is typically an employment contract, often stating that all intellectual property rights pertaining to source code written by an employee are property of the employer. This provides the employer with complete control of the rights granted of licensed software. In OSS development, you have a situation where multiple developers are contributing to a software system. Without copyright assignment or a CLAs, changing a software license requires the consent of every contributor to that system. We observed this situation in the case of the Linux kernel, which suggested that without a CLA, license evolution can become inhibited or prevented as the number of contributors, each with differing values and objectives, increases. To understand how changes in software licenses affect software development processes, we must also investigate changes in CLAs.

We address these issues with two case studies. The first examines the creation and deployment of the Apache Software License, Version 2.0. The second looks at an update to the contributor license agreement in the NetBeans project.

2 Background Work

Legal scholars, such as St. Laurent [3] and Larry Rosen [4], former general counsel and secretary for the Open Source Initiative (OSI), have written extensively on license selection. They note that quite often, the choice of license is somewhat outside the control of a particular developer. This is certainly the case for code that is inherited or dependent on code that is either reciprocally licensed, or at the very least, requires a certain license for the sake of compatibility. However, outside such cases, both St. Laurent and Rosen advocate for the use of existing and well-tested, well-understood licenses as opposed to the practice of creating new licenses. Such license proliferation is seen as a source of confusion among users and is often unnecessary given the extensive set of licenses that already exist for a diverse set of purposes. Lerner and Tirole [5] observe specific determinant factors in license selection.   Of the 40,000 Sourceforge projects studied, projects geared towards end-users tended towards more restrictive license terms, while projects directed towards software developers tended towards less restrictive licenses. Highly restrictive licenses were also found more common in consumer software (e.g., games) but less common for software on consumer-oriented platforms (e.g., Microsoft Windows) as compared to non-consumer-oriented platforms.   Meanwhile, Rosen specifically addresses the issue of relicensing, commenting that license changes made by fiat are likely to fracture the community. This case of relicensing is exactly the focus of our case studies here.

The drafting and release of the GNU General Public License, Version 3.0 was done in a public fashion, inviting many prominent members of the OSS community to participate in the process. In fact, we even see a sort of prescriptive process specification outlining, at a high level, how the new license was to be created. This license revision process is interesting from the perspective that the license in question is not used by one project or one foundation, but rather is an update of the most commonly used open source license in practice. As such the process of its update and impact of its revision on software development is both wide ranging and widely discussed.

Di Penta, et al. [6], examined changes to license headers in source code files in several major open source projects. Their three primary research questions sought to understand how frequently licensing statements in source code files change, the extent of the changes, and how copyright years change in source code files. Their work shows that most of the changes observed to source code files are small, though even small changes could signify a migration to a different license. The authors also note that little research available speaks to license evolution, pointing to the need for greater understanding in this area.

Lindman, et al., [2] examine how companies perceive open source licenses and what major factors contribute to license choice in companies releasing open source software. The study reveals a tight connection between business model, patent potential, the motivation for community members to participate in development, control of project direction, company size, and network externalities (compatibility with other systems) and licensing choice.

Lindman, et al., provide a model of a software company, its developers, and users in the context of an OSS system developed and released from a corporate environment [2]. However, few systems are developed in complete isolation. Rather, they leverage existing libraries, components, and other systems developed by third parties. Moreover, as Goldman and Gabriel point out, open source is more than just source code in a public place released under an OSS license [7]; communities matter. Fig. 1 shows the production and consumption of open source software, highlighting the impact of software licenses and contributor license agreements.

Going a step further, Oreizy [8] describes a canonical high-level software customization process for systems and components, highlighting intra-organizational software development processes and resource flow between a system application developer, an add-on developer, a system integrator, and an end user.

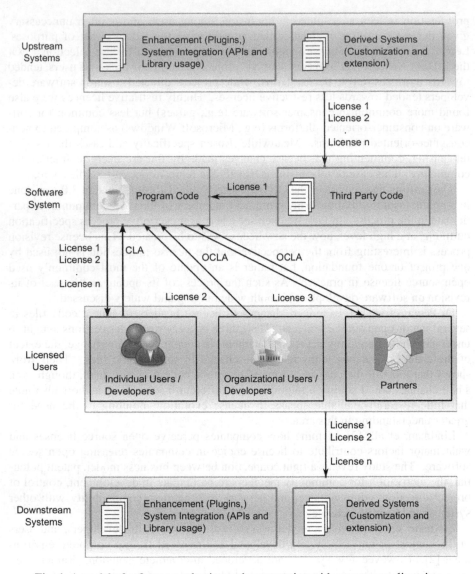

Fig. 1. A model of software production and consumption with open source licensing

Similarly, we have examined such concepts in the context of software ecosystems [9] in the context of process interaction. Software license change can precipitate integrative forms of process interaction in the case of dual and multi-licensing by enabling new opportunities for use of software systems upstream of a project to provide added functionality or support, as well as projects downstream vis a vis use as a library, plugin development, support tool development, and via customization and extension. In such cases, software source becomes a resource flowing between interacting projects. However, license change can also trigger interproject process conflict if new license terms render two systems incompatible. At that point, the resource flow between projects can be cut off, when downstream consumers of software source code no longer receive updates. A more common example with non-OSS is license expiration. License-based interproject process conflicts can also manifest as unmet dependencies in software builds or an inability to fix defects or add enhancements to software, resulting in process breakdown, and failing recover, project failure. OSS licenses, however, guarantee that even when conflict occurs, recovery is possible because the source is available and can be forked.

3 Methodology

The case studies in this report are part of an ongoing, multi-year research project discovering and modeling open source software processes. Our research methodology is ethnographically informed, applying a grounded theory to the analysis of artifacts found in OSS projects. The primary data sources in this study come from mailing list archives of the Apache and NetBeans projects.

Our primary data sources were mailing list messages. However, we also found supplementary documentation on each project's websites that served to inform our study. These supplementary documents were often, though not always referenced by the messages in the mailing list. Cases regarding the NetBeans project all took place between April and June of 2003, involving over 300 email messages, whereas the Apache cases were spread over several discrete time periods and consisted of more than 350 messages.

Case selection happened in two ways. For NetBeans, the cases arose during our study of requirements and release processes, having stood out as prominent issues facing the community during the time period studied. Although we observed additional incidents appropriate for discussion, the three cases selected fit together nicely as a cohesive story. This approach was also used in the study of the Apache project. However, due to a lower incident frequency, we expanded our study over a longer time period to find incidents that proved substantial. As a testament to the nature of project interaction, issues raised in mailing list discussions proved to be short-lived, either because they were resolved quickly or because the conversation simply ceased. It is possible to suggest this is the normal behavior pattern for both projects. A few issues proved outliers, having more focused discussions, and these were selected for further study. We also observed a tendency for discussions to play out in a series of short-lived discussions sessions. A topic would be raised, receiving little or no attention. Then, at a later time, it would be raised again. The JCA discussion in NetBeans and Subversion migration discussion in the Apache project demonstrated such conversational resurgence. We observed, in general, that discussion topics carry certain conversational momentum. Topics with a high degree of momentum tended to have lengthier discussion periods or frequent discussion sessions until fully resolved or abandoned while topics with a low degree of momentum were addressed quickly or simply died off. The causes and factors affecting changes in momentum were not investigated as they laid too far afield from the focus of this study. We do note that although *consensus by attrition* has been cited in other communities (e.g., [10 and 11]), we did not observe it in effect in any of the cases studied, but rather that the primary participants in discussions remained active in their respective projects for several months following the reported incidents. The creation of the Apache License, version 2.0 was directed to us by a colleague familiar with the project. Data for the Apache licensing case was gathered from email messages sent to a mailing list established for the purpose of discussing the proposed changes.

Considering the difficulties we experienced with building our own search engine to support process discovery, we still faced the challenge of keeping track of process data once we found it as we were building our models. Up until this point, our approach to providing process traceability was simply to include links to project artifacts in our models. However, this strategy did not help us build the models, themselves. We returned the search problem back to the projects, themselves using their own search engines to locate process data, looking for more lightweight support for discovery.

Our current strategy for providing computer support for process discovery returns to using each project's own search engine to locate process information. We have operationalized the reference model as an OWL ontology with the Protégé ontology editor [12], using only the OWL class and individual constructs to store process concepts and their associated search queries respectively. Secondly, we built a Firefox plugin, Ontology [13], to display the reference model ontology in the Firefox web browser. Next, we enlisted the Zotero citation database Firefox plugin [14] to store process evidence elicited from project data, integrating the two plugins such that each datum added to the citation database from a project artifact is automatically tagged with the selected reference model entities.

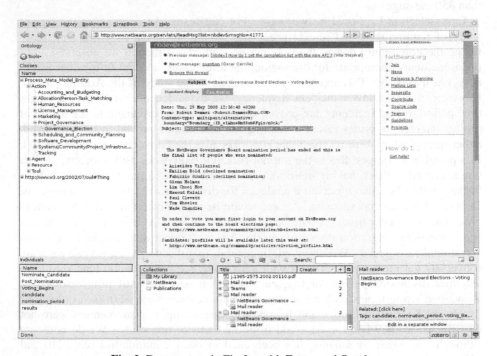

Fig. 2. Data capture in Firefox with Zotero and Ontology

The use of a citation database as a research data repository may seem unintuitive. Zotero, however, has proven well suited for our needs. Like many Firefox plugins, Zotero can create records simply from highlighted sections of a web document, though the creation of arbitrary entries (not gleaned from document text selections) is also possible. It can also save a snapshot of the entire document for later review, which is useful given the high frequency of changes of some web documents- changes that evidence steps in a software processes. The tag, note, and date fields for each entry are useful for recording reference model associations and memos about the entry for use in constructing process steps and ascertaining their order. A screenshot of Zotero with Ontology appears in Fig. 2.

The plugin integration greatly facilitates the coding of process evidence and provides traceability from raw research data to analyzed process models. As the tool set is browser-based, it is not limited to analysis of a particular data set, whether local or remote. Moreover, the tool set does not limit users to a single ontology or Zotero database, thereby allowing users to construct research models using multiple ontologies describing other (e.g. non-OSS process) phenomenon and reuse the tool set for analysis of additional data sets. Thus, it may be easily appropriated for grounded theory research in other fields of study.

The elicitation of process evidence is still search driven. Rather than use one highly customized search engine for all examined data repositories, the search task has been shifted back to the organizations of study. This decision has several implications in comparison with the previous approach, both positive and negative. Using an organization's own search engine limits our ability to extract document-type specific metadata, however among the organizations we have studied, their search tools provide greater coverage of document and artifact types than Lucene handled at that time. Furthermore, this approach does not suffer the data set limitations imposed by web crawler exclusion rules. The ability to query the data set in a scripted fashion has been lost, yet some scientists would see this as a gain. The use of computer-assisted qualitative data analysis software (CAQDAS) historically has put into question the validity of both the research method and results [15,16].

This tool was still quite unfinished as we began governance process discovery and modeling. As we added functionality, we had to return to some of our data sources and recapture it. Although we have high hope to use the integrated timeline feature to assist in process activity composition and sequencing, the time and date support within Zotero's native date format was insufficiently precise. With provisions only for year, month, and day, there is no ability to capture action sequences that happen on the same day. After adding support for greater date and time, we found having to enter the date and time for every piece of data we captured rather tedious. Eventually we have had to prioritize completion of discovery and modeling ahead of computer-support for process discovery, and we had to disable the time and date entry. Unable to utilize Zotero to our intended effect in discovery and modeling, our efforts with Zotero remain in progress, pending usability improvements.

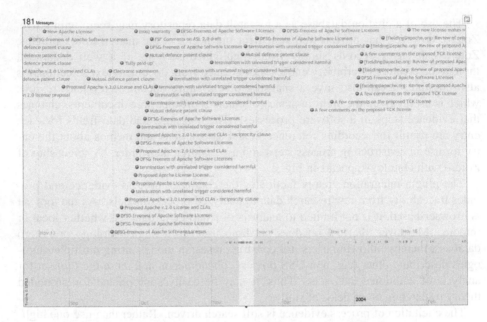

Fig. 3. Timeline of the Review and Approval of the Apache License, Version 2.0

4   Creation and Migration to the Apache License, Version 2.0

The Apache Software Foundation created a new version of their license in the end of 2003 and beginning of 2004. Roy Fielding, then director of the ASF, announced the license proposal on 8 November 2003 [17], inviting review and discussion on a mailing list set up specifically for said purpose. Per Roy's message, the motivations for the proposed license included

- Reducing the number of frequently asked questions about the Apache License.
- Allowing the license to be usable by any (including non-Apache) projects
- Requiring a patent license on contributions that necessarily infringe the contributor's own patents
- Moving the full text of the license and specific conditions outside the source code
- Roy further indicated a desire to have a license compatible with other OSS licenses, notably the GPL.

As you can see from Fig. 3, most of the discussion took place in mid November of 2003. In fact, given that the ApacheCon conference that ran from 16-19 November, we can see a high message density in the days leading up to ApacheCon, with a steady rate continuing on for a few days afterward. Beyond this, the frequency becomes sparse. An update to the proposed license was announced on 24 December 2003, after some internal review, a part of the process that is not publicly visible. This update prompted a brief discussion. A second active time period is observable in January 2004, when Fielding announces a final update (20 January 2004) and that the final version of the license has been approved by the board [18 and 19] (21 January 2004).

The primary discussion point of the creation and migration to the 2.0 version of the Apache License centered around a patent clause in the proposed license. According to Brian Behlendorf, who was serving on the ASF board of directors at the time, the ASF's patent-related goals were to "prevent a company from sneaking code into the codebase covered by their own patent and then seeking royalties from either the ASF or end-users" [20]. The clause in question read:

> *5. Reciprocity. If You institute patent litigation against a Contributor with respect to a patent applicable to software (including a cross-claim or counterclaim in a lawsuit), then any patent licenses granted by that Contributor to You under this License shall terminate as of the date such litigation is filed. In addition, if You institute patent litigation against any entity (including a cross-claim or counterclaim in a lawsuit) alleging that the Work itself (excluding combinations of the Work with other software or hardware) infringes Your patent(s), then any patent licenses granted to You under this License for that Work shall terminate as of the date such litigation is filed. [21]*

Consequences of this clause sparked discussion in a few areas, mainly surrounding the first sentence of the clause regarding license termination. Legal representatives from industry stated objections to losing usage rights for patent litigation regarding any software, even software unrelated to that covered by the license [22], proposing alternative wordings to achieve the stated license goals but restricting the trigger to litigation pertaining to patents covered by the ASF licensed code [23]. Uncertainty regarding the roles of people in the license revision process [24] and proposed changes [25] created additional confusion regarding the patent reciprocity stance.

Eben Moglen, General Counsel for the Free Software Foundation (FSF), adds that the first sentence of the license clause carries great risk for unintended and serious consequences, and is an inappropriate vehicle for protecting free software against patent litigation [26]. As such, the FSF has deemed the clause causes the license to be incompatible with version 2 of the GPL, failing one of the goals of the proposed Apache License.

Brian Carlson reports that the Debian community's consensus is that the proposed license does not meet the criteria for *Free Software Licenses* under the Debian Free Software Guidelines [27]. Consequently, code licensed as such would sandboxed into the non-free archive, and therefore, not automatically built for Debian distributions, nor receive quality assurance attention. Again, the license termination aspect of the reciprocity clause is cited as the critical sticking point [28], with several members of the Debian community arguing that free software licenses should only restrict modification and distribution, but not usage of free software.

The patent reciprocity clause was not entirely rejected. There was support for extending it to provide mutual defense against patent litigation attacks against all open source software [29]. The idea was quickly nixed on the grounds that it could lead to users being attacked and unable to defend themselves if someone were to maliciously violate a user's patent on an unrelated piece of software and create an open source version. In such a scenario, the user would have to choose between using Apache licensed software and losing all their patents [30].

On 18 November, Fielding indicates that there have been "several iterations on the patent sentences, mostly to deal with derivative work" [24], mentioning he will probably include the suggested changes in the patent language recommended by one of the legal representatives from industry. Fielding notes that he has been in contact with representatives from other organizations, among them Apple, Sun, the OSI, Mozilla, and a few independent attorneys, although the details of these portions of the process remain hidden.

The next milestone in the process occurs on 24 December, when Fielding mentions that a second draft, version 1.23, has been prepared after internal review due to extensive changes [31], and has been posted to the proposed licenses website [32] and the mailing list. The new proposed license [33] incorporates many of the proposed changes, including the removal of the contested first sentence of the patent reciprocity clause, leaving the generally agreed upon patent termination condition:

If You institute patent litigation against any entity (including a cross-claim or counterclaim in a lawsuit) alleging that the Work or a Contribution incorporated within the Work constitutes direct or contributory patent infringement, then any patent licenses granted to You under this License for that Work shall terminate as of the date such litigation is filed.

The 1.23 version of the license received little feedback on the license discussion mailing list. Aside from definition clarifications, there was an inquiry about GPL compatibility. Behlendorf commented that Moglen's suggestions had been incorporated to address the two issues with GPL compliance, but he had been contacted earlier in the week to take a look at the current draft [34]. As a result, Behlendorf (on 7 January 2004) offers that the issues presented have been addressed to his satisfaction and is willing to propose the license to the board at the January 2004 meeting [35]. However, before the board meeting, Fielding announces a version 1.24, featuring a change to the definition of "Contributor" [36] and a 1.25 version very shortly thereafter to address the way "Copyright" is represented due to various laws and the use of "(C)" to indicate copyright [37]. Finally, the Apache License, Version 2.0 was approved by the ASF board by a unanimous vote on 20 January 2004 [18] and announced to the mailing list by Fielding the following day [19]. Per the board meeting minutes:

> *WHEREAS, the foundation membership has expressed a strong desire for an update to the license under which Apache software is released,*
> *WHEREAS, proposed text for the new license has been reworked and refined for many, many months, based on feedback from the membership and other parties outside the ASF,*
> *NOW, THEREFORE, BE IT RESOLVED, that the proposed license found at http://www.apache.org/licenses/proposed/LICENSE-2.0.txt is officially named the Apache Software License 2.0. To grant a sufficient transition time, this license is to be used for all software releases from the Foundation after the date of March 1st, 2004.*

The conversation continued on, briefly, to address two points. Firstly, a return to the GPL compatibility discussion. Don Armstrong requested verification as to whether Moglen/the FSF has identified the license as GPL compatible (Fielding's announcement claimed it was) [38]. Fielding responds, saying Moglen sent a private communication commenting on the license compatibility, and furthermore, that it was the belief of the ASF that "a derivative work consisting of both Apache Licensed code and GPL code can be distributed under the GPL," and, as such, there wasn't anything further to consider, as far as the ASF was concerned [39]. Incidentally, the FSF standing is that due to the patent issue, the Apache license 2.0 is GPL3 compatible but not GPL2 compatible [40]. Secondly, Vincent Massol requested information about moving his Apache sub-project to the ASL2 license and what file license headers should be used [41], to which Behlendorf responds [42]. A flow graph of the License creation and migration process appears in Fig. 4.

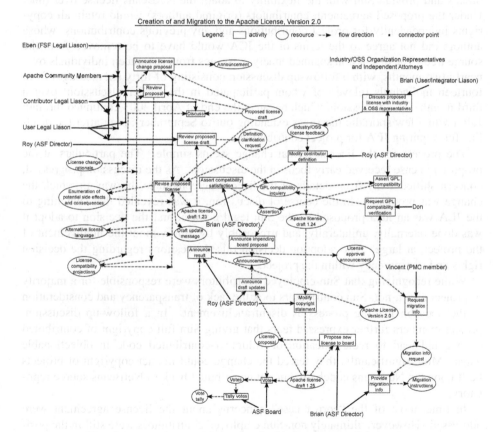

Fig. 4. Process flow graph for Apache License Version 2.0 creation

5 Introduction of the Joint License Agreement

Rosen [4] suggests that copyright assignment is sought for two purposes:

1. So the project can defend itself in court without the participation and approval of its contributors.
2. To give the project (and not the contributor) the right to make licensing decisions, such as relicensing, about the software

The NetBeans case is interesting because it is not simple copyright assignment, but rather affords both the contributor and the project (Sun Microsystems, specifically) equal and independent copyright to contributed source.

The Joint License Agreement (JLA) was introduced to the NetBeans project on 28 April 2003 by Evan Adams, a prominent project participant working for Sun Microsystems [43]. Adams states that the JLA was being introduced in response to observations by Sun's legal team of Mozilla and other open source projects and believed that Sun required full copyright authority to protect the NetBeans project from legal threats and provide Sun with the flexibility to adapt the NetBeans license over time. Under the proposed agreement, contributors (original authors) would retain all copyrights independently for project contributions and any previous contributions whose authors did not agree to the terms of the JCA would have to be removed from the source tree. The discussion spanned ninety messages from seventeen individuals over nearly two months, with a follow-up discussion consisting of forty six messages from fourteen individuals (eleven of whom participated in the earlier discussion) over a third month. The discussion, which began at the end of April 2003 continued through July (with a few sporadic messages extending out to September), long after the deadline for requiring JLA for project contributions.

The process for the license format change seems simple. The particulars of the proposed license received early focus in the discussion. As the discussion progressed, concern shifted away from details of the license agreement to the way in which the change was proposed. In the course of discussion, it was revealed that switching to the JLA was an idea proposed by the Sun legal counsel and the decision to adopt it was done internally, unilaterally, and irrevocably by Sun without the involvement of the project, at large. The adoption decision raised questions regarding the decision rights and transparency within the project.

While recognizing that Sun-employed contributors were responsible for a majority of project effort, non-Sun contributors took the lack of transparency and consideration in the decision making process as disenfranchisement. In a follow-up discussion, project members further expressed fears that giving Sun full copyright of contributed code could lead to reclassification of volunteer-contributed code in objectionable ways. More significantly, they feared the change could impact copyright of projects built upon the NetBeans codebase, but not contributed back to NetBeans source repository.

In time, most of the "corner case" concerns about the license agreement were addressed. However, ultimately non-Sun employed contributors were still in the position of having to trust Sun to act in an acceptable manner with a grant of full copyright. Moreover, the discussion drew out larger concerns regarding Sun's role position of leadership and control of the project, and regarding transparency in decision making. A flow graph of the JCA introduction process appears in Fig. 5.

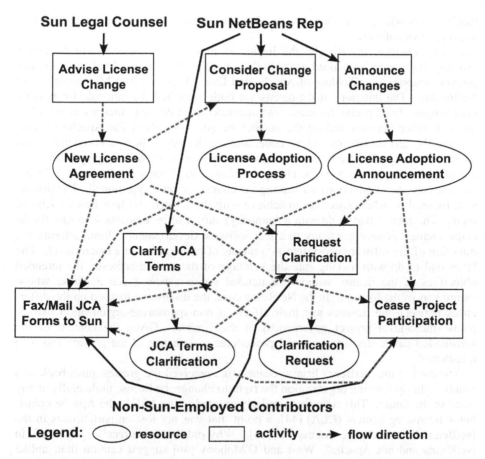

Fig. 5. NetBeans JCA introduction process flow graph

6 Discussion and Conclusions

The two cases presented are not directly comparable. The Apache study looks at the process of creating a new license, to be used by all projects under the domain of the Apache Software Foundation. The NetBeans study focuses on the adoption of a new license agreement for contributors to the NetBeans IDE and platform. Software source licenses govern the rights and responsibilities of software consumers to (among other things) use, modify, and distribute software. Contributor license agreements (CLAs), on the other hand, govern the rights and responsibilities to (among other things) use, modify, and distribute contributions of the organization to which the contributions are submitted, and those retained by the contributor. The new CLA stated that copyright of project contributions would be jointly owned by the originating contributors, as well as the project's benefactor, Sun Microsystems. Code contribution agreements may not be of interest to end users of software executables. However, the OSS movement is known for its tendency towards user-contributors;

that is, users who contribute to the development of the software and developers who use their own software.

If we consider, specifically, the license changes in the Apache and NetBeans projects, both were introduced as inevitable changes by persons of authority in each project (founder Roy Fielding of Apache and Evan Adams of Sun Microsystems for NetBeans). The initiators of the discussion both presented the rationale for making the changes. For Apache, the move was motivated by a desire to increase compatibility with other licenses, reduce the number of questions about the Apache license, moving the text outside the source code, and require patent license on contributions where necessary. For NetBeans, the motivations were to protect the project from legal threats and provide Sun the ability to change the license in the future. In the Apache case, the motivations for making the changes went unquestioned. The discussion focused on what objectives to achieve with the change and how best to achieve them. The former had to do with a (minority) subset of participants who saw the license change as an opportunity to affect software development culture, altering the direction of the software ecosystem as a means of governance on a macro level. The latter had to do with making sure the verbiage of the license achieved the intended objectives of the license without unintended consequences (such as those whose nature was of the former). In the NetBeans case, the discussion focused on the differences between the licenses and their affect on non-sponsoring-organization participants (meso-level project governance) of the license. Given the context of the surrounded cases, the structural and procedural governance of the project was also questioned.

The area of the NetBeans license change that received the greatest push-back was granting the sponsoring organization the right to change the license unilaterally at any point in the future. This right was similarly granted to the ASF in the Apache contributor license agreement (CLA) [44], a point that was not lost on participants in the NetBeans license change discussions [45]. Why did this issue receive push-back in NetBeans and not Apache? West and O'Mahony [46] suggest caution that, unlike community-initiated projects, sponsored OSS projects must achieve a balance between establishing pre-emptive governance design (as we saw here) and establishing boundaries between commercial and community ownership and control. The surrounding cases served to create an atmosphere of distrust within the project. The distrust led to fears that contributions from the sponsoring organization would become closed off from the community, perhaps saved for the organization's commercial version of the product, leaving the sponsoring organization as free-riders [47 and 48] profiting off of the efforts of others without giving back [49] or otherwise limit what project participants can do with project code.

Perhaps the most striking difference in the way the two license changes were introduced is that the Apache case invited project participants (as well as the software ecosystem and the public, at large) to be a part of the change, whereas the NetBeans case did not. Participants in the NetBeans project were left without a sense of *transparency in the decision-making process* in that the change was put on them without any warning before the decision was made. Moreover, they were left without *representation in the decision-making process* in that they did not participate in determining the outcome of a decision that had a large impact on them. This is not to say that the Apache case was entirely transparent. There are clear indications from the messages on the list that conversations were held off-list. Likewise, there were misconceptions over what roles participants played and participant affiliation. However, the process was not questioned, nor the result.

In conclusion, we have taken a first step to understanding how license change processes impact software development processes by discovering and modeling the update process for the Apache License and the update to the contributor license agreement in the NetBeans project. We observed how differences in the processes in introducing change intent influenced response to the changes. To put these cases into context, NetBeans underwent two license changes since the events described above, neither of which received significant push-back from the community. The first shifted the license to the CDDL. The second was a move to dual license NetBeans under the GPLv2. This second licensing shift was considered by Sun "at the request from the community" [50]. Unlike the introduction of the JCA, the GPL shift was presented to the community by Sun for feedback (in August 2007) as an added option (rather than a complete relicensing) before the change was made. Thus, we can clearly see further change in the processes used to govern the community in a way that directly addressed the defects in the project's governance processes circa 2003. Shah [51] echoes these concerns, observing that code ownership by firms creates the possibility that non-firm-employed contributors will be denied future access to project code. In other projects, these threats can lead to forking of the source, as happened when the MySQL corporation was purchased by Sun Microsystems, which, in turn, has recently been acquired by Oracle.

Acknowledgements. The research described in this report is supported by grants from the Center for Edge Power at the Naval Postgraduate School, and the National Science Foundation, #0534771 and #0808783. No endorsement implied.

References

1. Scacchi, W., Alspaugh, T., Asuncion, H.: The Role of Software Licenses in Open Architecture Ecosystems. In: Intern. Workshop on Software Ecosystems, Intern. Conf. Software, Reuse, Falls Church, VA (September 2009)
2. Lindman, J., Paajanen, A., Rossi, M.: Choosing an Open Source Software License in Commercial Context: A Managerial Perspective. In: 2010 36th EUROMICRO Conference on Software Engineering and Advanced Applications Software Engineering and Advanced Applications, Euromicro Conference, pp. 237–244 (2010)
3. Laurent, A.M.: Understanding Open Source and Free Software Licensing. Reilly Media, Inc., Sebastopol (2004)
4. Rosen, L.: Open Source Licensing: Software Freedom and Intellectual Property Law. Prentice-Hall, Englewood Cliffs (2005)
5. Lerner, J., Tirole, J.: The Scope of Open Source Licensing. The Journal of Law, Economics & Organization 21(1), 20–56 (2005)
6. Di Penta, M., German, D., Guéhéneuc, Y., Antoniol, G.: An exploratory study of the evolution of software licensing. In: Proceedings of the 32nd ACM/IEEE International Conference on Software Engineering (ICSE 2010), vol. 1, pp. 145–154. ACM, New York (2010)

7. Goldman, R., Gabriel, R.: Innovation Happens Elsewhere: How and Why a Com-pany should Participate in Open Source. Morgan Kaufmann Publishers Inc., San Francisco (2004)
8. Oreizy, P.: Open Architecture Software: A Flexible Approach to Decentralized Software Evolution. Ph.D. Information and Computer Sciences, Irvine, CA, University of California, Irvine (2000)
9. Jensen, C., Scacchi, W.: Process Modeling Across the Web Information Infrastructure. Software Process: Improvement and Practice 10(3), 255–272 (2005)
10. .Hedhman, N.: Mailing list message 07:18:55 -0000 "Re: [ANN] Avalon Closed (December 16, 2004),
    http://www.mail-archive.com/communityapache.org/msg03889.html
    (last accessed September 15, 2009)
11. Dailey, D.: Mailing list message 10:38:26 -0400 "Re: Support Existing Content / consensus through attrition? (May 02, 2007),
    http://lists.w3.org/Archives/Public/public-html/2007May/
    0214.html (last accessed September 15, 2009)
12. The Protégé Ontology Editor Project, http://protege.stanford.edu/ (last accessed June 23, 2008)
13. The Firefox Ontology Plugin project,
    http://rotterdam.ics.uci.edu/development/padme/browser/ontol
    ogy (last accessed June 23, 2008)
14. The Zotero Project, http://www.zotero.org/ (last accessed June 23, 2008)
15. Bringer, J.D., Johnston, L.H., Brackenridge, C.H.: Using Computer-Assisted Qualitative Data Analysis. Software to Develop a Grounded Theory Project Field Methods 18(3), 245–266 (2006)
16. Kelle, U.: Theory Building in Qualitative Research and Computer Programs for the Management of Textual Data. Sociological Research Online 2(2) (1997),
    http://www.socresonline.org.uk/socresonline/2/2/1.html (last accessed June 23, 2008)
17. Fielding, R.: Message 02:39:09 GMT "Review of pro-posed Apache License, version 2.0 (November 08, 2003),
    http://mail-archives.apache.org/mod/_mbox/archive-license/
    200311.mbox/%3cBAAB287A-1194-11D8-842D-
    000393753936apache.org/%3e (last acccessed August 14, 2009)
18. Board meeting minutes of The Apache Software Foundation (January 2004),
    http://apache.org/foundation/records/minutes/2004/board_minu
    tes_2004_01_21.txt (last accessed August 13, 2009)
19. Fielding, R.: Mailing list message 01:34:36 GMT "Apache License, Version 2.0 (January 24, 2004),
    http://mail-archives.apache.org/mod_mbox/archive-license/
    200401.mbox/%3C781EEF08-4E0D-11D8-915D-
    000393753936apache.org%3E (last accessed August 13, 2009)
20. Behlendorf, B.: Mailing list message 07:31:40 GMT "RE: termination with unrelated trigger considered harmful (November 22, 2003),
    http://mail-archives.apache.org/mod_mbox/archive-license/
    200311.mbox/%3C20031121232552.X38821fez.hyperreal.org%3E   (last accessed August 13, 2009)

21. Carlson, B. M.: Mailing list message 10:03:55 +0000 "Re: [fielding@apache.org: Review of proposed Apache License, version 2.0] (November 8, 2003), http://lists.debian.org/debian-legal/2003/11/msg00053.html (last accessed August 12, 2009)

22. Peterson, S.K.: Mailing list message 14:52:54 GMT "ter-mination with unrelated trigger considered harmful (November 14, 2003), http://mailarchives.apache.org/mod_mbox/archivelicense/20031 1.mbox/%3C6D6463F31027B14FB3B1FB094F2C744704A11176tayexc17.americas.cpqcorp.net%3E last accessed August 13, 2009)

23. Machovec, J.: Mailing list message 16:49:09 GMT "Re: termination with unrelated trigger considered harmful (November 14, 2003), http://mailarchives.apache.org/mod_mbox/archivelicense/20031 1.mbox/%3C3FB50785.7010801@golux.com%3E (last accessed August 13, 2009)

24. Fielding, R.: Mailing list message 02:10:27 GMT "Re: [fielding@apache.org: Review of proposed Apache License, version 2.0] (November 18, 2003), http://mail-archives.apache.org/mod_mbox/archive-license/200311.mbox/%3c60AEF3C1-196C-11D8-A8F4-000393753936apache.org%3e (last accessed August 13, 2009)

25. Engelfriet, A.: Mailing list message, 20:59:53 GMT Re: [fielding@apache.org: Review of proposed Apache License, version 2.0] (November 17, 2003), http://mail-archives.apache.org/mod_mbox/archive-license/200311.mbox/%3c20031117205953.GA95846stack.nl%3e (last accessed August 13, 2009)

26. Moglen, E.: Mailing list message, 21:28:32 GMT FSF Comments on ASL 2.0 draft (November 14, 2003), http://mailarchives.apache.org/mod_mbox/archivelicense/20031 1.mbox/%3c16309.18688.540989.283163@new.law.columbia.edu%3e (last accessed August 13, 2009)

27. Carlson, B. M.: Mailing list message 05:39:49 GMT DFSG-freeness of Apache Software Licenses (November 13, 2003), http://mailarchives.apache.org/mod_mbox/archivelicense/20031 1.mbox/%3c20031113053949.GD23250@stonewall%3e (last accessed August 13 2009)

28. Armstrong, D.: Mailing list message, GMT Re: DFSG-freeness of Apache Software Licenses (November 14, 2003), http://mailarchives.apache.org/mod_mbox/archivelicense/20031 1.mbox/%3C20031114043950.GM2707@donarmstrong.com%3E, (last accessed August 13, 2009)

29. Johnson, P.: Mailing list message GMT Mu-tual defence patent clause (November 12, 2003), http://mailarchives.apache.org/mod_mbox/archivelicense/20031 1.mbox/%3C003d01c3a8c1$f9b55170$c6ba400c@protocol.com%3E (last accessed August 12, 2009)

30. Behlendorf, B.: Mailing list message 21:09:32 GMT Re: Mutual defence patent clause (November 12, 2003), http://mailarchives.apache.org/mod_mbox/archivelicense/200311.mbox/%3C2003111213 0508.H497@fez.hyperreal.org%3E, last accessed 13 August 2009

31. Fielding, R.: Mailing list message. Re: Review of proposed Apache License, version 2.0 (December 24, 2003), `http://mail-archives.apache.org/mod_mbox/archive-license/ 200312.mbox/%3c464B4006-3604-11D8-9A9F- 000393753936@apache.org%3e` (last accessed, August 12, 2009)

32. Apache License Proposal Website, `http://www.apache.org/licenses/proposed/` (last accessed August 13, 2009)

33. Apache License, Version 1.23, `http://mail-archives.apache.org/mod_mbox/archive-license/ 00312.mbox` (accessed August 13, 2009)

34. Behlendorf, B.: Mailing list message 22:42:52 GMT Re: Review of proposed Apache License, version 2.0 (January 09, 2004), `http://mail-archives.apache.org/mod_mbox/archive-license/ 200401.mbox/%3c20040109143803.G31301fez.hyperreal.org%3e`, (last accessed August 13, 2009)

35. Behlendorf, B.: Mailing list message. Re: Review of proposed Apache License, version 2.0 (January 07, 2004), `http://mail-archives.apache.org/mod_mbox/archive-license/ 200401.mbox/%3c20040107140658.A23429@fez.hyperreal.org%3e` (last accessed August 13, 2009)

36. Fielding, R.: Mailing list message Re: Review of proposed Apache License, version 2.0 (January 14, 2004), `http://mail-archives.apache.org/mod_mbox/archive-license/ 200401.mbox/%3cD81EA136-46CF-11D8-B08A- 000393753936@apache.org%3e` (last accessed August 13 2009)

37. Fielding, R.: Mailing list message Re: Review of proposed Apache License, version 2.0 (January 14 , 2004), `http://mail-archives.apache.org/mod_mbox/archive-license/ 200401.mbox/%3cD6DB9454-46D3-11D8-B08A- 000393753936@apache.org%3e` (last accessed August 13, 2009)

38. Armstrong, D.: Mailing list message. Re: Apache License, Version 2.0 (January 24, 2004), `http://mailarchives.apache.org/mod_mbox/archivelicense/20040 1.mbox%3C20040124021350.GG306@0archimedes.ucr.edu%3E` (last accessed August 13, 2009)

39. Fielding, R.: Mailing list message Re: Apache License, Version 2.0 (January 24, 2004), `http://mail-archives.apache.org/mod_mbox/archive- license/200401.mbox/%3C23385101-4E15-11D8-915D- 000393753936@apache.org%3E` (last accessed August 13, 2009)

40. Free Software Foundation Licenses webpage, `http://www.fsf.org/licensing/licenses/index_html#GPLCompatib leLicenses`, (last accessed August 14, 2009)

41. Massol, V.: Mailing list message dated Sun, How to use the 2.0 license? (January 25 2004), `http://mailarchives.apache.org/mod_mbox/archivelicense/20040 1.mbox/%3C012f01c3e35c$78e229d0$2502a8c@0vma%3E` (last accessed August 13, 2009)

42. Behlendorf, B.: Mailing list message Re: How to use the 2.0 license? (January 25, 2004), http://mailarchives.apache.org/mod_mbox/archivelicense/20040 1.mbox/%3C20040125121456.H396@fez.hyperreal.org%3E (last accessed August 13, 2009)

43. Adams, E.: NBDiscuss mailing list message:Joint Copyright Assignment, http://www.netbeans.org/servlets/ReadMsg?list=nbdiscuss&msgN o-=2228 (last accessed August 6, 2009)

44. The Apache Software Foundation Individual Contributor License Agreement, Version 2.0, http://www.apache.org/licenses/icla.txt (last accessed October 20, 2009)

45. Brabant, V.: mailing list message [nbdis-cuss] Re: licenses and trees (July 15, 2003), http://www.netbeans.org/servlets/ReadMsg?listName=nbdiscuss& msgNo=2547 (last accessed October 20, 2009)

46. West, J., O'Mahony, S.: Contrasting Community Building in Sponsored and Com-munity Founded Open Source Projects. In: Proceedings of the Proceedings of the 38th Annual Hawaii International Conference on System Sciences, HICSS, vol. 07, p. 196.3. IEEE Computer Society, Washington, DC (2005)

47. Lerner, J., Tirole, J.: The simple economics of open source. NBER Working paper series, WP 7600. Harvard University, Cambridge (2000)

48. von Hippel, E., von Krogh, G.: Open source software and the private-collective innovation model: Issues for organizational science. Organization Science 14(2), 209–223 (2003)

49. Hedhman, N.: mailing list message dated Sun, 29 Jun 2003 13:31:48 +0800 "[nbdiscuss] Re: licenses and trees (was: Anti-Sun Animosity)," available online at http://www.netbeans.org/servlets/ReadMsg?listName=nbdiscuss&msgNo=2578, last accessed 21 October 2009.

50. NBDiscuss mailing list message, http://www.netbeans.org/servlets/ReadMsg?list=nbdiscuss&msgN o=3784 (last accessed February 28 , 2009)

51. Shah, S.K.: Motivation, governance and the viability of hybrid forms in open source soft-ware development. Management Science 52(7), 1000–1014 (2006)

A Historical Account of the Value of Free and Open Source Software: From Software Commune to Commercial Commons

Magnus Bergquist, Jan Ljungberg, and Bertil Rolandsson

University of Gothenburg
{magnus.bergquist,jan.ljungberg,bertil.rolandsson}@gu.se

Abstract. Free and open source software has transformed from what has been characterized as a resistance movement against proprietary software to become a commercially viable form of software development, integrated in various forms with proprietary software business. In this paper we explain this development as a dependence on historical formations, shaped by different ways of justifying the use of open source during different periods of time. These formations are described as arrangements of different justificatory logics within a certain time frame or a certain group of actors motivating the use of free and open source software by referring to different potentialities. The justificatory arrangements change over time, and tracing these changes makes it easier to understand how the cultural, economic and social practices of open source movements are currently being absorbed and adopted in a commercial context.

Keywords: Free and Open Source Software, Justification, Historical approach.

1 Introduction

Over the last decade, free and open source software (FOSS) have transformed from being an ideologically driven movement, organizing resistance against proprietary software development, to a means for revitalizing the way firms produce software and make business. This change can be described as a development, where open source gradually have been recognized by different actors outside the open source communities and incorporated in corporate software development contexts. While this development has been described by different authors e.g. [4, 5, 8, 9, 10, 13, 24, 38], the conditions and circumstances enabling the transformation of FOSS has rarely been addressed.

The purpose of this paper is to articulate this transformation, by analyzing how the use of FOSS has been justified during different periods of time. These justifications are analyzed by using Boltanski and Thevévenot's [3] framework of "justificatory regimes" that enable actors in various settings to justify different means and initiatives. We claim that these historically formed arrangements condition the

S.A. Hissam et al. (Eds.): OSS 2011, IFIP AICT 365, pp. 196–207, 2011.

adoption of FOSS software and methods and its cultural, economic and social practices in a commercial context.

2 Logics of Justification

FOSS has during its development been motivated or justified with different arguments ranging from a moral non-utilitarian stance to a pragmatic, utilitarian stance [36]. In this paper justification refers to how actors make use of different logics to embrace ideas of change or novelties. Logics of justification are ways by which actors make the changes legitimate through ongoing processes of valuation.

The analysis focus on how the importance of FOSS is recognized through combinations of logics of justification, reshaped over time. In order to distinguish more than e.g. the importance of a market value, we will then talk about different types of worth [39]. An important start point is the theoretical elaboration on the concept of justification done by Luc Boltanski together with Laurent Thevénot [3], in which different social and moral aspects are considered as important to how change is justified. More precise, they depict six different logics or "justificatory regimes" that enable actors in various settings to justify means and initiatives. Boltanski and Chiapello also used this framework to map the "spirit of capitalism", i.e. the justifications of people's commitment to capitalism in a certain era. Here we apply it to the development of FOSS and changes in the dominant value system. Thus, this is an analysis of "the spirit of open source" or the ideology that justifies people's commitment to FOSS in a certain time period. The six logics of justification suggested by Boltanski and Thevénot [3] are:

- An *inspirational logic* is founded on a principle of grace or artistry serving what is perceived as authentic qualities of life.
- A *domestic logic* can be traced when an established hierarchy made out of personal interdependencies, with a patriarch or guru on top, is justified as natural by referring to a stable social order or tradition. An example could be a conservative family organization, ruled by an authoritarian father or elder.
- In a *popular logic* justification is reached through importance of being renowned, i.e. by being granted credit and esteem in the opinion of others. The achieved worth becomes dependent on identification and fame.
- Within a *civic logic*, justification relies on being representative and on acting in accordance with a collective will. Worth is created through the capacity to mobilize collectives around common interests. In this process, moral claims, and definition of identity become important.
- Justification within the *market logic* depends on individuals and their ability to possess and compete. The worth would then be related to individuals' selling and buying goods and services. This can be perceived as an egoistic practice. However, the right to possess and seize market opportunities is related to a claim that, if done fairly common good will emerge out of market transactions.
- An *industrial logic* justifies actions and initiatives by referring to efficiency and the scale of abilities. Contrary to the market logic, the industrial logic focuses on whether functionality and productivity is organized in a reliable way.

To Boltanski and Thevénot [3] these logics describe how the worth of initiatives can be perceived differently, but also that any justification relies on claims that are based on socioeconomic conditions as well as some sort of moral order. Thus, the ability to make trustworthy references to both a general fairness and social order will be necessary for a justifying logic to emerge.

In our analysis, the various logics that Boltanski and Thevénot identify are shaping the justifying arrangements emerging within a certain time frame or a certain group of actors. The tensions between FOSS practices and proprietary practices, creates uncertainties about the future impact of adopting FOSS software. Thus, different actors struggle to justify the use of FOSS by referring to different potentialities. These justifying arrangements can work as integrative forces, but they are also associated with tensions between e.g. social order and moral claims. The analysis concerns how old perspectives on FOSS software development become active components in new circumstances, justified according to principles that involved actors in different organizational contexts can agree and act upon [3]. Hence, we will follow how new and old discourses overlap and form interpretative arrangements, guiding how means and measures are motivated and common principles articulated [2].

3 Method

In order to understand how perceptions of FOSS software have developed over time, we have traced justifying arrangements that historically have been used to define the value of FOSS. This has been done by going through canonical texts and previous research [3]; i.e. we look at research and publications considered to have had a major impact on the perception of FOSS. Typical for this archeology of knowledge [14] is to compare series of sources over time in order to capture changes in dominant modes of thinking, acting and organizing.

We have been looking at three time periods, were we claim that certain arrangements of justification logic is constituted. Certain events function as approximate starting points of these time periods. The formulation of the free software definition and constitution of the Free Software Foundation is the starting point for the first time period (early 1980s), constituting the first justificatory arrangement. Here analysis is based on texts that evolved around the Free Software Foundation (FSF) and the front figure Richard Stallman (e.g. [15, 17, 36]). The starting point of the second time period is the formulation of the open source definition and constitution of the Open Source Initiative (late 1990s). This is based on texts related to the Open Source Initiative (OSI) and the front figure Eric Raymond and his seminal and much referred texts that were later published as the book *The Cathedral and the Bazaar* [29]. The starting point for the third time period is the emergence of public sector policy documents regarding FOSS, created by policy making bodies, advocacy groups and governments representing public sector interests (early 2000s). Some of these documents, e.g. national reports and policy documents, constitute the basis for the analysis of the justificatory arrangement that we call public commons e.g. [27, 28, 32, 33].

4 Arrangements of Ideological Justification in the History of FOSS

4.1 First Arrangement: Software Commune

When the free software movement started to mobilize during the eighties it was a reaction against the emerging software industry, and it was organized as an ideologically framed commune. Earlier no software industry or market for software did exist because software was developed directly for specific hardware [6]. Since intellectual property for software was a non-issue, the programmers were used to share solutions, knowledge and the source code itself. They took pride in being skilled programmers, and were eager to help fellow programmers. However, when the market for software took off, the programmers' old practices of sharing were abandoned, and the source code became a private company property to be carefully protected. This provoked some developers to take action in the shape of a politically driven movement. One of the key persons in this process was Richard Stallman (RMS) who still plays an important role in the movement. Stallman's work on the text editor Emacs is a good example of the spirit of the movement. Emacs was given away by Stallman on the condition that other programmers should "give back all extensions they made, so as to help Emacs improve. I called this arrangement the Emacs commune" [Stallman in 25, p.   416]. The emerging copyright protected software development practices faced Stallman with what he describes as a stark moral choice:

> "The easy choice was to join the proprietary software world, signing nondisclosure agreements and promising not to help my fellow hacker." [15, p. 17].

Stallman chose another route and facilitated a number of initiatives that institutionalized the resistance to proprietary software, such as the GNU project, the Free Software definition [17], the Free Software Foundation, and the GNU General Public License (GPL) [15, 36] that was designed to ensure that the rights of the free software definition were preserved (i.e. an inscription of the free software definition in copyright law). The "viral" character of GPL, i.e. that other software that is bundled with a GPL-licensed software must also be released under GPL, created tensions with proprietary software.

Here, justification was based on a *civic logic* based on principles and rules defining free software as a common good. Software code must be made available for anyone to use, alter and redistribute to secure future development of the ideas that the code entails. The proprietary development was a threat against the programmers' freedom: "The fundamental act of friendship among programmers is the sharing of programs; marketing arrangements now typically used essentially forbid programmers to treat others as friends." [15]. It was also threatening a more general public interest in the freedom of information.

Besides the civic logic, an *inspirational logic* could be identified, emanating from the roots of the free software movement in the hacker culture of the early sixties. This was mainly formed around several MIT research groups who were experimenting with new technologies (e.g. TX-0 computer, MIT AI Lab and the Unix operating system). Levy [25] described this culture as:

"a new way of life, with a philosophy, an ethic and a dream.[...] hackers that by devoting their technical abilities to computing with a devotion rarely seen outside monasteries they were the vanguard of a daring symbiosis between man and machine." [25, p. 39].

Here hacking and playing with technology were justified as the authentic values of life and the true motivational force for programmers' engagement. The activity of programming itself is often referred to as an art [11, 22], e.g. as Donald Knuth has formulated it "The chief goal of my work as educator and author is to help people learn how to write *beautiful programs*" [22, p. 6]. The word hack and hacking changed over time from "a spirit of harmless, creative fun" to "acquire a sharper, more rebellious edge" [36]. Still, the hacker concept is deeply linked to the ability to solve difficult problems for its own sake, as the definition Stallman gives to it: "Playfully doing something difficult, whether useful or not, that is hacking." [15].

Also a *popular logic* was visible, since the reputation of being a skilled hacker is at the heart of the very concept. To become a hacker is not something that individuals decide by themselves, it is something they earn by getting respect from the community. Public opinion itself establishes the worth of FOSS initiatives and actors, in the sense that popular and famous projects or persons will attract many contributors. There is even a special word in the hacker dictionary for the most admired programmers with an exceptional reputation - demigood: "A hacker with years of experience, a world-wide reputation, and a major role in the development of at least one design, tool, or game used by or known to more than half of the hacker community." (Jargon-file 4.3.1).

Furthermore, the tight community with its' closed clan-like hierarchy of personal interdependencies and patriarchic governance, resembles a *domestic logic*. This is what Raymond in his book "the Cathedral and the Bazaar" criticized as the cathedral-building style of development, even though Raymond himself was a former believer:

"I believed that the most important software (operating systems and really large tools like Emacs) needed to be built as cathedrals, carefully crafted by individual wizards or small bands of mages working in splendid isolation, with no beta released before its time." [29].

We have labeled the justifying arrangement of the first time period, *commune*, due to its nature of a tight community, kept together by strong common hacker values. As has been shown, this arrangement drew mainly on a technically driven civic *community logic* rooted in the hacker movement, demanding free access to information and source code while fighting against proprietary commercial interests in software development. In addition, the arrangement also relied upon an *inspirational logic* stressing the importance of authentic grace or technical artistry. This encourages the developers to independently realize personal creativity, and thereby improve their status within the community. This leads also to a *popular logic* stressing the importance of reputation and fame. Also, a domestic logic follows from the nature of the movement as a closed tight community were highly respected developers took on roles representing hierarchical superiority typical for the domestic logic.

4.2   Second Arrangement: The Bazaar

During the mid nineties a new approach to justify FOSS can be identified. Eric Raymond, saw the earlier movement's hostile attitude to commercial software as a big problem:

> "It seemed clear to us in retrospect that the term 'free software' had done our movement tremendous damage over the years. Part of this stemmed from the well-known 'free speech/free-beer' ambiguity. Most of it came from something worse -- the strong association of the term 'free software' with hostility to intellectual property rights, communism, and other ideas hardly likely to endear themselves to an MIS manager." [29].

In order to avoid these connotations, the term open source was coined, indicating that open source is viewed as a means to an end of producing software of high quality. The Open Source definition is similar to the Free Software definition, but it explicitly states that an open source license must not contaminate other software (as the GPL-license), claiming that this would hamper commercial use of open source. A plethora of more permissive licenses [35], were used to make it easier for open source and proprietary software to coexist. The Open Source Initiative (OSI) was founded in 1998 to support the new focus on technology rather than ideology. This more pragmatic nature of the movement, downplayed some of the most ideological parts of the value system, but also contributed to a wider diffusion of free and open source software. The movement grew substantially, and included both large traditional software companies (e.g. IBM, HP) and small companies that were founded on open source business models (e.g. Red Hat, Mandrake).

In this time period the dominating technically driven civic imperative is replaced by clearly visible *market logic*:

> "RMS's best propaganda has always been his hacking. So it is for all of us; to the rest of the world outside our little tribe, the excellence of our software is a far more persuasive argument for openness and freedom than any amount of highfalutin appeal to abstract principles. So the next time RMS, or anybody else, urges you to "talk about freedom", I urge you to reply "Shut up and show them the code." [30]

It is not philosophical or political principles, but the excellence of the software that should convince. The excellence of the software also points to quality ideals that are often found in an *industrial logic*. This also stresses the importance of the code itself and its accessibility, as a key to the arrangement. As an alternative to the domestically oriented cathedral style, where wizards were leading a tight, closed tribe of skilled hackers, Raymond proposed the "Linus Torvalds's style or the bazaar style of development - release early and often, delegate everything you can, be open to the point of promiscuity" [29, p, 30].

However, the *inspirational* and *popular logic* are still visible in the bazaar. Inspirational worth of open source software would depend on spontaneous and passionate initiatives, like Linus Torvalds' initiative to write the Linux system in order to learn how operating systems work [34], and software development close to artistry were still appreciated. The *popular logic* was strengthened in the bazaar where skilled programmers could gain reputation and fame if they succeeded to pass the peer review system [1]. They could gain reputation among an even larger crowd of developers, due to the open character of the bazaar. Also a *domestic logic* could be detected, e.g. in the coordinating model often referred to as "benevolent dictatorship" with Linus Torvalds as the prime example [29], and the informal hierarchies resulting from differences in status and skill within FOSS communities.

We have labeled the justifying arrangement of this time period, the *bazaar*, in accordance with Raymond's metaphor. Here, FOSS and proprietary software will coexist, and anyone is free to choose what is considered the best solution (*market logic*). FOSS is viewed as a better, more efficient method for developing software of good quality (*industrial logic*). Hacker values are still emphasized claiming that free access to code would improve developers' opportunities to do innovative and artistic programming (*inspirational logic*). The spread of the movement make opportunities to get reputation and fame among peers for making good contributions even more attractive (*popular logic*). However, despite its strong market component, open source software was still associated with a *civic logic* where freedom of information became important.

4.3 Third Arrangement: The Public Commons

In the beginning of 2000, a pragmatic version of FOSS started to be appropriated by large user groups outside FOSS communities. Especially governments and public sector organizations found an interesting potential in keeping computing costs down by using FOSS software. The domination of FOSS applications in the horizontal domain of infrastructural software (e.g. operating systems, web servers, and data bases) was complemented with an increasing use of vertical software as desktop and enterprise systems [13]. With a growing number of FOSS users that were not producers or experts but "general end-users", FOSS moved out of the pure hacker domain.

The growing use of FOSS in this context revitalized the former *civic logic*. Manifested in a number of policy texts, it was reinterpreted by public sector organizations and advocacy groups representing public sector interests (e.g. governments, municipalities, FSF and OSI representatives). The incorporation of FOSS in public sector was seen as an appraisal of values associated with democracy, citizenship and the relationship between citizens and public sector. On the one hand FOSS became attractive to public authorities due to new demands and needs dictated by changes in their own organizations and in society. Economically it was a way to cut costs in the public sector, and get value for taxpayer money. Ideologically FOSS was seen as an expression of the principle of the commons, a way to promote ideas associated with public sector organizations' role in a democracy. It was presented as a radical alternative that could liberate the public sector by getting rid of bureaucratic and expensive non-democratic historical burdens; the public sector would serve the people while standing free from partial interests on the market. On the other hand the FOSS movement took the opportunity to influence public authorities by lobbying activities in order to gain a widespread impact of their goals. FOSS provided the public sector with the ideology and examples it needed to make its point.

A set of policy documents formulated around year 2000 [20, 21, 26, 27, 28, 32, 33, see also 18] give a more detailed view of basic arguments about the use of FOSS in public sector. One argued advantage was cost reasons. FOSS was often made available at a low acquisition cost, and without licensing costs. By adopting FOSS solutions it was also argued that different public actors could develop shareable solutions that would decrease development costs. The cost argument is an efficiency argument, relating to an *industrial logic*, but also to a public governance version of the *civic logic* demanding transparency in how taxpayer money is used. Other arguments aligned to an industrial logic are related to the supposed quality and reliability of FOSS. The same holds for common arguments related to security, transparency and privacy. The free availability of the source code supposedly offers better protection against malware, meaning better protection for the citizens' integrity. These arguments also relates to how an *industrial logic* of efficiency is combined with a public version of the *civic logic*, i.e. how the civic mission best could be implemented in an efficient way.

Other arguments relates to market *logic*. FOSS devotion to promoting open standards and interoperability secures that systems ensure access to government data without possible barriers posed by proprietary software and data formats. This would lead to a situation where lock-in effects of proprietary companies' software could be avoided. By promoting open standards and interoperability in its own systems, public sector contributes actively to well functioning software markets, minimizing monopolies and lock-in effects. Another argument related to a *market logic* is that regional software industry was supposed to prosper as a consequence of public sector interventions in FOSS. Local programmers were to be engaged in flexible adaptation to specific needs that were not supported by global commercial actors, and create new niche markets to be exploited by local entrepreneurs rather than by global software firms.

Finally, a set of arguments more directly related to a *civic logic* could be found. Arguments related to political reasons, claimed the advantages of national-wide FOSS based IT-infrastructure in developing countries. Here post-colonial arguments were raised, highlighting the possible independence from Western software houses controlling the IT-development: "If South Africa chooses the open route [...] South Africa can break dependence on foreign companies, and potentially become a player in the world of software development and software services markets" [32]. Other arguments directly linked to a civic logic stress *freedom* and *democracy* as basic values inherent in FOSS.

The justifying arrangement of the third time period is labeled *public commons*. This is the first justifying arrangement that takes shape outside the movement. This becomes evident while looking at how the *civic logic* is reinterpreted from a public sector perspective; open source is seen as a mean for enhancing democracy and making the public sector free in relation to commercial interest. The civic logic promotes the public sector as a service provider for citizens, which calls for certain moral claims regarding loyalty to the public who elects officials to represent them. This justifies claims on honest and transparent development of software made for the citizens by using FOSS. The public sector is given a mission based on the *market logic* and the *industrial logic*, directing attention towards issues of cost efficiency, reliability and quality. Furthermore, the promotion of open source in public sector is supposed to contribute to a well functioning software market.

5   Emerging Justificatory Logics of Contemporary FOSS

The historically based arrangements presented above show how FOSS has gone from being justified as a community driven software development endeavor with the developer at the center of attention, to become more motivated by external interests. As shown, these are emerging arrangements in which the content of identified justifying logics continuously have changed. The transformation of FOSS into a commercially viable form has been described by different authors as OSS 2.0, progressive open source, corporate code, professional open source etc. [12, 13, 16]. Here, we view the characteristics of all these phenomena as parts of an emerging arrangement, partly overlapping in time with the mentioned public commons arrangement. The arrangements presented also reveal how e.g. initial civic and inspirational logics are reinterpreted over time. We will now discuss how these logics play out in the formation of a new emerging justifying arrangement, and what this may mean for the adoption of open source in a corporate context.

According to Boltanski and Thevénot [3] justification through the *industrial logic* is achieved by making claims on efficiency, expertise and the scale of abilities. Focus also lays upon whether technological innovations and functionality is organized in a scientifically controlled and predictable way. In accordance, many descriptions of FOSS have always focused on what can be described as an industrial logic [12, 13, 16], and today is further emphasized in descriptions of how e.g. the voluntary nature of FOSS is substituted by strategic planning, bulletin board like product support becomes professional, the open access to source code is challenged by giving controlled access only to specific business partners, or only internally behind corporate fire walls [12].

The basic idea within market logic is that justification is based on individuals' ability to possess and compete [3]. The worth of such a justification is created when as many as possible are able to sell and buy goods and services. One of the main arguments for FOSS in public commons was to maximize the positive effects of taxpayers' money by making them operate on a more open and transparent market. Hereby the market could be used for reaching a higher cause and thus legitimize public sector civic claims. This way of reasoning, supporting the civic emphasis on honest markets, is also found in firms built around FOSS today. However, these firms also struggle to find ways of combining FOSS with proprietary code [19, 31].

The *civic justification logic* has undergone an interesting development. In the commune and the bazaar, civic justification was the nexus of the FOSS movement's ideological agenda. The aim was to strengthen democracy and free access to information and source code by the help of software communities supporting a common right to independently control software. In the public commons arrangement FOSS was then justified as a mean helping the public sector to become independent from private companies' proprietary standards and lock-in strategies. This also paved the way for a customer and user perspective, supporting a market logic that emphasizes common good rather than proprietary strategies. Open source software becomes a way to improve honest, flexible and efficient relations with customers and end-users. Now, similar claims on challenging the idea of possessing software is found in contemporary pure play firms built around FOSS business models [31], where competition with free and open standards is said to be a more honest approach to customers and users.

Boltanski and Thevénot [3] describe *inspirational logic* as a type of justification that refers to principles such as grace and artistry serving authentic qualities of life. These principles lies at the heart of the initial hacker culture and were prominent both in the commune and bazaar arrangements, but could not be identified in the public commons arrangement. The inspirational logic traced in firms today seems to be reinterpretcd from a business perspective, by being less associated with contributions to a higher cause and more associated with being engaged in work. Movement driven inspiration is replaced by professional inspiration associated with a hobby or a scientific quiz triggering lust for work by making professional developers free to access and manipulate the code [13, 31]. In addition, the inspirational and industrial logic do then also support each other. This type of inspirational logic resembles FOSS research on intrinsic motivation, stressing that open source may connect the professional world of software development with the exercise of a hobby [4].

According to Boltanski and Thevénot [3] justification through the *popular logic* is reached through reputation; i.e. being granted credit and esteem in the opinion of others is a goal in itself. The popular logic identified in the commune and in the bazaar arrangement, related to the reputation that was gained when programmers succeeded to get their contributions of code accepted by peer-reviewers and introduced into the code base [4, 1]. Highest reputation was attached to the visionaries and ideological leaders, gaining reputation through developing widely used FOSS programs. Companies that struggle to attract the best FOSS-programmers, indicates that this logic is re-articulated in firms today. The majority of contributions appear to be rejected, and accepted contributions are still manifestations of programming skills and status. However, focus rather lay on the status of the project than the individual programmer; i.e. the success of attracting contributors to a company owned FOSS project is main issue [31].

In the *domestic logic* Boltanski and Thevénot [3] describe how justification is reached through the stability of conventions or traditions, revolving around a family like organization and its ruler. In the commune, with its tight tribe of developers and highly reputed ideological leaders, a domestic logic could be sensed. The bazaar arrangement then challenged this closed commune, and in the public commons arrangement the domestic logic was not visible at all. The domestic logic is hard to trace in contemporary FOSS in corporate settings. However, it is potentially inherent in many FOSS practices, where developers are part of an informal hierarchy due to skill and reputation, and were the most respected developers is dedicated to roles that clearly points out a hierarchical superiority. Also, FOSS still appears as heavily male dominated, indicating a stable gender structure that could be investigated further in terms of a domestic logic [7, 23].

6 Conclusions

By focusing on the value accredited to FOSS by different groups, and how the justifications of FOSS have been formed in different time periods, we may move beyond the established distinction between the initial movement driven approach and the current business driven OSS 2.0. It becomes possible to describe how different justifying logics are re-articulated in the intersection of FOSS movement and corporations. While the industrial and the market logics emerge as major justificatory means in contemporary commercial FOSS, it is important to notice that core driving forces as the inspirational logic and the civic logic still could be considered as important parts of the FOSS bandwagon. Even in a commercial context marked by industrial and markets logics, these logics still makes it possible to justify the use of FOSS by referring to potentialities that could inspire and empower programmers developing open source, as well as to contribute to the society as a whole.

References

1. Bergquist, M., Ljungberg, J.: The Power of Gifts: Organizing Social Relationships in Open Source Communities. Information Systems Journal 11, 305–320 (2001)
2. Boltanski, L., Chiapello, E.: The New Spirit of Capitalism. Verso, London (2005)
3. Boltanski, L., Thévenot, L.: On Justification: Economies of Worth. Princeton University Press, Princeton (2006)
4. Bonaccorsi, A., Rossi, C.: Comparing Motivations of Individual Programmers and Firms to Take Part in the Open Source Movement: From Community to Business. Knowledge, Technology and Policy 18(4), 40–64 (2006)
5. Bonaccorsi, A., Rossi, C.: Why Open Source Software can Succeed. Research Policy 32(7), 1243–1258 (2003)
6. Campbell-Kelly, M.: From Airline Reservations to Sonic the Hedgehog: A History of the Software Industry. MIT Press, Cambridge (2003)
7. Cuckier, W.: Constructing the IT Skills Shortage in Canada: The Implications of Institutional Discouse and Practices for the Participation of Women. In: SIGMIS Conference Copyright 2003. ACM, Philadelphia (2003)
8. Dahlander, L., Magnusson, M.G.: Relationships between open source software companies and communities: Observations from Nordic firms. Research Policy 34(4) (2005)
9. Dahlander, L., Magnusson, M.G.: How do Firms Make Use of Open Source Communities? Long Range Planning 41 (6) (2008)
10. Demil, B., Lecocq, X.: Neither Market, nor Hierarchy or Network: The Emergence of Bazaar Governance. Organization Studies 27(10), 1447–1466 (2006)
11. Dijkstra, E.W.: EWD316: A Short Introduction to the Art of Programming. T. H. Eindhoven, The Netherlands (1971)
12. Dinkelacker, J., Garg, P.K., Miller, R., Nelson, D.: Progressive Open Source. In: Proceedings of ICSE 2002, Orlando, May 19-25 (2002)
13. Fitzgerald, B.: The Transformation of Open Source Software. MIS Quarterly 30(3), 587–598 (2006)
14. Focault, M.: The Archaeology of Knowledge. Routledge Classics, London (2002)
15. Gay, J. (ed.): Free Software, Free Society: Selected Essays of Richard M. Stallman. GNU Press, Boston (2002)
16. Gurbani, V.K., Garvert, A., Herbsleb, J.D.: A case study of open source tools and practices in a commercial setting. In: Proceedings of the 3rd IFIP Working Group 2.13 International Conference on Open Source Software (OSS 2007), Limerick, Ireland, June 11-14., vol. 234. Springer, Heidelberg (2007)
17. GNU's Bulletin 1 (1): 8, http://www.gnu.org/bulletins/bull1 (accessed March 5, 2010)

18. Hahn, R.W. (ed.): Government Policy toward Open Source Software. AEI-Brookings Joint Center for Regulatory Studies, Washington DC (2002)
19. Höst, M., Orucević-Alagić, A.: A systematic review of research on open source software in commercial software product development. Inform. Softw. Technol. (2011), doi:doi:10.1016/j.infsof
20. IDA Study: Study into the use of Open Source Software in the Public Sector Part 1, http://europa.eu.int/ISPO/ida/export/files/en/840.pdf; Part 3, http://europa.eu.int/ISPO/ida/export/files/en/835.pdf (2001)
21. International Institute of Infonomics Free/Libre and Open Source Software: Survey and Study (2002), http://www.infonomics.nl/FLOSS/report/ [040401]
22. Knuth, D.: Computer Programming as an Art. Communications of the ACM (December 1974)
23. Lakhani, K.R., Wolf, R.G.: Why Hackers Do What They Do: Understanding Motivation and Effort in Free/Open Source Software Projects. In: Feller, J., Fitzgerald, B., Hissam, S., Lakhani, K. (eds.) Perspectives on Free and Open Source Software. MIT Press, Cambridge (2005)
24. Lerner, J., Tirole, J.: Some Simple Economics of Open Source. Journal of Industrial Economics 50(2), 197–234 (2002)
25. Levy, S.: Hackers: Heroes of the Computer Revolution. Anchor Press/Doubleday, New York (1984)
26. LinuxToday, LinuxPR: Munich Goes with Open Source Software (May 28, 2003), http://linuxtoday.com/infrastructure/2003052802126NWDTPB.
27. MIMOS Berhad; Worldwide Open Source Policy: National Summaries (2003), http://community.asiaosc.org/~iwsmith/policy/ [040401]
28. MITRE Corporation Use of Free and Open-Source Software (FOSS) in the U.S. Department of Defense (2002), http://www.egovos.org/rawmedia_repository/588347ad_c97c_48b9 _a63d_821cb0e8422d?/document.pdf
29. Raymond, E.S.: The Cathedral and the Bazaar: Musings on Linux and Open Source by an Accidental Revolutionary. O'Reilly and Associates, Sebastopol (1999)
30. Raymond, E.S.: Shut Up and Show them the Code, Linux Today (June 28, 1999b)
31. Rolandsson, B., Bergquist, M., Ljungberg, J.: Open source in the firm: Opening up professional practices of software development. Research Policy 40(3), 576–587 (2011)
32. South African National Advisory Council on Innovation: Open Software & Open Standards in South Africa: A Critical Issue for Addressing the Digital Divide (2002), http://www.naci.org.za/pdfs/opensource.pdf [040401].
33. Statskontoret: Free and Open Source Software – a feasibility study, Appendix 1: Extensive survey (2003), http://www.statskontoret.se/pdf/200308eng.pdf [040401]
34. Torvalds, L., Diamond, D.: Just For Fun: The Story of an Accidental Revolutionary. HarperCollins, New York (2001)
35. Välimäki, M.: The Rise of Open Source Licensing: A Challenge to the Use of Intellectual Property in the Software Industry. Helsinki, Turre Publishing. versity Press (2005)
36. Williams, S.: Free as in Freedom: Richard Stallman's Crusade for Free Software. O'Reilly Media, Sebastopol (2002)
37. Zittrain, J.: Normative Principles for Evaluating Free and Proprietary Software. University of Chicago Law Review 71(1) (2004)
38. Ågerfalk, P.J., Fitzgerald, B.: Outsorcing to an unknown workforce: exploring opensourcing as a global sourcing strategy. MIS Quarterly 32(2), 385–409 (2008)
39. Stark, D.: The Sense of Dissonance. Accounts of Worth in Economic Life. Princeton University Press, Princeton (2009)

Framing the *Conundrum* of Total Cost of Ownership of Open Source Software

Maha Shaikh and Tony Cornford

London School of Economics and Political Science, Information Systems
and Innovation Group, Houghton Street, London WC2A 2AE, UK
{m.i.shaikh,t.cornford}@lse.ac.uk
http://personal.lse.ac.uk/shaikh/,
http://personal.lse.ac.uk/cornford

Abstract. This paper reflects the results of phase I of our study on the total cost of ownership (TCO) of open source software adoption. Not only have we found TCO to be an intriguing issue but it is contentious, baffling and each company approaches it in a distinctive manner (and sometimes not at all). In effect it is a conundrum that needs unpacking before it can be explained and understood. Our paper discusses the components of TCO as total cost of ownership and total cost of acquisition (and besides). Using this broad dichotomy and its various components we then analyze our data to make sense of procurement decisions in relation to open source software in the public sector and private companies.

Keywords: open source software, total cost of ownership, benefits, exit costs, software adoption.

1 Introduction

Total cost of ownership (TCO) is considered to be a fundamental issue when making software procurement decisions [1-4] in organizations yet this is an area that has received limited attention. In this paper we are concerned with TCO but more specifically in relation to open source software (OSS) adoption decisions by organizations[1]. This adds yet another layer of complexity because the assessment of open source software procurement is not exactly the same as that for proprietary software [5]. Indeed, we find that by unpacking the idea of open source TCO we become more aware of the taken for granted in proprietary software procurement decisions. TCO has been defined as an understanding of 'the "true cost" of doing business with a particular supplier for a good or service' [6]. The idea of a 'true cost' and the ability to be able to assess it accurately, however is something most academics and practitioners would agree is not straightforward [7]. Thus we prefer a definition of TCO offered by Lerner and Schankerman [8] which distinguishes between different costs, and TCO is

---

[1] Our research is funded by the UK Cabinet Office and the OpenForum Europe. The aim of this research is to assess the various costs and value of open source adoption by the public sector and private organizations.

S.A. Hissam et al. (Eds.): OSS 2011, IFIP AICT 365, pp. 208–219, 2011.

understood as the total cost of providing a functionality using one program. The proper accounting of cost should include total costs of procurement, management and support, associated hardware costs, and when one is thinking of changing software solutions, migration costs' (p107). In this definition we have a range of costs mentioned, all of which need attention before any true grasp of TCO of software can be reached.

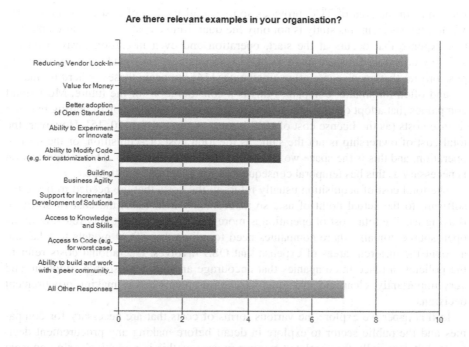

Fig. 1. This figure shows the various factors companies in our pro-forma considered when making a decision to move to open source, and the relative weight of each factor

Some authors claim that there is no such thing as 'the cost of software' [8] implying that cost is a more multi-faceted issue which needs to be understood better. Thus, like all evaluation decisions TCO has a quantitative and qualitative element. In section 3 below we detail the various elements of each type of cost that we have noted from literature and from our data. We found this to be a recognized issue amongst practitioners in the public sector and the private. Most decisions taken on procurement in either sector are understandably based on cost but this is not the only factor and with open source software, we found that this is not even the most relevant. The larger concern for companies eager to adopt open source software was reduced vendor lock-in, and what companies understood as 'value for money' (see Fig 1).

Lerner and Schankerman [8], with their distinctions in costs, indicate the conundrum of TCO. Companies are becoming more aware of these issues but the smaller ones do not have the resources to actually carry out a full detailed TCO study. How does a company assess the softer costs [9] surrounding software procurement espe-cially when the software is open source (a relatively less familiar category for many companies)? It is important to make sense of the categories of cost and exactly what they entail to better manage them and make better informed decisions.

1.1 Conundrum of TCO of OSS

Literature in the area of TCO provides some useful models of cost evaluation [10]. What interests us in this study is not only the quantifiable costs but the more amorph-ous expense that occurs at the start, operation and even migration away from the software adoption. Business analysts [11-13] in this area have made note of the many possibilities that open source offers companies [14] but again these are hard to quanti-fy, and often for smaller companies this is a consideration for the future. Most small companies that adopt open source software do so with the intention of cutting back on license costs (as the license cost of OSS is zero or close to it) [10, 15]. However, the total cost of ownership is not the same as the total cost of acquisition, or the cost of operation, and this is the space we want to explore with our study. This categorization is necessary as this has temporal consequences for adoption of software.

The total cost of acquisition usually refers to the costs that are needed to bring the software to the actual point of use, so theoretically it includes the cost of software (buying it). The total cost of operation is more nuanced a cost but very relevant in the open source domain where companies need to adjust their TCO models to take into account the different areas of expense that OSS involves. Operational costs refer to the policies in place in companies that encourage and aid open source adoption, and very importantly, clear and structured TCO assessments before making procurement decisions.

In this paper we explore the various forms of costs that are necessary for compa-nies and the public sector to explore in detail before making any procurement deci-sions, but especially those related to open source as this is a relatively new concept for them and different in that the license cost is very low yet other costs seem to mount (but are often ignored and not understood).

2 Methodology

This study is structured to take place in two phases. This paper reflects some of the results from phase I which includes a data collection pro-forma of twenty-five ques-tions. These questions cover the basics of the company size, name and focus, but then go on to ask some very detailed questions about the various applications that are open source, why they were chosen, if they replaced proprietary software, was any differ-ence in cost experienced, and finally, what prompted this change or need to adopt open source software. The responses to the pro-forma, especially to the last question will be fed into an interview guide. This then takes us to phase II where we aim to conduct between 35-40 in-depth interviews. Access in a number of organizations that responded to the pro-forma has already been negotiated and phase II will take place over the summer of 2011.

2.1 Phase I – Pro-Forma

Phase 1 involved the creation of the pro-forma which was based on literature and documentation that helped to understand TCO models used in companies. The pro-forma was used as an early and simple data collection device rather than a fully structured and detailed survey. Our study is more qualitative in approach so the pro-forma was meant simply to gather material to help us set-up interviews, gain access and get some early comments to amend our interview guide. The pro-forma does not ask for facts, and figures on TCO for each organization, instead we ask the respondents to reply in relation to a Likert scale of 1-5.

The pro-forma was set up for access in two ways, document form (available in odt, pdf and doc formats) and an online version set up in SurveyMonkey. The aim of this study is to make sense of adoption of open source software by both private companies and public sector organizations. Though funded by the UK Cabinet Office to assess and evaluate the costs and issues involved in open source adoption by government agencies we decided that a more sound methodology would involve a balanced mix of commercial and public sector organizations. Public sector organizations are not profit orientated yet there is much to learn from private companies and their manner of dealing with open source. The larger idea here is the level of experience and comfort that private companies bring to open source adoption which is sorely lacking in the public sector. There are some exemplary cases of open source adoption by the public sector like the Extremadura case in Spain [16, 17] but there are far more 'success' stories of open source adoption by commercial companies [18-21]. The factors that encourage private companies to adopt open source software, especially considering most business models of such adoption indicate that the software itself does not lead to value creation or capture directly [22-25], make some of the lessons translatable across both sectors.

The pro-forma was put online for a period of two months (and is still online but for the sake of this paper we only took into account the pro-formas completed in the first two months) and we received twenty-four responses. We also received seven paper based pro-formas sent back to us as scanned documents via email. This made a total of thirty-one pro-formas. We had set the pro-forma to ensure that details of the respondent was a required category. This was done to be able to filter out any responses that were biased, duplicate or simply not completed with any seriousness. Of the 24 online pro-formas two were filled in by people calling themselves 'test' and 'anonymous'. We discounted the results from both these pro-formas. We also had two incomplete pro-formas online. Incomplete pro-formas were those where some questions were skipped. As this exercise was carried out as a precursor and data gathering exercise more for the interest of creating a strong and clear interview guide for phase II we accepted the results of the incomplete pro-formas. Phase II is where the researchers involved in this study hope to gain a more detailed understanding of TCO models and the decision-making process in organizations so it was felt that so long as the pro-formas were recognized to be valid (not anonymous or biased) and useful (filled in 75% of the pro-forma and added some non-mandatory comments that helped us to evaluate the experience of the company with open source) we would include the results to help shape the interview guide for phase II.

The pro-forma has four sections. The aim of section one is to ask for simple information like the name and affiliation of the respondent. This includes the size and name of the company/local authority. Section two prompts for the sort of OSS used by the organization and the time span of use. The aim of section three is to gather details on strategic drivers that lead to OSS adoption, and section 4 is concerned with eliciting the TCO models used.

2.2 Phase I Leading to Phase II

Of the total pro-formas we received the majority of them were filled in by small to medium sized private companies (44%). Small to medium sized enterprises included all those with a number of employees ranging between 1-100. We had 24% of the pro-formas completed by employees of large, and in many cases global companies (employees ranging from 101 and above). Public sector replies made 32% of the total. In phase II we intend to cover a larger portion of the public sector.

The pro-forma had a number of questions where respondents were asked to add comments or spell out the category of 'other' in more detail. Responses to such questions gave rise to some very interesting issues which will become a part of the interview guide and informed the researchers involved. The respondents for the pro-forma were asked for their contact details and phase II will draw us back to the those that made very intriguing comments. Phase II will involve in-depth interviews focused on 5-7 case studies. Key personnel involved in making procurement decisions and strategy of open source use in the organization will be interviewed.

The cases will be chosen on the basis of whether there has been involvement with open source adoption, use and/or redistribution for at least a period of two years. This is to ensure that there has been time enough for reflection on the process and there are some indicators that show 'success' or 'failure' – more simply, is open source still being used, has the use of it increased over time, and has it spread up the stack. We also want an even mix of public and private organizations to make it possible to reflect across the cases and build on lessons learned for an exchange of ideas. The organizations and their experience with open source needs to have been fruitful at some level but we are equally aware that we need to have examples of less successful implementations as this will enrich our work and understanding of the concerns with open source that can be faced. Indicators for the less successful cases include a return to proprietary software use, move to outsourcing their software development and a shrinkage in their in-house IT department. And lastly, our cases will include local authorities in the UK and other European government as well such as the Municipality of Munich, Municipality of Andalucía.

3   Analysis and Discussion

The pro-forma results are very interesting and we only have space to share some of the key ideas that emerged. These ideas include the importance of liberty [15] and flexibility (reduced vendor lock-in) provided by open source to companies and the public sector, that long term costs vary far more across companies considering their size and experience with open source, short term costs are slightly higher, that most companies choose a combination of open source and proprietary software where their decisions are based on pragmatism and need rather than questions of openness.

In this section we take these broad themes and frame them in relation to another interesting dimension that we noted from our data, that of a more fine-tuned TCO categorization than has been offered so far by other studies. We found that the cost categories were not limited to two broad ones, cost of acquisition and cost of operation, but instead we recognized two other very key cost factors that companies are beginning to take very seriously in relation to open source software procurement decisions – cost of adoption and exit costs (see Table 1).

The cost of operation and software are more quantifiable and thus easier to measure and evaluate. Cost of *operation* includes the expense of conducting a TCO study before making a decision of implementation. This is similar for proprietary software so in some respects this is not specific to open source, but rather to software decisions in general. The cost of *software* includes the cost of the license, setting up costs, and other costs which are similar to the cost of operation in their lending themselves to be quantified. However, the cost of adoption and exit costs for open source pose some interesting challenges but also opportunities.

3.1   Cost of Adoption

The cost of adoption, we found, concerns all the relevant but more qualitative expenses involved with the broad idea of adoption such as the learning necessary when you adopt open source for the first time or for a new part of the stack (see Table 2). Very importantly it also includes interoperability costs which many companies surprisingly ignore even though this is a feature of proprietary software as well. The difference with open source is that some respondents stated that they feel they can make the

Table 1. Different Categories of TCO, and what each involves

Categories of TCO			
Cost of operation	**Cost of software**	**Cost of adoption**	**Exit costs**
Formal TCO assessment	Initial purchase price	Learning	Migration costs
TCO policy	Monetary costs of set-up	Interoperability	Re-training
Cost of evaluating software (tinkering)	Customization expense	Support services	Switching costs
	Software scaling cost	Training	
		Access to upgrades	

necessary adjustments because the code is open, yet as we are becoming more aware, there is a steep learning curve with all software not created in-house. Upgrades are a growing concern for companies with open source because most open source software tends to adapt and be changed more frequently than proprietary. Of course the choice to upgrade is with the user yet interoperability can also become a problem if one software is upgraded but other applications and infrastructure are not.

3.2 Exit Costs

Exit costs are yet a more intriguing idea. Respondents agreed that this aspect was the most ignored and yet it formed a very positive aspect of the overall TCO of open source adoption. Table 2 outlines the areas where the costs with open source were considered to be the lowest but take note of the 'other' category in Table 2 and Figure 2. Upon reading the comments added by the respondents it was evident that though open source saved the organization money there were costs that had been ignored and ill-understood.

Exit costs include all the expenses of switching from one software to another, various interoperability expenses, costs related to legacy systems, retraining staff and initial teething issue costs. This is an area where at least at present companies feel that open source costs are higher and not clear at all. The low license costs with open source software, according to the respondents though very real can become misleading because companies simply begin to base their decisions on that cost alone and dismiss any other factor.

Table 2. Areas and applications of OSS which Saved Money for Organizations

Which OSS saved your organization money?			
	Agree	Agree some-what	Completely disagree	No change in expense - same
Applications:	75.0%	12.5%	0.0%	6.3%
Enterprise systems	71.4%	0.0%	0.0%	0.0%
Vertical/line of business	58.3%	0.0%	0.0%	0.0%
Desktop	84.6%	7.7%	0.0%	0.0%
System's De-velopment	71.4%	7.1%	0.0%	7.1%
Infrastructure:	84.6%	7.7%	0.0%	0.0%
OS Platforms	88.2%	11.8%	0.0%	0.0%
Application Servers	84.6%	0.0%	0.0%	7.7%
Web services	93.3%	0.0%	0.0%	0.0%
Networking	86.7%	6.7%	0.0%	0.0%
Database	86.7%	0.0%	0.0%	13.3%
Other	72.6%	0.0%	0.0%	0.0%

However, when migrating between one open source product to another the migration costs are then lower because open source is based on open standards and there is greater interoperability. Most companies that had a more long term understanding of their software adoption added comments in the pro-forma to the effect that the migration costs (exit costs) were more favourable for open source and so this became one of the deciding factors in favour of OSS. Simply put, migration costs between one proprietary application to another are always considerable as neither products are open and it is thus difficult to manage the necessary interfacing and interoperability changes across all the software applications that need to 'speak to each other'.

3.3 Vendor Lock-in and Lock-Out

Vendor lock-in though a real problem with proprietary software is less so with open source. This may well be a real consideration yet what we note from our study so far is that expertise of the software (be it open source or otherwise) and a lack of good documentation which is a problem with open source often becomes a *lock-out*. Companies feel discouraged from adopting any software they cannot control [29], but also cannot obtain comprehensive services and troubleshooting. The idea of reaching out to an unknown community [26] has a romantic appeal but is not practical.

Indeed, such promises spread FUD about open source adoption and lead to lock-out because companies avoid anything they are not familiar with. It is easy enough in theory to take code and customize it yet as many respondents noted this is not so in practice. They are forced to hire experts and look for support to communities outside the company. This is a drain on their resources and an expense that was not considered by the decision takers, even if a TCO assessment was carried out before procurement decisions were taken. This would not be such an issue if the software is vendor-supported, however, we have found that many companies and local authorities are drawn to open source because they wish to control the software source and make changes to it both in-house and with the help of a strong external community.

3.4 Temporal Element of TCO

Another key theme that arose from our pro-forma data was that of cost temporality. Of the four costs outlined in Table 1 the cost of adoption and exit costs are relatively quite high for open source software. This is even more marked with the added complexity of the size and experience of the organization. If the company is large *and* experienced with open source then these two costs are often well-understood and thus less expensive. Large companies can diversify and absorb costs better than smaller companies and this is largely true for the public sector as well. Smaller organizations however usually jump on the open source bandwagon with the naive idea that this will prove cheaper. They have also not undertaken a TCO analysis and if they are not experienced with open source it was then found that the expense of open source surprised many. In some cases, especially in the public sector (coupled with issues of poor interoperability) we have seen a return to proprietary software. Phase II of our study will include this local authority as a full case study to better explore the problems, issues and dilemmas that forced it to return to closed source software adoption.

3.5 Pragmatism and the Idea of Value

Most companies, like software developers and hackers work on the basis of pragmatism. If something is good enough and not broken then it will be continued to be used. Open source requires some experience and practice and very key, the licenses involved with open source need good skills and expertise, something most small companies do not have the resources for and the public sector simply does not consider. In effect if something is good enough then change is considered problematic and unnecessary. In the public sector in the UK we have found that local authorities are beginning to gravitate towards open source simply because of the lower costs promised by open source (due to the recession), however a better understanding of the benefits of

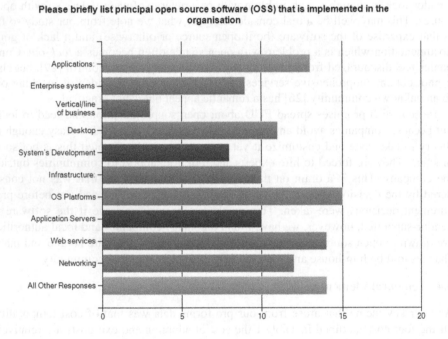

Fig. 2. This figure shows the various open source software concentration in implementation in the different organizations

open source is needed. The idea of where the *value* lies in open source software adoption is the needed. And by value we do not mean a monetary value (though it does include this category) we refer to the softer side of benefits – that of flexibility, openness, freedoms, ability to tweak and customize and along with open standards and open data – a far more open and accessible software environment. This is where the true benefits and cost reductions will come in the future.

The real value in open source adoption is clearly the collaborative co-creative ideology and spirit it encourages. This in turn leads to value creation, innovation, and a stronger ecosystem [27].

4 Conclusion

This paper maps the initial stages of our study of the total cost of ownership of open source software. The analysis provides some interesting answers to broad questions about cost but more importantly it raises relevant questions about the rather enigmatic quality of assessing the total cost of software, especially open source software. It is a conundrum as most companies lack experience with open source and more to the point, a slightly different understanding of 'value for money' is needed. The actual monetary cost of open source software adoption in many cases is not unlike that of proprietary software but it is the liberties, flexibility and control that it provides that draws companies and the public sector towards it.

Phase I of our study has given rise to some very interesting findings. The main ideas the respondents focused on included the lack of maturity level of open source software, license confusions and lack of knowledge about the implications of various open source licenses. Other ideas which arose were somewhat more surprising, most organizations do not even attempt a TCO study before making procurement decisions because of the expense such studies involve. The models used to assess TCO are also more suited for proprietary software and companies are not comfortable or skilled to tweak them for open source. And lastly, there is no policy in most companies for open source adoption. These decisions are made more ad-hoc and usually based on pragmatic decisions of use and need rather than cost.

However, there is much detail yet to be teased out in this amorphous area, and during the course of this study we hope to be able to show more clearly where the large part of the costs lie with open source software adoption, exactly where and how they are distinctive in comparison to proprietary software, what strategies and practices the public sector and companies can employ to make more effective use of the unique qualities of open source so that the software can yield a stronger feeling of 'value for money'. Indeed, can we go so far as to suggest that open source software adoption is an idea that the public sector not only needs to explore more seriously but in fact it will prove more effective, valuable, cheaper and necessary in the future? We aim to be able to provide a more nuanced answer to this and other questions at the end of our study, and encourage other researchers to explore such aspects because we feel that public sector adoption of open source has the potential to have real influence on open government strategies [28], open standards, and open data.

References

1. Ellram, L.M.: A Taxonomy of Total Cost of Ownership Model. Journal of Business Logistics 15, 171–191 (1994)
2. Ellram, L.M.: Total cost of ownership an analysis approach for purchasing. Journal of Physical Distribution and Logistics 25, 4–23 (1995)
3. Ellram, L.M., Siferd, S.P.: Total Cost of Ownership: A Key Concept in Strategic Cost Management Decisions. Journal of Business Logistics 19, 55–84 (1998)
4. Hurkens, K., van der Valk, W., Wynstra, F.: Total Cost of Ownership in the Services Sector: A Case Study. The Journal of Supply Chain Management 42, 27–37 (2006)

5. MacCormack, A.: Evaluating Total Cost of Ownership for Software Platforms: Comparing Apples, Oranges, and Cucumbers. AEI-Brookings Joint Center for Regulatory Studies Series (2003)
6. Ellram, L.M.: A Framework for Total Cost of Ownership. The International Journal of Logistics Management 4, 46–60 (1993)
7. Wouters, M.J.F., Anderson, J.C., Wynstra, F.: The Adoption of Total Cost of Ownership for Sourcing Decisions - A Structural Equations Analysis. Accounting, Organizations and Society 30, 167–191 (2005)
8. Lerner, J., Schankerman, M.: The Comingled Code: Open Source and Economic Development. MIT Press, Hong Kong (2010)
9. Carr, L.P., Ittner, C.D.: Measuring the Cost of Ownership. Journal of Cost Management 6, 42–51 (1992)
10. Russo, B., Succi, G.: A Cost Model of Open Source Software Adoption. In: IJOSSP, pp. 60–82 (2009)
11. Sutor, R.: Managing open source adoption in your IT organization (2009), http://www.sutor.com/newsite/blog-open/?p=3260
12. Galoppini, R.: Open Source TCO: Total Cost of Ownership and the Fermat's Theorem (2009), http://robertogaloppini.net/2009/01/08/open-source-tco-total-cost-of-ownership-and-the-fermats-theorem/
13. Burkhardt, R.: Seven Predictions for Open Source in (2009), http://it.syscon.com/node/797241
14. Wheeler, D.A.: Why Open Source Software / Free Software (OSS/FS, FLOSS, or FOSS)? Look at the Numbers (2007), http://www.dwheeler.com/oss_fs_why.html
15. Phipps, S.: Open Source Procurement: Subscriptions. In: Computer World, UK (2011)
16. Zuliani, P., Succi, G.: Migrating public administrations to open source software. In: E-society IADIS International Conference, Avila, Spain, pp. 829–832 (2004)
17. Zuliani, P., Succi, G.: An Experience of Transition to Open Source Software in Local Authorities. In: e-Challenges on Software Engineering, Vienna, Austria (2004)
18. Dinkelacker, J., Garg, P., Miller, R., Nelson, D.: Progressive Open Source, Hewlett-Packard, Palo Alto, California HPL-2001-233 (September 28, 2001)
19. Dahlander, L.: Penguin in a newsuit: a tale of how de novo entrants emerged to harness free and open source software communities. Industrial and Corporate Change 16, 913–943 (2007)
20. Fitzgerald, B.: The Transformation of Open Source Software. MIS Quarterly 30, 587–598 (2006)
21. O'Mahony, S., Diaz, F.C., Mamas, E.: IBM and Eclipse (A), Harvard Business Review Case Study, December 16 (2005)
22. West, J., Gallagher, S.: Challenges of Open Innovation: The Paradox of Firm Investment in Open Source Software. R&D Management 36, 315–328 (2006)
23. West, J.: How Open is Open Enough? Melding Proprietary and Open Source Platform Strategies. Research Policy 32, 1259–1285 (2003)
24. Osterwalder, A., Pigneur, Y., Tucci, C.L.: Clarifying Business Models: Origins, Present, and Future of the Concept. Communications of the Association for Information Systems 15, 1–40 (2005)
25. Vargo, S.L., Lusch, R.F.: Evolving to a New Dominant Logic for Marketing. Journal of Marketing 68, 1–17 (2004)
26. Agerfalk, P., Fitzgerald, B.: Outsourcing to an Unknown Workforce: Exploring Opensourcing as a Global Sourcing Strategy. MIS Quarterly 32, 385–400 (2008)

27. Shaikh, M., Cornford, T.: Understanding Commercial Open Source as Product and Service Innovation. In: 2011 Academy of Management Annual Meeting, San Antonio, Texas, USA (2011)
28. Noveck, B.S.: Defining Open Government (2011),
    http://cairns.typepad.com/blog/2011/04/whats-in-a-name-open-gov-we-gov-gov-20-collaborative-government.html
29. Shaikh, M., Cornford, T.: Letting Go of Control to Embrace Open Source: Implications for Company and Community. In: The Hawaii International Conference on System Sciences (HICSS), Koloa, Kauai, Hawaii, USA, vol. 43 (2010)

Libre Software as an Innovation Enabler in India
Experiences of a Bangalorian Software SME

Katja Henttonen

VTT Technical Research Centre of Finland
Oulu, Finland
katja.henttonen@vtt.fi
www.vtt.fi

Abstract. Free/Libre and open source software (FLOSS) has been advocated for its presumed capacity to support native software industries in developing countries. It is said to create new spaces for exploration and to lower entry barriers to mature software markets, for example. However, little empirical research has been conducted concerning FLOSS business in a developing country setting and, thus, there is not much evidence to support or refute these claims. This paper presents a business case study conducted in India, a country branded as a 'software powerhouse' of the developing world. The findings show how FLOSS has opened up significant opportunities for the case company, especially in terms of improving its innovative capability and upgrading in the software value chain. On the other hand, they also highlight some challenges to FLOSS involvement that rise specifically from the Indian context.

Keywords: Open source, innovation, India, free software, software business.

1 Introduction

Free/Libre and open source software (FLOSS) has been widely advocated [e.g. 1-4] as a way to promote endogenous software innovation in developing countries. The developmental opportunities created by the FLOSS phenomenon have been noticed both by international development institutions (e.g. World Bank and UNDP) and many of the developing countries themselves [1,3,4]. However, despite the enthusiasm, there remains very little empirical research on how developing country companies could successfully integrate FLOSS efforts into their internal innovative activities. Studies on commercially-motivated FLOSS in the US and Europe abound, but the results may not be directly applicable to the diverse innovation environments in the global South. This paper presents some key results of a qualitative case study [5] conducted in India, the country with the most well-known software industry in the developing world. The aim is to understand FLOSS-created opportunities and challenges from the viewpoint of an indigenous software SME.

The focus of the study is on the impacts of FLOSS on the innovativeness and profitability of the case company. Herein, innovativeness means the ability to create and implement new ideas which generate commercial value [cf. 6]. This can entail

S.A. Hissam et al. (Eds.): OSS 2011, IFIP AICT 365, pp. 220–232, 2011.

improvements to products, internal operations or a mix of markets. The study concerns modest incremental innovations, which an SME can generate on a regular basis.

The rest of the paper is structured as follows. The second chapter is divided into two sections: the first summarizes theoretical concepts underlying the study and the second one briefly introduces the current debate on whether and how Indian primary software sector could benefit from FLOSS.[1] The third chapter describes the research approach and methods employed in this study, and also very briefly introduces the case company. The fourth chapter presents the actual case study results; it is organized in three sections reflecting three different approaches to open innovation (more on these below). The fifth chapter discusses the meaning of some findings for further research. Conclusions close the paper.

2 Background

2.1 FLOSS as Open innovation: Three Archetypes

This study builds on the Chesbrough's [7] Open innovation theory, which describes the recent tendency of companies to 'open up' their innovation processes. In open innovation, not all good ideas need to be developed internally, and not all ideas should necessarily be further developed within firm's boundaries [8]. Chesbrough and Crowther [9, cf. 10] distinguish two archetypes of open innovation: inbound and outbound. In the case of inbound open innovation, companies monitor the surrounding environment of the firm to find technology and knowledge to complement in-house R&D. In the case of outbound open innovation, companies are looking for external organizations to take internally developed technology into new markets. An additional approach to openness is an interactive value co-creation in strategic partnerships [11, cf. 10] Here, the focus is on innovating together rather than on bringing resources over company borders (inside or outside) [8].

From a perspective of a private company, FLOSS involvement becomes open innovation when it is combined with a sustainable business model [12]. The aforementioned 'subtypes' of open innovation can be used to categorize how primary software companies engage with FLOSS [5, cf. 12,13]. In inbound open innovation, a company sources free-of-charge intellectual property (IP) from FLOSS communities and uses it to produce commercial software products or services. Typically, the main goal is to save own R&D expenses and/or achieve faster time to market[2]. The outbound open innovation entails what West and Gallagher [12] call "open source spin-out": a company brings internally developed IP into FLOSS domain. It may aim to to create demand for associated commercial offerings or advance strategic goals such as standards creation, for example. OSS communities can also be platforms for open value co-creation where diverse stakeholders join forces to achieve a common R&D goal and pooled contributions are made available to all [cf. 12].

---

[1] The focus is on introducing the points put forwards in the development literature; the discourse is somewhat different in the FLOSS business literature. For the comparison of discussions in the two disciplines, please see [5].

[2] This does not necessarily equal to a 'parasite approach': a company may motivate external innovation, e.g. by financially sponsoring FLOSS development [5, cf. 13,14].

2.2   FLOSS-Based Innovation in the Indian Context

While Indian software exports have grown exponentially over the past two decades [15,16], many observers have pointed out that the industry's innovative capability has remained relatively low [15,17,18]. The vast majority of Indian software exports consists of low-value-adding off-shoring services such as maintenance of legacy systems [15,17,18]. Due to barriers such as heavy financial constraints, 'late-comer disadvantage' and geographical distance from key customers, many Indian software entrepreneurs struggle to upgrade in the software value chain [15,17]. Meanwhile, 'FLOSS debate' is getting heated: academics and policy makers are arguing [e.g.4,19-22] on whether FLOSS could help some Indian software companies, especially SMEs, to increase innovativeness, add more value and capture more returns.

The proponents point out that sourcing technology from FLOSS communities (i.e. inbound open innovation) saves R&D time and costs and can thereby help Indian companies to overcome financial constraints and 'catch-up' to older players on the global software markets [3,4,23]. Another key argument relates to inter-organizational learning through gradually deepening FLOSS participation (in open co-creation). Unlike off-shoring parent companies, who often have a strong incentive to prevent knowledge spill-overs, FLOSS communities are very motivated to share knowledge across organizational and geographical boundaries [24,25]. This is said to offer valuable learning opportunities to Indian and other Southern companies [2,3,22]. Interestingly, the possible benefits of outbound open innovation has not been discussed much in the development literature, perhaps reflecting a tacit assumption that relevant IP and technical knowledge flows 'from the West to the Rest' rather than vice versa.

Some critics have argued that any competitive advantage derived from FLOSS-enabled cost and time savings is mitigated by GPL-like licensing terms [19,26]. As these licenses make it difficult to sell mass-distributed packaged software, they are said to deprive Southern software companies from the opportunity to benefit from the 'economies of repetition' [19]. Others have pointed out that 'price parity' with pirated software is shirking the markets for FLOSS in the South [21,27]. It also widely acknowledged that the cultural and linguistic barriers may hinder learning trough participative process in FLOSS communities [20,28].There are also significant differences between FLOSS communities on how they draw the boundaries of peripheral participation: some are highly inclusive, while others welcome only very advanced programmers [28,29]. Further, open co-creation and outbound open innovation both require significant investments in non-(directly) revenue generating activities [13] and because Indian companies typically face heavier financial constraints than their Western counterparts, affordability can become a major problem [20]. Launching an own FLOSS project is considered particularly costly and challenging human resource wise [30-32].

Somewhat surprisingly, despite the lively debate, empirical studies on FLOSS activities of primary software companies in India are almost non-existent. Some authors [e.g. 19] even dismiss the subject by saying that FLOSS plays no role in the Indian software industry. However, an international survey [33-35] indicates that, while commercially-motivated FLOSS involvement seems to be a relatively weak phenomenon in India (e.g. in comparison to Europe or Brazil), many small FLOSS companies are still 'out there' and FLOSS experience is also highly appreciated by recruiters in more 'mainstream' software companies. The survey [34,35] also suggests that most Indian companies limit themselves to inbound open innovation as far as FLOSS is concerned. Outbound open innovation seems particularly rare, only three Indian companies were found to author their own FLOSS projects [33]. Mahammodan and De [36] also analysed FLOSS reuse by six proprietary software producers in India. While these organizations reportedly attained significant cost savings by using FLOSS components as 'black box', their developers often lacked sufficient time or skills to even read the source code, leave alone contribute back.

3   Research Approach and Methodology

The paper presents results from a single case study conducted in a company called Mahiti Infotech Private Limited[3] (in short 'Mahiti') which is headquartered in Bangalore and employs 70-90 people. The company employs a customized product development model [37]: it develops 'semi-finished' products, often co-creatively with FLOSS communities, and later adds value by customizing them to the needs of individual end-clients. The tailored products go to market either as bespoke software or through the application service provision (ASP) model. Technical consulting provides additional revenue streams.

While planning to conduct more case studies in the future, the author believes that findings from this one case study alone may be valuable for the research community. Especially so, because, to the knowledge of the author, no previous academic study has aimed to 'give a say' to FLOSS entrepreneurs in India. Further, even though the case cannot be argued to be perfectly 'revelatory' nor 'exemplifying' in a strict sense [cf. 38], there are certainly many interesting characteristics to it. For example, despite its relatively small size, Mahiti has an extremely visible role in the Indian FLOSS scene. It can also be regarded as a notable example of an SME which has successfully used FLOSS strategically in order to upgrade in the software value chain. The case company also integrates elements from all the three archetypes of open innovation, thereby allowing to analyse outbound/inbound open innovation and open co-creation within the same organization.

The primary method of data collection was semi-structured interviews of the case company personnel. Three directors, the company's chief executive officer (CEO), chief technical officer (CTO) and marketing director were interviewed along with few senior developers. Two other sources of evidence, documentation (e.g websites and mailing list archives) and unobtrusive observation (mostly of employee interaction on FLOSS related IRC channels) were used to collaborate and augment evidence collected in the interviews. In order to cross-check data further [cf.39,40], some questions were also made to representatives of partner organizations. Most interviews were recorded and transcribed; in few cases, it was necessary to rely on note taking instead. A qualitative method called Template Analysis [41] was employed to thematically analyse the interview transcripts and, to a much smaller extent, some documentary evidence. In short, this means that a coding template was developed

---

[3] Researchers have argued both for and against disclosing the organization's name in case studies, see [47] for an overview. In this study, the company directors were given a choice and they selected recognition over anonymity.

iteratively whilst the analytical process moved forwards. The final template served as a basis for interpreting the data and writing up the findings. In addition to the thematic coding, some aspects of the Value Network Analysis [42] approach were used. The role of this method was complementary and it is not elaborated herein.

This study aims to confirm to the criteria that Guba [43,44] suggests for qualitative research: credibility (a parallel of internal validity), dependability (a parallel of reliability) and transferability (a parallel of external validity). To improve *credibility*, the study relies on several data sources and two different analysis methods as explained earlier [cf. 40]. The results report has also been shown to key informants for confirmation [cf.39,45]. To ensure *dependability*, complete records have been kept of the collected raw data (a case study database) so that other researchers can check them per request [cf, 22,50]. As for *transferability,* the results from a single case study are not generalisable to other situations, but they can still contribute to the understanding of the target phenomena and thereby provide valuable leads for future research [40,46]. Further, a longer research report available online [5] provides additional contextual information which can help others to make judgements on the transferability of the findings to other settings [cf. 38].

4   Case Study Results

4.1   Experiences in Inbound Open Innovation

In order to save costs, Mahiti intensively encourages the use of FLOSS code and components in all of their software projects. One of the founders gauged that an average Indian software company pays approximately 15% of their profits back in licensing fees, an expense they avoided. However, the cost savings and their profitability implications varied a great deal in practice as illustrated by two recent customer projects (see Fig.1). In the first case, the company only needed to make minor modifications to an existing FLOSS product, but could still charge a 'premium price', higher than that of all proprietary software vendors participating in the bid. This is because

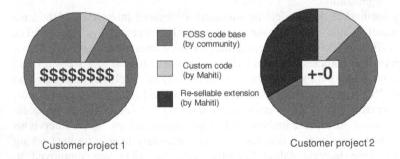

Fig. 1. Proportions of FLOSS and 'own' code in two projects [5]

the FLOSS product in question met well with the needs of the customer as such and Mahiti could offer the fastest lead-time. The profit margin was very high. In the second case, the company had to build almost half of the source code by itself before customer requirements were met, but could still charge a much lower price. The project was not profitable alone, but was still worth doing because the extension developed in this project was expected to be resold to several other customers over time.

FLOSS also brings cost savings to customers and, according to Mahiti's experience, this is helping to expand bespoke software markets in India and other developing countries. Interestingly, unlike most Indian software SMEs [15,18], Mahiti has highly diversified export markets with customers in countries such as Mongolia, the Bahamas, Brazil and Tajikistan. They believed this is partially because FLOSS based solutions are more affordable to Southern customizers, though it is obvious that many other factors are also at play. Nevertheless, it is noteworthy that, while the ease of piracy diminishes the cost advantage of FLOSS on the realm of mass software, the impact is not the same on the bespoke software markets. Pirated mass products can be customized to a certain extent (e.g. Microsoft Excel with macros), but such possibilities are very limited.

FLOSS licensing did not cause any fundamental changes to the company's revenue models[4]: instead of tailoring proprietary software packages, they were customizing FLOSS solutions. The latter allowed them to add-more value in-house, thanks to the low 'purchase price' and unlimited customization options. Profiting from the 'economies of repetition' through closed-source product development was seen infeasible due to financial constraints and highly mature markets: "basically, the curve to recover the funds is very high and this kind of [business] model is not viable for a company like ours". To the question of whether FLOSS licensing terms limit profit-making, the CEO replied:

— "Yes, if you choose to build your product with open source, you will most probably not become Bill Gates or Steve Jobs. But this is like any career choice, well, you can become a mortgage banker or a broker. [...] Microsoft is what it is today because they have spent money on every single line of code that they wrote. But I cannot start from scratch and build an operating system, I cannot achieve anything like that unless I do it with open source. And when I benefit from the efforts of others, I cannot expect to keep all of the profits alone."

When asked about the main difficulties in FLOSS reuse, directors pointed to difficulties in finding recruits with any previous experience on FLOSS technologies and development practices. This was seen as stemming from the tendency of local engineering education to ignore the skills needs of FLOSS companies and from the relatively small number of volunteer FLOSS developers [cf. 33] in the country. There was a recognition of some recent positive developments on the field of education. However, while some FLOSS technologies were slowly making their way to the engineering curricula, general code reuse skills were reportedly not. Consequently, the vast majority of new recruits were totally unfamiliar with the whole concept code reuse, only vaguely associating it with 'cut and paste'. They had to be taught 'by hand' which tended to bring up training expenses. As a strategy to address the skills gap, the company has started to offer free-of-charge lectures on FLOSS skills to local engineering colleges.

---

[4] For more information on FLOSS licensing issues in the case context, please see [5].

For the case company, another concern is that, as the vast majority of FLOSS projects originate from the global North [cf. 48], they do not always address regional needs as well as locally created software could. For example, the directors pointed out that there are practically no FLOSS applications addressing non-urban development needs in India, such as monitoring the quality of water or coordinating rural health-care. "All of these are possible with FLOSS, but there are very few projects moving despite a huge demand", the CTO said. He added that, while many FLOSS applications are relatively easy to localize in terms of language or metric systems, there are also more fundamental differences in software requirements between countries and regions. Referring to the cultural diversity within India, he continued: "this is a vast country and on the way from Bombay to Delhi the requirements change also... so no matter how much FLOSS there is in a market, it is not enough.".

4.2 Experiences in Outbound Open Innovation

Mahiti is one of the very few Indian software companies [cf. 33] to author its own FLOSS development project. Recently launched OurBank (www.ourbank.in) is a micro-finance software community which has attracted dozens of volunteer develop-ers, mostly Indian engineering students, and NGOs from as far as Brazil have contrib-uted localization effort. Based on their positive experience, Mahiti's directors are convinced that it is feasible for a resource-constraint SME to launch its own FLOSS community. Success on this arena was seen to depend on "energy and passion" as well as certain key capabilities (e.g. social networking skills) rather than large mone-tary investments. On the other hand, the CEO did admit that capturing returns from FLOSS spin-outs can be difficult: "Creating a product, architecting it, developing it, convincing other people that it is good and building a community - it is a painful thing to do, but it is sustainable in the long run. However, it does not provide you with re-turns like these [FLOSS customization projects]."

The mentioned profit-making challenges exist despite some institutional donations (e.g. from EuropeAid) towards the development of OurBank. However, most diffi-culties were believed to relate to the incubation phase. In the long run, Mahiti plans to step aside from leading the community and become just one of the many contributors. Such partial 'hand-off' was seen necessary so that the community can "evolve natu-rally". Time will show how the transition will work out in practise.

Apart from the spin-out described above, Mahiti has a longer history in doing re-leases which could be called 'spin-offs'. Whenever they have a piece of 'surplus' source code, which has reached the end of it's life cycle, they put it freely available on SourceForge or similar OSS platform. Because nothing is invested in community building or even making a website, the cost of open-sourcing is very low in this case. The company reported concrete and significant benefits once the IP got 'new life' in FLOSS domain. For example, they once open sourced a very small business applica-tion, a leave management system, which was only meant to be used in-house. Later on, they were contacted by a large foundation, who had downloaded the software and wanted to have it extended. The company gained a very important customer in this way, but the marketing effort only consisted of a few mouse clicks.

4.3 Experiences in Open Value Co-creation

Mahiti also plays a globally important role in the development of some FLOSS products such as the Plone content management system. When asked about business gains from strategic FLOSS participation, the global marketing benefits were typically mentioned first. The company does not need to engage in conventional marketing, directors said, because "FLOSS gives us complete visibility". Being listed as a partner on the Plone website was alone considered to be a major advantage. Further visibility resulted from employees' contributions, which they were always advised to do under the name of the company, and from co-organizing Plone conferences. However, it was not only about having one's name visible but, more importantly, about being seen as a 'shaper' of the technology: "They [customers] come to us because they see us as people who vision the [Plone] product and not only as people having [third-party] expertise on it", explains the marketing director. In addition, FLOSS communities are specialized social networks were 'word-of-mouth' travels quickly. Recommendations from other FLOSS community members brought in many customers. To exemplify such discussions, a UK-based member recommends Mahiti to another organization on a Plone mailing list saying "I've been told Mahiti has very good Plone/Zope skills and also knows the server side".

Somewhat expectedly, another group of reported benefits related to inter-organizational learning. The employee training at Mahiti is closely integrated with FLOSS participation. New employees started by following discussions on FLOSS forums and they were encouraged to gradually deepen their participation, very much in line with the classic 'onion model' [49] frequently stated in FLOSS research. The interviewed employees seemed genuinely enthusiastic about this training method. One said that while engineering education had only taught her to complete specific tasks, FLOSS participation had taught her to find solutions independently. From the management viewpoint, there are cost advantages because new employees are coached free-of-charge[5] by external experts. Some drawbacks were also mentioned, for example, FLOSS project administrators did not always explain why they rejected a contribution, which obviously constrained what an employee could learn from the experience.

As to other forms of inter-organizational learning, the company had benefited from adopting process innovations from FLOSS communities. For example, FLOSS involvement had prompted the company to adopt and improve the practice of end-user co-development. As a result, intense collaboration with domestic end-customers, who paid below-market prices in return of participating in R&D and beta-testing, had become a key part of their innovation process. Further, as a result of their FLOSS activities, the company has become geared towards writing well-commented and highly modular source code which is easy to reuse internally. They have even introduced an 'internal source forge', a repository where developers search for reusable source code developed in previous customer projects. These new development practices have enabled the case company to move away from one-time-project development towards developing 'semi-packages' out of reusable modules. In this way, FLOSS had clearly become a booster rather than a barrier to the 'economies of repetition' discussed earlier.

---

[5] Alternatively, the coaching can be understood as a social return from investments which the company makes to foster its relationship with FLOSS communities [cf. 13]. One developer said that Mahiti's 'good reputation' in communities helped her to get assistance.

The challenges discussed in the context of inbound open innovation also have an impact on open co-creation. In addition, directors acknowledged there are economic barriers to making FLOSS contributions. However, Mahiti has found several ways to keep expenses reasonable. Making minor contributions like bug fixes is integrated into employee training as explained earlier. The company also intermediates contributions made by others, for example, it facilitated local Myanmarian refugees to translate Plone into Burmese and put their contribution online. Or, as in the Hecker's [50] "sell-it-free-it model", large FLOSS contributions often consist of source code that has already been sold to one or more customers. The later model is not only an issue of affordability though; co-operation with end-customers was also seen crucial for ensuring the quality of the contribution. The CTO explains: "You cannot release something [to a FLOSS community] and expect miracles, unless you have tested it and the only way to test a product is to test with a customer...once it's a stable product only then the masses [of FLOSS users] can use it". Reportedly, most of the company's customers do not mind a contractual clause saying that the source code developed for them might be open source later.

Interestingly, developers said that they had not experienced any language barriers to FLOSS participation. Tertiary education had left them with a decent command of English and, besides, they felt that only technical terms are needed to talk on FLOSS related IRC channels. This is not an ethnographic study and it was not possible to detect how more 'subtle' cultural or linguistic issues may shape their identity building as FLOSS developers and effect their sense of belonging to a FLOSS community. On the surface, however, the employees seemed to feel sincerely proud of being well-recognized and respected members of the FLOSS communities where they contributed. For example, they very positively recalled that Joel Burton from the Plone Foundation had visited Mahiti and socialized with them. This was understood to be evidence that their participation is highly appreciated. "If we worked with Microsoft, Bill Gates would not come to chat with us", compared one.

5 Discussion

This paper does not aim to advocate Mahiti's experiences as a success model, which Indian software SMEs in masses could imitate. Firstly, it is appreciated that the study has succeeded to identify more opportunities than challenges. Despite cross-checking information from different sources (including non-company ones), the study still essentially relies on what the informants decided to share. Most people prefer to talk about their successes rather than their failures and the informants were supposedly not free from this common human tendency. Secondly, the case company seems to possess unique capabilities and also has a different market position than most Indian software SMEs. To exemplify the latter, Mahiti serves direct end-customers, over half of which are non-profit organizations. As such customers often agree to open-sourcing the code, which they have already paid for, the company can benefit from 'the sell-it-free-it' model. The scene is supposedly very different for most Indian SMEs, which do subcontracting work for multinational ICT companies.

The paper has 'scanned' several opportunities and challenges faced by the case company and none of these could be discussed in great depth. However, the author hopes that the paper has helped to highlight the wide range of perspectives, which one should take into consideration when discussing FLOSS business in India, or possibly other Southern contexts. For example, some development writers [e.g. 19] argue that FLOSS business models are 'less profitable' without discussing what are the likely alternatives for software companies in that particular country/region. Or, on the other side of the debate, the 'endogenous' nature of FLOSS is often strongly advocated [e.g. 1,4] without discussing the challenges that Southern organizations face when trying to launch their own FLOSS projects.

Most prior work on FLOSS-enabled learning, especially in the development context, focuses on technological knowledge transfer [e.g.2,25,29]. However, this study points to significant benefits from learning new development practices on customer co-development and code reuse. The study also suggests that FLOSS can have mixed impacts on the costs of employee training. These are both interesting subjects for further research, especially considering that the low level of code reuse (often below 5%) and high training expenses are often mentioned among key factors hindering profitability of Indian software SMEs [15,17,18]. Other topic, which deserves more attention, is the potential ability of FLOSS to expand low-cost markets for bespoke software in the South. The strong emphasis, which the interviewees placed on the global marketing benefits of FLOSS participation, is also noteworthy. Very hypothetically, this could related to the cost of international marketing (e.g. adverts in international magazines) being proportionally higher than the cost of R&D labour (i.e. FLOSS participation) for Indian companies.

From the viewpoint of the Open Innovation theory, Mahiti's experiences in uploading 'surplus' source code to the Internet are particularly interesting. Their habit strongly reflects one of the Chesbrough's [7] main "ethos": one should never 'sit' on the surplus intellectual property. The case study hints that SourgeForge-like platforms might provide a low-cost route for releasing IP which is no longer creating value in-house. If the released IP creates value elsewhere, there is a chance to claim a portion of that value. While getting theoretical support from Open innovation researchers [e.g. 51], this idea conflicts with many prior studies [13,30,32], which suggest that any commercially-motivated FLOSS release should be supported by significant investments in marketing and infrastructure building.

6 Conclusion

The study illustrates how FLOSS can blur the boundary between software vendor and third-party service provider, thereby opening up new opportunities for companies who lack resources to develop own products from 'scratch'. FLOSS co-creation has helped the case company to develop 'vendor-like' in-depth expertise and build an image as a co-creator of certain technologies. Due to the availability of source code and the absence of licensing fees, they can also add more value to FLOSS products than a non-vendor can typically add on proprietary products. In some cases, FLOSS releases have even helped to open up routes to new markets. Meanwhile, the case company continues to face many challenges such as the poor availability of new recruits with FLOSS competences in India. More research is needed to understand how the findings may apply beyond the single case setting and whether FLOSS has any potential to transform the Indian software sector at large.

Acknowledgements. I want to thank Mr. Pasi Pussinen, a research scientist at VTT, who gave some excellent feedback on a draft of this paper and all the informants, who spent their valuable time being interviewed. As to financing, this work has been supported by the research project ITEI-VTT, which is co-funded by VTT and Finnish National Technology Agency (Tekes).

References

1. Dravis, P.: Open source software. Perspectives for development. Global information and Communication Technologies Department, the World Bank, Washington (2003)
2. Tapia, A., Maldonado, E.: An ICT Skills Cascade: Government-Mandated Open Source Policy as a Potential Driver for ICT Skills Transfer. Information Technologies and International Development 5(2), 31–51 (2009)
3. Weerawarana, S., Weeratunge, J.: Open Source in Developing Countries. SIDA, Stockholm (2004)
4. Wong, K.: Free/open source software: government policy, UNDP Asia Pacific Development Information Program in cooperation with Elsevier, New Delhi (2004)
5. Henttonen K.: Open source as an innovation enabler: A Case Study of an Indian Software SME. Dissertation, The University of Manchester. Institute for Development Policy and Management (2011),
   http://opensource.erve.vtt.fi/publications/henttonendisserta
   tion.pdf
6. Subramaniam, M., Youndt, S.: The influence of intellectual capital on the type of innovative capabilities. Academy of Management Journal 48(3), 450–463 (2005)
7. Chesbrough, H.: Open innovation: the new imperative for creating and profiting from technology. Harvard University Press, Boston (2003)
8. Koskela, K., Koivumäki, T., Näkki, P.: Art of openness. In: Pikkarainen, M., Codenie, W., Boucart, N., Heredia, J. (eds.) The Art of Software Innovation. Eight Practice Areas to Inspire your Business. Springer, Heidelberg (2011) (to be published)
9. Chesbrough, H., Crowther, A.: Beyond high tech: early adopters of open innovation in other industries. R&D Management 36, 229–236 (2006)
10. Gassmann, O., Enkel, E.: Towards a theory of Open innovation: three core process archetypes. In: The Proceedings of the R&D Management Conference, Sesimbra, Portugal (2004)
11. Piller, F., Ihl, C.: Open Innovation with Customers – Foundations, Competences and International Trends. Trend Study within the BMBF Project. International Monitoring', RWTH Aachen University, Aachen (2009)
12. West, J., Gallagher, S.: Patterns of Open innovation in open source software development. In: Chesbrough, H., Vanhaverbeke, W., West, J. (eds.) Open Innovation:Researching a New Paradigm, pp. 82–106. Oxford University Press, Oxford (2006)
13. Goldman, G., Gabriel, R.: Innovation happens elsewhere. Open source as business strategy. Elsivier, San Fransisco (2005)
14. Dahlander, L., Magnusson, M.: Relationships between open source software companies and communities: Observations from Nordic firms. Research Policy 34(4), 481–493 (2005)
15. Arora, A.: The Indian software industry and its prospects. In: Bhagwati, J., Calomiris, C. (eds.) Sustaining India's growth miracle, pp. 166–215. Columbia University Press, New York (2008)

16. Athreye, S.: The Indian software industry. In: Arora, A., Gambardella (eds.) From Under-dogs to Tigers: The Rise and Growth of the Software Industry in Brazil, China, India, Ireland, and Israel, pp. 7–14. Oxford University Press, Oxford (2005)

17. D'Costa, A.: Export Growth and Path Dependence: Locking Innovations. Software Industry, Science, Technology and Society 7(1), 51–81 (2002)

18. Nirjar, A., Tylecote, A.: Breaking out of lock-in: Insights from case studies into ways up the value ladder for Indian software SMEs. Information Resources Management Journal 18(4), 40–61 (2005)

19. Debroy, B., Morris, J.: Open to development: Open-Source software and economic development. International Policy Network, London (2004)

20. O'Donnell, C.: A case for Indian outsourcing: open source interests in IT jobs. First Monday 9(11) (2004)

21. Sharma, A., Adkins, R.: OSS in India. In: Dibona, C., Cooper, D., Stone, M. (eds.) Open sources: the Continuing Evolution, O'Reilly, Sebastopol (2005)

22. Suman, A., Bhardwaj, K.: Open Source Software and Growth of Linux: The Indian Perspective. DESIDOC Bulletin of Information Technology 23(6), 9–16 (2003)

23. May, C.: The FLOSS alternative: TRIPs, non-proprietary software and development. Knowledge, Technology, and Policy 18(4), 142–163 (2006)

24. Krogh, G., Spaeth, S., Lakhani, K.: Community, joining, and specialization in open source software innovation: a case study. Research Policy 32(7), 1217–1230 (2003)

25. Staring, K., TitleStad, O.: Development as a Free Software: Extending Commons Based Peer Production to the South. In: The Proceedings of the Twenty Ninth International Conference on Information Systems (ICIS 2008), Paris (2008)

26. Reddy, B., Evans, D.: Government Preferences for Promoting Open-Source Software: A Solution in Search of a Problem. Social Science Research Network (2002)

27. Heeks, R.: Free and Open Source Software: A Blind Alley for Developing Countries? IDPM Development Informatics Briefing Paper, Institute of Development Policy and Management. The University of Manchester (2005)

28. Vaden, T., Vainio, N.: Free and Open Source Software Strategies for Sustainable Information Society. In: O. Hietanen (ed.) University Partnerships for International Development: Finnish Development Knowledge, Finland Futures Research Centre, Turku (2005)

29. Wernberg-Tougaard, C., Schmitz, P., Herning, K., Gøtze, J.: (Evaluating Open Source in Government: Methodological Considerations in strategizing the Use of Open Source in the Public Sector. In: Lytras, M., Naeve, A. (eds.) Open Source for Knowledge and Learning Management: Strategies Beyond Tools, Idea Group Publishing, London (2007)

30. Henttonen, K., Matinlassi, M.: Contributing to Eclipse - a case study. In: Proceedings of the Software Engineering 2007 Conference (SE 2007), Hamburg, Germany (2007)

31. Järvensivu, J., Mikkonen, T.: Forging A Community? Not: Experiences On Establishing An Open Source Project. In: Russo, B., Damiani, E., Hissam, S., Lundell, B., Succi, G. (eds.) Open Source Development Communities and Quality, IFIP Working Group 2.13 on Open Source Software Systems (OSS 2008), Milano, Italy (2008)

32. West, J., O'Mahony, S.: Contrasting Community Building in Sponsored and Community Founded Open Source Projects. In: The Proceedings of the 38th Annual Hawaii International Conference on System Science. IEEE Computer Society, Los Alamitos (2005)

33. UNU-MERIT Free/Libre and Open Source Software: Worldwide Impact Study. D31: Track 1 International Report. Skills Study. United Nations University, Maastricht (2007)

34. UNU-MERIT Free/Libre and Open Source Software: Worldwide Impact Study. D7: Track 1 Survey Report - India. Skills Study. United Nations University, Maastricht (2007)

35. UNU-MERIT Free/Libre and Open Source Software: Worldwide Impact Study. D7: Track 2 Survey Report - India. Software Study. United Nations University, Maastricht (2007)
36. Madanmohan, T., De, R.: Open source reuse in commercial firms. IEEE Software 21(6), 62–69 (2004)
37. Codenie, W., Pikkarainen, M., Boucart, N., Deleu, J.: Software innovation in different companies. In: Pikkarainen, M., Codenie, W., Boucart, N., Heredia, J. (eds.) The Art of Software Innovation. Springer, Heidelberg (2011) (to be published)
38. Bryman, A.: Social Research Methods, 3rd edn. Oxford University Press, New York (2008)
39. Chetty, S.: The case study method for research in small and medium sized firms. International Small Business Journal 15(1), 73–85 (1996)
40. Yin, R.: Case Study Research: Design and Methods, 4th edn. Sage Publications, California (2009)
41. King, N.: Template analysis. In: Symon, G., Cassell, C. (eds.) Qualitative Methods and Analysis in Organizational Research: A Practical Guide, pp. 118–134. Sage Publications, California (1998)
42. Allee, V.: Value Network Analysis and Value Conversion of Tangible and Intangible Assets. Journal of Intellectual Capital 9(1), 5–24 (2008)
43. Guba, E.: Criteria for assessing the trustworthiness of naturalistic inquiries. Educational Technology Research and Development 29(2), 75–91 (1981)
44. Guba, E.G., Lincoln, Y.S.: Competing paradigms in qualitative research. In: Denzin, N., Lincoln, Y. (eds.) Handbook of Qualitative Research, pp. 163–194. Sage Publications, London (1994)
45. Rowley, J.: Using case studies in research. Management Research News 25(1), 16–27 (2002)
46. Flyvbjerg, B.: Five Misunderstandings About Case-Study Research. Qualitative Inquiry 12(2), 219–245 (2006)
47. Walsham, G.: Doing interpretive research. European Journal of Information Systems 15(3), 320–330 (2006)
48. Robles, G., Gonzalez-Barahona, J.M.: Geographic location of developers at SourceForge. In: The Proceedings of the 2006 International Workshop on Mining Software Repositories, Shanghai, China, pp. 144–150 (2006)
49. Ye, Y., Kishida, K.: Toward an understanding of the motivation of open source software developers. In: The Proceedings of the 25th International Conference on Software Engineering, ICSE 2003 (2003)
50. Hecker: Setting up shop: the business of Open-Source software. IEEE Software 16(1), 45–51 (1999)
51. Henkel, J.: Selective revealing in Open innovation process: The case of embedded Linux. Research Policy 35, 953–969 (2006)

Adoption of OSS Development Practices by the Software Industry: A Survey

Etiel Petrinja, Alberto Sillitti, and Giancarlo Succi

CASE - Center for Applied Software Engineering
Free University of Bozen-Bolzano
Piazza Domenicani 3, 39100 Bolzano, Italy
{etiel.petrinja,alberto.sillitti,
giancarlo.succi}@unibz.it
http://www.case.unibz.it

Abstract. The paper presents a survey of aspects related to the adoption of Open Source Software by the software industry. The aim of this study was to collect data related to practices and elements in the development process of companies that influence the trust in the quality of the product by potential adopters. The work is part of the research done inside the QualiPSo project and was carried out using a qualitative study based on a structured questionnaire focused on perceptions of experts and development practices used by companies involved in the Open Source Software industry. The results of the survey confirm intuitive concerns related to the adoption of Open Source Software as: the selection of the license, the quality issues addressed, and the development process tasks inside Open Source Software projects. The study uncovered specific aspects related to trust and trustworthiness of the Open Source Software development process that we did not find in previous studies as: the standards implemented by the OSS project, the project's roadmap is respected, and the communication channels that are available.

1 Introduction

The Open Source Software (OSS) industry is continuously growing and it is influencing also practices that are part of the traditional software development process [4, 17, 20]. OSS development is supported by software companies and OSS projects are not interesting only to enthusiasts and volunteers [6]. Similar initiatives are not limited to the software area alone, but they are encompassing also areas as book publishing, scientific publishing, and other familiar areas [8]. The benefits that OSS brings are an interesting research area [16] and many studies have been conducted in the past decade. An important research initiative is focused in the definition of measures that based on empirical values characterise the OSS, and allow standard measurement of its quality. The grouping of these measures in a form of an assessment methodology provides to the software industry a convenient tool that can be used before adopting a new OSS product.

The number of OSS projects implementing similar functions is large. The quality of those projects depend on the development process used inside the project and the

S.A. Hissam et al. (Eds.): OSS 2011, IFIP AICT 365, pp. 233–243, 2011.
© IFIP International Federation for Information Processing 2011

skills of participants in the community that are participating to the development. Companies interested in reusing OSS components should be able to select the OSS product that best suits their needs. This is usually a challenging task that requires assessment tools and skills but it also takes time to be performed. Several assessment methods for OSS have been proposed and new ones are under development; some of the already available are:

• the QualiPSo OpenSource Maturity Model (OMM) (2009) [14],
• the Open Business Readiness Rating (OpenBRR) (2005) [24],
• the Open Source Maturity Model (OSMM) from Cap Gemini (2003) [7],
• the Open Source Maturity Model (OSMM) from Navica (2004) [12],
• the Open Business Quality Rating (Open BQR) (2007) [21], and
• the Methodology of Qualification and Selection of Open Source software (QSOS) (2004) [1].

The aim of the study was to understand which are the key OSS development activities perceived as important by experts from the software industry. Those activities will influence the creation of a new assessment method that is addressing the OSS development process. The methods listed (except QualiPSo OMM) are focused in the OSS product and not in the development process. The development process has an important role when deciding to adopt an OSS product. Product characteristics alone are usually not a sufficient indicator of quality for stakeholders that are interested to actively participate in the OSS project. Specially for software integrators that plan to create a new software product reusing OSS components.

Adoption and integration of a OSS product is strongly related to further modifications of the product, therefore a good interaction with the community developing the product is essential. Process aspects have to be identified and assessed; the best known process assessment methodology in the software domain is the Capability Maturity Model Integrated (CMM/CMMI) [9, 19]. However, the CMMI is not appropriate for assessing OSS projects. It is complex, it does not focus on single projects but on the software company, and it does not address OSS specific aspects as: contribution level, reputation of the project, licenses used, etc. The aim of the new assessment methodology is to propose a CMMI-like method able to address OSS specific aspects. Software companies are knowledgeable in CMMI, therefore we planed to leverage on this skills of experts working in software companies.

The development of a OSS development process assessment methodology is one of the results of the EU funded project QualiPSo [15]. The basic task for defining the methodology was to conduct the survey, presented in this paper, and collect data related to the practices and trust elements in the development process of surveyed companies. Even if a higher quality of OSS products in comparison with closed source software products was demonstrated in several case studies [5, 10, 11], OSS is generally still perceived of low quality. The trustworthiness of OSS is an aspect we consider critical for a larger usage of OSS.

We included in our study more than 50 professionals working for 20 European software companies including: Siemens (Germany), Engineering Ingegneria Informatica (Italy), Bull (France), Atos (Spain), IBM, Mandriva (France), Thales (France), and others. A similar survey focused on OSS communities was published in 2008 [13].

The results of the study provided a set of best practices related to the OSS development process and a set of elements that bring trust to participants of the survey. Trustworthy elements are elements that when addressed during the OSS development process, guarantee to developers, users, and software integrators that the project is of good quality.

The paper is organized as follows: in Section 2 we describe the survey with the presentation of the methodology used. Section 3 is dedicated to the results of the survey, highlighting key aspects of the OSS development process. The last section contains conclusions and indications of possible future work.

2 The Survey

The European software industry is interested in OSS because it is an expanding market alternative to the closed source software market where non-EU, mainly US, software companies have a strong position [3]. OSS is an area where European software companies can offer their expertise and implement new business models [22]. The study presented in this paper is part of a large research conducted with the aim to improve the adoption of OSS by the software industry.

2.1 Interviews with European Software Companies

The qualitative study gives an overview of OSS related development practices used in European software companies, the adoption of new OSS development procedures, and the usage of OSS products. Interviewees were asked which factors influence their perception of the quality of the OSS process and what is their opinion about a wider adoption of OSS in the company. The questionnaire was developed by defining specific topics and identifying possible questions. We decided which questions are important for the OSS development process and which aspects characteristics of the final OSS product should be also collected to understand better the development process. The complete questionnaire is available as part of the results of the QualiPSo project at www.qualipso.org.

2.2 Methodology

The design of the research was based on the approach proposed by Silverman [18]. The approach requires the design of a structured and formal research involving two basic and partially related decisions:

• The method, that is, whether performing a quantitative or a qualitative investigation.
• The methodology and the specific technique for gathering data (with an interview, a questionnaire, with observations, etc.).

The decision is based on the evaluation of major goals of the research and the type of information required. As far as opinions on OSS products were concerned, our goal was to investigate those OSS product factors that influence the trustworthiness of the OSS development process. Following the approach proposed by Silverman, this kind of information requires a qualitative investigation. Our research methodology was based on a semi-structured questionnaire where some of the questions were closed (offering a limited list of possible answers) and other questions were open allowing to provide any answer from the interviewee. The questionnaire has been filled in during face-to-face or telephone interviews. The interviewees were employees of European software companies. The first set of interviewees was selected by the companies that were involved in the QualiPSo project (19 partners) [15]. An additional set of companies was selected based on their involvement in OSS projects. The total number of interviewees was 53.

During the second phase, we conducted a quantitative study by creating a shorter questionnaire on key elements and submitting the questionnaire to mailing lists, conferences, and using it during personal meetings. We designed a web version of the questionnaire and we collected most of the answers through a web form. The second questionnaire reused many questions from the first version and additionally (randomly) listed answers obtained during the first iteration of the research. We closed most of the questions (providing possible answers) aiming to identify the priority of answers compared between them. We excluded outliers and incomplete answers from the first iteration. The results of the on-line questionnaire are just one part of the results of the study presented in this paper. We compared the results obtained with personal interviews and results obtained with the survey; the syntheses of the results of the two iterations is presented in this paper.

The overall structure of the research was based on the Goal Question Metric (GQM) approach [2]:

1. **Goal:** Identify trust related issues in the adoption of OSS in the European software industry.
2. **Question:** The questionnaire includes 53 questions with additional sub-questions. They were developed to ask the interviewee on specific goal related issues.
3. **Metric:** Metrics about the level of adoption and the trust in OSS process are selected to be able to measure answers to questions defined in the precedent step.

The final form of the questionnaire was achieved through several iterations of drafts. The questionnaire was divided into sections covering different topics related to the OSS development process. The three main sections focused on:

1. trust and quality related aspects;
2. stakeholders related aspects as the roles and responsibilities used inside the development process; and
3. aspects related to the technology used.

The data gathering process was organized as follows:

• The respondents were contacted to determine their general interest in the study.
• The questionnaire was sent to the respondents to verify the actual availability.
• Data were collected by personal or telephone interviews in English.
• The results of the interviews were recorded and the interviewees were asked for a final check.

The majority of interviews was conducted at the interweave working place. We travelled to meet the majority of interviewees. Only in few cases the interview was done through a conference call. The duration of the interviews was not specially recorded and varied between different interviews. However the interviews lasted approximately one hour. We were not able to measure the duration of the surveys performed on the web. However, we have tested the survey questionnaire with colleagues before publishing it on the web and the mailing lists and the time necessary to complete the questionnaire varied between thirty and forty-five minutes. There were always two interviewers present during interviews. One was usually reading questions and the other was writing the answers. Two persons were able to collect more detailed responses from the interviewed person and limited the duration of interviews. Soon after the end of the interview, the answers recorded in the protocol were sent back to the interviewed person. Only upon a positive feedback from the interviewee, the questionnaire was considered accepted and the data were processed.

Participants were guaranteed anonymity and the information reported was reviewed so that no single person or company can be identified. The number of individuals interviewed from each company varied from one up to six employees per company. The people interviewed were developers and managers in companies included in the study.

3 Results

We present the results about three types of aspects in separate subsections, to summarize the results of: quality related aspects, OSS development stakeholder related aspects, and technology aspects. Some of the 53 collected questionnaires were not filled completely; however, the key questions that were focused on topics presented in this paper were filled in more than 90% of the questionnaires. During the second phase of the study, we obtained 56 filled questionnaires. The subjects involved in the two phases were different, nevertheless the results of the second phase confirmed the results obtained during the first phase.

3.1 Trust and Quality Related Aspects

The interviewees spontaneously mentioned various factors that influence the trust they have in the OSS process. The most frequently mentioned criteria were:

- the availability and the quality of the documentation about the OSS product,
- the number of downloads and the number of potential users of the OSS product, and
- open standards used for the development of the product.

A list of the most important characteristics and the percentage of respondents that think that a characteristic influences trust is presented in Fig. 1. For example, the popularity of the product was listed as important by 81% of the respondents. Identified characteristics confirm the results of previous studies; however, some unexpected aspects have been reported. Interviewers asked the respondents only what they consider important when they think about trustworthiness of the OSS development process. This focus allowed to discover some aspects that are considered important specially for the trustworthiness. Some additional interesting aspects recorded were: the importance of companies and the industry that is sponsoring the OSS project, the presence of an independent body that checks the product and the development process used. We were especially interested in the perception of the importance of the development process. Half of the surveyed participants were working for companies that are concerned about the development process. The other half reported that they are almost completely uninterested in the process and the interest was merely focused on the characteristics of the OSS product. The companies that just wanted to use the product were in general not interested in the development process. On contrary, the companies that wish to further develop the product were interested also in the development process.

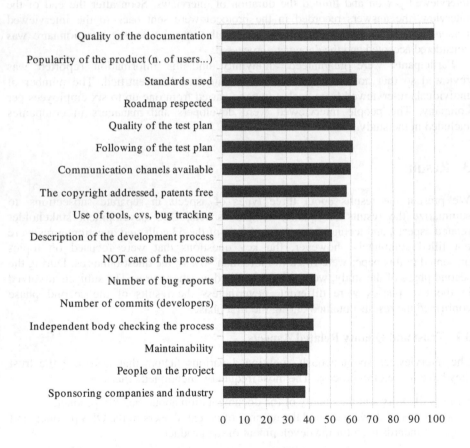

Fig. 1. Trustworthy factors related to the OSS development process

The cost of the license was one of the characteristics addressed specifically by a question in our questionnaire. We considered this factor important because it is one of the most influential aspect emerging from the results of previous OSS studies [23]. Interviewees stressed the importance of the license cost of OSS in comparison with proprietary software. An additional question was if the total cost of the ownership (TCO) affects the importance of the cost of the license; and the answer was unexpectedly negative. The absence of an initial license cost allows users to test and experiment code modifications; this is a positive aspect of OSS in comparison with proprietary software. The absence of license costs contributes strongly to an easier and larger adoption of OSS products. Subsequent training and maintenance costs encountered are considered less important than the benefits users have at the beginning of the use of a OSS product.

The survey confirmed that testing is an important part of the adoption process of a new OSS product. Surveyed companies test OSS using manual and automatic tests. Manual tests are usually unstructured and often done ad-hoc. Automatic tests are structured and standardized. Details of the testing procedure depends on the importance the new OSS product has for the company.

Despite the availability of some assessment methodologies as QSOS and OpenBRR, the surveyed companies usually use their own set of criteria to test the quality of a OSS product. Many of these criteria are already part of the listed methodologies but the methodologies are not yet well established and known by the industry and their use is still limited.

Aspects that companies consider important when testing OSS vary considerably across companies and the summary described in this paper presents characteristics mentioned by a large percentage of interviewees. The testing process is usually done informally by developers of the company (80%). Testing is also done by the OSS community that is using the product and that reports bugs and proposes new features and improvements. OSS communities are a good short-cut for companies when they need to choose which OSS product to adopt, as communities can provide volunteers for the testing process. If they need additional test results related to each company's specific requirements, they conduct in-house tests carried on either by a group of specialized developers, by the project manager, or sometimes also by external teams.

Two frequently mentioned characteristics considered important for assessing the quality of OSS are: results of in-house or external tests of the product, and the size of the user community. Both were mentioned by all the participants to the study (as evident from Fig. 2). The number of users that already use a OSS product is an indication of the variability of opinions and comments about the product that can offer insight in aspects of the OSS project. Another important aspect is the type of standards implemented inside the OSS project (73%). This aspect is often related to technologies, programming languages, and frameworks used and supported by projects. Documentation is also considered important by three-fourths of the companies. Satisfaction of user expectations reported on forums and mailing lists by the community of users of the product are considered important by half of the surveyed interviewees. Just half of the interviewees consider important the process followed (either the Rational Unified Process - RUP, a process assessed according to the Capability Maturity Model - CMM, an IT Infrastructure Library - ITIL, benchmarked process or others) for the development of the product. Less frequently mentioned aspects are: the availability and use of measures such as bug reports, the size of the components, the certification of the quality of the product by a third-party company, and others. The answers are graphically presented in Fig. 2.

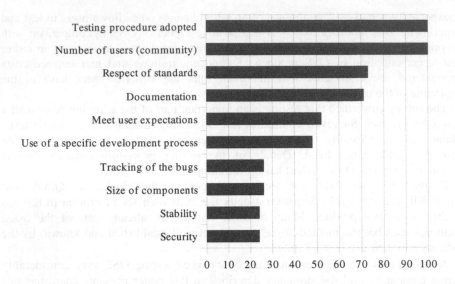

Fig. 2. Criteria for assessing the quality of OSS

3.2 Aspects Related to Stakeholders

Roles and responsibilities of people inside companies participating in our research play an important role when considering the adoption and development of OSS. Surveyed companies use OSS products but in some cases they also develop OSS. There is a correlation between the development of OSS and its usage. Part of the participants only use OSS products, however the interviewees that participate in the development of OSS products usually also use the product. This behaviour was expected and it was reported in previous studies. We have asked how are the OSS projects managed inside the companies and which are the responsibilities related to OSS. For most of the companies, OSS development is only one of their development activities. The majority of companies involved in the survey, is still exploring the OSS development and their current OSS development approach is a mix set of activities typical of a proprietary software development and of OSS development (as the one adopted by OSS communities as Linux Kernel, Apache, and others). The number of developers and the size of OSS communities formed by the companies is still rather small. They are typically composed of up to 100 developers and contributors and only rarely reach 1000 contributors. A limited OSS development approach is evident from the responsibilities that members of the OSS community have. Important decisions and responsibilities are assigned to employees of the company. This way, companies still maintain the leading role and decide the development directions of the OSS project. This way companies lose important advantages that a more democratic OSS development process offers, such as a larger amount of source code contributions, the development based on meritocracy, and better motivation for the involvement in the OSS development process.

As for the roles and responsibilities, also the definition of requirements and implementation of new features is influenced strongly by the business strategy of the companies. After the implementation of critical features that are imposed by architectural and design decisions; key contributors of new features are customers of surveyed companies. More than 80% of interviewees consider more important the suggestions from their regular customers than suggestions coming from the OSS community. Decisions of which features are implemented first are usually taken by the coordinator of the project that is almost always an employee of the company.

3.3 Technology Aspects

The technology used inside OSS projects is an aspect that strongly influences the whole OSS project and specially the development process. Interviewees answered in details on questions related to technology related aspects and showed a high interest on these characteristics. We dedicate this subsection to answers of interviewees to some strongly perceived technology aspects.

Surveyed companies use different operating systems for developing OSS; Linux is the most frequently used. Some companies also use the Windows operating system and sometimes Solaris, in addition to Linux. Fig. 3 presents the percentage of use of each operating system.

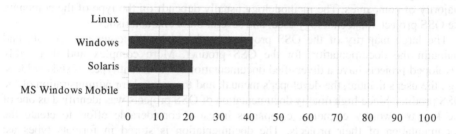

Fig. 3. Operating System used by interviewees

The most frequently used programming language is Java which is regularly used by all the companies. C++ is the second most frequently used programming language. The surveyed companies use also other languages as: C, Python, PHP, Perl, some proprietary domain languages, JavaScript, MS proposed languages, and others. Languages and technologies usually influence the development methodologies used and also the efficiency of specific development activities. The frequent use of Java and Linux confirmed our expectations related to the language and the platform used by OSS communities. Fig. 4 presents the frequency of the use of programming languages.

The surveyed companies that develop OSS, do it incrementally (80%) with small exploration projects, starting with some basic features and releasing the first versions of the product. New features are added later according to new requirements collected first by the customers and then by the community. Such development is similar to the process followed in OSS communities that are not strongly influenced by software companies. The results of the survey also show that in some cases, the core part of the project is entirely developed and implemented inside the company and only then it is disclosed to the community and opened for new suggestions.

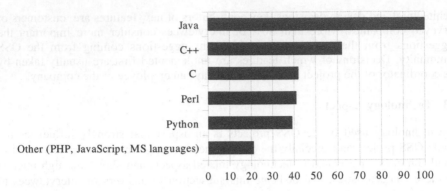

Fig. 4. Programming language used by interviewees

Two important aspects of OSS projects that are supported by the surveyed companies are: the use of open standards, and the modularization of the system architecture. From the survey, we could not identify a OSS development strategy that was used by the majority of companies. The methodology usually depends on the type of the company, the OSS project, the technology used, and several other characteristics.

The large majority of the OSS projects developed by the companies, create and maintain the documentation for the OSS product. More complex and thoroughly developed projects have a diversified documentation for different types of stakeholders, e.g., the user's manual, the developer's manual, and the manual for administration of the OSS product. Since high quality documentation of OSS projects was identified as one of the key trustworthy elements, companies invest a considerable effort to create the documentation of their projects. The documentation is stored in formats types as: readme files, wikis, web pages, user guides, as comments to the source code, and in other forms. Developers are responsible to write the documentation related to their part of the project. Only occasionally the documentation is written by specialized in-house or external groups of experts.

4 Conclusions

The study presented in this paper has confirmed the interest of European software companies in OSS in general and also in the OSS development process. The majority of the surveyed companies are trying to participate on the OSS market starting with small, exploration projects and then incrementally extending them. The majority of companies participating in the survey started mixed proprietary-OSS projects. OSS products developed are free to be used and improved by everybody and the source code is available; the development process, the management responsibilities and evolution of the project are, however, coordinated by the company that started the project. The quality of the OSS development process and the lack of a high quality assessment process, that can be used to measure characteristics of the development process, is one of the main limitation to a wider diffusion of OSS. The key issue is related to the perceived quality of OSS and the trust people have in its development process. The survey was focused on the usage of OSS in the European software industry and the results sheds some light on this growing market.

Interviewees report that companies, that are releasing some of the management and planning aspects of the project to the OSS community, are observing growing contributions from volunteers. Benefits provided by OSS communities allow faster improvement of OSS projects. Interviewees reported a perceivable improvement of the business in companies that have started to use OSS and that have decided to use the OSS development process for some of their projects. The key result of the study was a set of characteristics that are considered important for improving the quality of the OSS development process. The most important one are: the quality of the documentation, the number of the users of the OSS product, and standards implemented by the OSS project. These characteristics were used for the definition of an assessment methodology that is now in the process of validation and refinement.

References

[1] Atos Origin, Method for Qualification and Selection of Open Source Software (QSOS) (2009), http://www.qsos.org (last visited: March 2011)

[2] Basili, V.R.: Software modeling and measurement: The Goal/Question/Metric paradigm. Technical Report CS-TR-2956, Department of Computer Science, University of Maryland, MD, USA (1992)

[3] Burzynski, O.R., Graeml, A.R., Balbinot, Z.: The internationalization of the software market: opportunities and challenges for brazilian companies. JISTEM J. Inf. Syst. Technol. Manag. 7(3), 499–516 (2010)

[4] Capiluppi, A., Feller, J., Fitzgerald, B., Hissam, S., Lakhani, K., Robles, G., Scacchi, W.: First International Workshop on Emerging Trends in FLOSS Research and Development. In: Proc. of ICSEcompanion, 29th International Conference on Software Engineering (ICSE 2007 Companion), pp. 135–136 (2007)

[5] DiBona, C., Ockman, S., Stone, M.: Open Sources: Voices from the Open Source Revolution, p. 280. O'Reilly, Sebastopol (1999)

[6] Dueñas, C.J., Parada, H.A., Cuadrado, G.F., Santillán, M., Ruiz, J.L.: Apache and Eclipse: Comparing Open Source Project Incubators. IEEE Software 24(6), 90–98 (2007)

[7] Duijnhouwer, F.-W., Widdows, C.: Capgemini Expert Letter Open Source Maturity Model, Capgemini (2003)

[8] Goth, G.: Sprinting toward Open Source Development. IEEE Software 24(1), 88–91 (2007)

[9] Humphrey, W.: Characterizing the software process: a maturity framework. IEEE Software 5(2), 73–79 (1988)

[10] Birendra, M., Ashutosh, P., Srinivasan, R.: Quality and Profits Under Open Source Versus Closed Source. In: ICIS, Proceedings. Paper 32 (2002)

Towards Improving OSS Products Selection – Matching Selectors and OSS Communities Perspectives

Claudia Ayala[1], Daniela S. Cruzes[2], Xavier Franch[1], and Reidar Conradi[2]

[1] Technical University of Catalunya
UPC-Campus Nord (Omega), 08034 Barcelona, Spain
{cayala,franch}@essi.upc.edu
[2] Norwegian University of Science and Technology
NTNU-Gløshaugen, Trondheim, Norway
{dcruzes,conradi}@idi.ntnu.no

Abstract. Adopting third-party software is becoming an economical and strategic need for today organizations. A fundamental part of its successful adoption is the informed selection of products that best fit the organization needs. One of the main current problems hampering selection, specially of OSS products is the vast amount of unstructured, incomplete, evolvable and widespread information about products that highly increases the risks of taking a wrong decision. In this paper, we aim to inform and provide evidence to OSS communities that help them to envisage improvements on their information rendering strategies to satisfy industrial OSS selectors' needs. Our results are from the matching between the informational needs of 23 OSS selectors from diverse software-intensive organizations, and the in-depth study of 9 OSS communities of different sizes and domains. The results evidenced specific areas of improvement that might help to enhance the industrial OSS selection practice.

Keywords: Open Source Software, selection, information rendering strategy, empirical study.

1 Introduction

Nowadays, the use of Open Source Software (OSS) provided by OSS communities is revolutionizing the software industry [1]. The fact that OSS products are freely available has influenced not only their significant adoption, but also the way that software is developed and commercialized [12]. Thus, fostering OSS adoption has been recognized as a crucial task for progressing towards improvements in a great variety of application areas [26].

The potential advantages of adopting OSS greatly depend on the ability to select the most suitable product for the task at hand [4]. Improper selection of an OSS product may result in wrong strategic decisions with subsequent economic loss and adverse effects on the business processes of the organizations [16].

In recent years there has been a plethora of proposals aimed to support software products selection, usually suggesting sets of evaluation criteria to evaluate and

S.A. Hissam et al. (Eds.): OSS 2011, IFIP AICT 365, pp. 244–258, 2011.

decide the most suitable alternative(s) (see [16], [18], [21], for comprehensive surveys). However, a recent survey about industrial OSS selection practices [2] shows that these proposals have not been greatly adopted in the industrial practice. Instead, in order to face time-to-market demands and reducing the potential risks, selectors (i.e., the person(s) in charge of the selection process) just base most of their decisions on their experience and tend to limit the use of OSS products to those that are already known and used by the development team. While the value of experience is important, the fact that it is currently considered as the most influential factor for selecting components is at the same time hampering the adoption and fully exploitation of the potential benefits of the high variety of OSS products in the marketplace. Furthermore, such study evidenced that one of the key problems is that selectors are struggling not only with the current diversity of OSS products available in the marketplace, but also with the great deal of widespread, incomplete, heterogeneous, and unstructured information describing each of them (e.g., formal/informal documentation, tutorials, comments in forums, internal experiences) that makes difficult to face a suitable selection process under time-to-market pressures [3], [4], [17]. In addition, the study emphasized that the main source for gathering OSS product information is the OSS community website.

In this context, in order to contribute to enable suitable OSS selection processes we need to envisage more pragmatic approaches than suggesting sets of evaluation criteria (especially when the evidence shows that the data to fill in these criteria is not usually available). Therefore, the goal of this study is to explore the following research questions:

— RQ1. *How much of the information required by OSS selectors for performing a suitable selection process is actually provided by OSS communities?*
— RQ2: *Are there OSS community characteristics that seem to influence its level of readiness for supporting OSS selection?*

With RQ1, we want to investigate the gap between the information "provided" by OSS communities on their OSS community websites and the information required by OSS selectors to perform an informed selection. RQ2 aims to explore whether some OSS community characteristic(s) seem to affect its readiness (i.e., the degree that the community covers the needs of OSS selectors). By answering these research questions, this paper aims to inform and provide evidence to OSS communities that help them to envisage improvements on their information rendering strategies. It is a first step to raise their awareness on areas that are required to improve the OSS selection industrial practice. This paper reports our results from a study about the matching of 9 OSS communities vs. the needs of 23 industrial OSS selectors.

The rest of the paper is organized as follows: Section 2 provides a brief background of the OSS literature and marketing principles that emphasize the importance of dealing with information rendering aspects and their influence on OSS products selection. Section 3 describes the methodological approach followed to perform the study. Section 4 presents the results obtained from the study, while Section 5 provides a discussion of main findings. Threats to validity are presented in Section 6. Section 7 summarizes the conclusions and future work.

2 Background

OSS research has largely ignored one interesting aspect that is becoming crucial for OSS projects: first-impression management [7]. Impression management theory refers to the process by which individuals or organizations try to control or manage the impressions that others form of them [28]. Due to the ever-increasing amount of information available on the Internet and the need to make quick choices among competing alternatives, first-impression management has been adopted as one of the main theoretical lenses in marketing literature. Choi et al [7] demonstrated that the OSS community website plays a critical role in attracting developers and users to the community. The mature status of well-known OSS projects likely attracts users given their greater activity and vitality. Furthermore some OSS products have become de facto standards. However this pathway is unavailable for most of the OSS projects and those newly initiated projects that struggle to attract users and contributors [6].

In the context of OSS selection if an OSS project is poorly presented and potential selectors feel that the community does not invest much care in providing the needed information for selection, they might formulate negative opinions about the project and fail to consider it as a candidate even if it might represent a promising alternative. Thus, poor first impressions not only impact on the rate of potential users in the short run, they can also produce negative externalities for the project in the long run. For instance, they might lose the synergies derived from collaborating with companies, such as greater project activity, higher user's base and popularity [7].

Several initiatives exist to develop a framework for the assessment of OSS products. Most of these initiatives suggest different kinds of criteria such as functionality, maturity, and the strategy of the organization around OSS. Relevant examples are: OpenBRR (Open Business Readiness Rating) [22], QSOS (Qualification and Selection of Open Source software) [25], or OSMM (Navica Open Source Maturity Model) [14]. The evaluation criteria are further explored by, for instance, Cruz et al. [9], the QualOSS Model Framework [8], and the QualiPSo model of OSS trustworthiness [10]. In addition, the factors that might attract developers to participate in OSS projects in order to sustain the vitality of the community have been also studied [5], [7], [11], [15]. However, as far as we know, there are not empirical studies that consider the needs of industrial OSS selectors as a way to improve first impression management.

Consolidated works from the marketing research have developed relevant tactics to help to influence first impression in a positive way [29]. We think that dealing with first impression management is important for OSS research for two main reasons: 1) OSS community contributors typically join OSS projects by first becoming selectors, subsequently as users and then evolving into contributors [7], therefore first impression management is a potential way to attract OSS potential users.   2) Marketing strategies are becoming crucial to pose OSS products into the marketplace given that nowadays the OSS phenomenon has evolved into a more commercially viable form, where both volunteers and commercial organizations collaborate in its ongoing development [1].

One of the initial grounds of first-impression management is to explore the needs of the potential users and to envisage the improvement tactics. Therefore, this study was designed to explore how the needs of OSS selectors are covered by OSS communities in order to raise observations that may serve to envisage suitable improvement tactics.

3   The Study

The study performed in this work was exploratory and aimed to investigate the research questions introduced above. Our main research strategy consisted of the in-depth study of 9 OSS community-based projects and exploring how these communities covered the informational needs of 23 industrial OSS selectors.

3.1 Sampling

The target population of the study was OSS community projects. Since the variety of OSS projects is quite wide not only regarding domain and size, but also regarding activity and popularity, we approached a stratified random sampling for improving its representativeness as well as the analysis of the results. We used the Ohloh.net directory as the reference directory for selecting OSS projects. We choose Ohloh as it is one of the largest and up-to-date OSS directories available, and has been widely used to create historical reports about the changing demographics of OSS projects.

From the 437,982 OSS projects referenced in Ohloh by February 2010, we ordered them with respect to their number of users and downloads. Then, we divided such a list into three equal parts (that were considered as stratums). Subsequently, we randomly selected 3 projects from each stratum. Table 1 summarizes the projects that fall into each stratum and provides a brief description of each of them.

3.2 Data Collection Instrument

In order to assess the OSS projects in a homogeneous way, we developed a data collection instrument based on the survey reported in [2]. This survey provides data about the information that is required in order to perform an informed OSS selection. 23 OSS selectors from 20 small and medium organizations in Spain, Norway and Luxembourg participated in the survey. It consisted of semi-structured interviews that were recorded in audio and then transcribed to text. We had access to the raw data from the respondents of such study (audio and text documents). From the responses of the selectors that participated in the study we elicited a total of 85 informational needs (i.e., specific information that they referred as needed for making informed decisions). We arranged similar answers using content analysis [19]. This process resulted in 21 informational needs. It is important to remark that similar efforts for establishing important evaluation criteria for selecting OSS products have been done in the literature as stated in section 2. Our primary goal was to assess if the information required by the selectors was provided by the OSS communities.

Table 1. Stratified random sampling

Stratum	Name	Description
1	Agilo for Scrum	It is one of the most widely used Scrum tools, offering many features to support Scrum and software development teams.
	Joomla	It is an award-winning content management system (CMS), which enables to build Web sites and powerful online applications.
	Subclipse	It is an Eclipse Team Provider plug-in providing support for Subversion within the Eclipse IDE.
2	Gimp	GIMP is an acronym for GNU Image Manipulation Program. It is a freely distributed program for such tasks as photo retouching, image composition and image authoring.
	GNU Grub	GRUB is the GRand Unified Bootloader for GNU
	IpTables	iptables is the user space command line program used to configure the Linux 2.4.x and 2.6.x IPv4 packet filtering ruleset. It is targeted towards system administrators.
3	Fluent NHibernate	A fluent API for simplifying the entity mapping of NHibernate. Add compile time safety, testability, and improved readability to NHibernate projects.
	MediaCoder	It is a free universal batch media transcoder, which integrates most popular audio/video codecs and tools into an all-in-one solution. New codecs and tools are added in constantly as well as support for new devices.
	StatusNet	StatusNet (formerly Laconica) is a microblogging service. Users post short (140 character) notices that are broadcast to their friends and fans using the Web, RSS, or instant messages.

In order to improve the quality of the data collection instrument, it was pre-tested with three researchers. Consequently, we decided to arrange the 21 informational needs in categories and subcategories that provided a more understandable, structured and informative way of collecting them. For instance, the informational need *Time of the product in the market* was grouped in the sub-category *History of the Product* which at its time was grouped in the category *To ensure technological stability and evolution of the OSS product and its provider*. This arrangement demonstrated to provide researchers with a better understanding of the informational needs and their contexts; it therefore enhanced the information gathering process. As a result, the 21 informational needs originally gathered were grouped into 8 categories and 3 sub-categories. The data collection instrument also gather information as: whether the informational need was provided or not by the OSS community; where the information was found, the time required to skim the OSS community website to find the information, and further comments from the researchers that performed and/or reviewed the OSS community. As it can be noted, our intention was not only to explore if the informational needs were available but also to have a first impression about how it was advertised, and how difficult it was to extract it. The resulting data collection instrument is shown in Table 2 and it also provides an example to illustrate the kind of information that was gathered.

Table 2. Data collection instrument. Example of the assessment of the Agilo for Scrum project.

Category/ Subcategory/ Informational Need	Results of the exploration		
	Status	Where	Further comments
1 Compliance with client's functional requirements			
List of main functional requirements of the OSS project.	√	(1)	The list of features provided seems quite comprehensive. If further information is required, there is an email available. We asked for further information about the product and our request was quickly processed.
2 To ensure technological stability and evolution of the OSS product and its provider			
Is it a commercial firm leading the community?	√		The community is lead by a single company called Agile42
List of companies/organizations collaborating in the community.	×		There is no information about any other company participating in the community.
History of the product.			
Time of the product in the market.	×		I navigated through the wiki and could not find this information
Versions of the product available.	×		
3 Evidence of successful OSS product usage			
Number of registered users	×		
List of companies using the product	√	(2)	There is documentation about success stories in companies as: ASDIS, eBuddy, be2, Ericsson, DHD24, Hypoport, Princenton Finantial systems and RES software among others.
Number of downloads	×		
Ratings and comments from users	√		Some textual comments from users as: *"The Agile project approach allowed be2 the ability to monitor the project's performance every two weeks, and to evaluate the performance and quality. agile42 did a perfect job in training and coaching distributed Scrum teams based in both Germany and Armenia for this project. agile42 has been an indispensable link for a successful company transformation."* **Dave Sharrock, Director IT, be2 S.à.r.l....**
4 Ease of OSS product integration			
Interoperability issues			
List of software system and subsystems required to ensure the correct functioning of the product.	√		Windows/MAC/Linux, andTrac 0.11. The information was widespread and it was not easy to gather it.
Hardware requirements	×		
Suitability of Code			
Well-commented code	√		Very well commented
Programming language	√		Python
5 Availability of support			
Free services	√	(3)	Very basic ones. Most of their services are not free. They have a commercial license where they provide further professional services. They also offer a blog, Google groups for commenting things about the product, and a Wiki for free.
Non-Free services	√	(4)	Personalized services. In addition there is an improved version of the product that is not OSS.
6 Availability of Tests Results			
Tests done by the OSS community	×		
Tests done by an external party	×		
7 Licensing terms			
Availability of detailed information about the licensing terms and explicitly state if they are listed by the OSS initiative	√	(5)	There are 2 licensing schemas. One that complies with the Apache Software License, and a non OSS (offering an improved version of the product)
8 Availability of documentation			
Documentation for final users	√	(6)	The documentation quality seems acceptable
Documentation for developers	√	(7)	The documentation for integrators is very basic and scarce
Available languages of the documentation	☑		Only English

General comments: The page is more oriented to business (the community is lead by a company called Agile42). A wiki is provided to report bugs and to inform about possible further involvement with the community as contributors.
Name of the researchers: TP + CA
Date of the assessment: 11/2010 – reviewed 02/2011
Mean time required for skimming the webpage: 4 hours

(1) http://www.agile42.com/cms/pages/features/; (2) http://www.agile42.com/cms/pages/references/;
(3) https://dev.agile42.com/wiki http://groups.google.com/group/agilo/topics http://agile 42.com/cms/blog/;
(4) http://agile42.com/cms/pages/support/; (5) http://agile42.com/cms/pages/agilo/;
(6) http://agile42.com/cms/pages/agilo-documentation/; (7) https://dev.agile42.com/wiki/agilo/dev

3.3 Study Procedures and Data Analysis

Each OSS project was assessed using the data collection instrument introduced above. Two different researchers were in charge of assessing each OSS community website. Subsequently, they discussed and agreed the observations. Once all OSS projects were explored and reviewed, the whole research team held discussion meetings to analyze the data and consolidate the results.

4 Results of the Study

Results are grouped in two subsections according to the research questions introduced above.

Table 3 shows a summary of the results from the analysis of the 9 OSS communities. In order to provide insights of the coverage of each OSS community to the selectors' needs, we assigned relative weights to each category of the data collection instrument as shown at the right side of each category in Table 3. Such assignment was based on the number of selectors' responses grouped into the category. For instance, the category *Compliance with client's functional requirements* had 18 similar responses; therefore its relative weight with respect to the 85 selector's responses resulted in 21.18%. The category *Availability and quality of the documentation* grouped 3 responses; therefore its weight was 3.53%. Based on such weights we calculated the percentage of coverage of each community to the categories of the data collection instrument. The last row of Table 3 shows the final coverage of each OSS community to the OSS selectors needs. These weights allow us to summarize our findings and provide useful insights to the reader to identify and understand the categories where there is a higher need of improvements.

4.1 How Much of the Information Required by Selectors is Provided by OSS Communities?

We found that the most important informational need belonging to the category *Compliance with client's functional requirements* was covered by all the analyzed projects. All of them show (with diverse levels of detail) a list of features of the OSS product.

The information required *To ensure technological stability and evolution of the OSS product and its provider* was poorly covered by most of the studied OSS communities. In addition, the coverage of informational needs belonging to this category was very diverse (see Standard Deviation in last column of Table 3). Most of the analyzed communities failed to clarify the kind of involvement of commercial firms. While in some cases it was clear that the leader of the community was a commercial firm and that several companies were also collaborating in the community under diverse schemas (coding, sponsoring, donating, etc.), in some other cases this information was not clear. The case of MediaCoder was outstanding as the project has radically changed its OSS nature by a purely commercial approach. At this respect, we found controversial comments in Ohloh claiming that MediaCoder should not be listed therein anymore mainly because the source code is not actually available and this violates one of the principles of OSS [23]. Other projects that did not offer clear information about the involvement of companies were FluentNHibernate and StatusNet, our observations regarding these projects led us to realize that such lack of clarity might come from the fact that these communities are currently in the process of defining a new business strategy by establishing commercial entities for making business around the products (e.g., selling expert support). Furthermore, basic informational needs as *Time of the product in the market* and *Versions of the product available* were not provided by several communities, especially those with a commercial orientation.

The informational needs grouped in the category *Evidence of Successful OSS product usage* were the ones that most communities failed to cover. None of the studied projects covered all the informational needs belonging to this category. Any of the OSS projects offered information about the number of downloads. Only two communities stated the number of registered users. Five projects stated a list of companies that have successfully used the product, and just one project offered comments from users of the product.

The informational needs belonging to the category *Ease of OSS integration* were mostly covered by the studied projects. The only informational need that was not successfully covered by most projects was related to the *Hardware requirements* needed to ensure the correct functioning of the OSS product. MediaCoder also failed to provide *well-commented* code and *programming language* (this is again due to the fact that it does not provide the source code of the product).

Regarding the category *Availability of support*, in all projects it was explicitly stated whether they provide non-free support, while free support was commonly characterized by wikis, email lists, IRC channels, and forums.

Most projects, excepting two (MediaCoder and StatusNet) offered clear information regarding *Licensing terms*. As mentioned above, these two projects were facing a business model change and therefore their licensing schemas were not clearly stated. Finally, regarding the *Availability of documentation*, almost all projects offered documentation for final users and for developers and most of them offered a variety of languages.

Summarizing, we found that the analyzed OSS projects cover the selectors' needs in a diverse degree. Such coverage ranges from 44.96% to 80.89%. Further discussions are provided in section 5.

4.2 Are There OSS Project Characteristics that Influence Its Level of Readiness for Supporting Selection?

The assessment of the 9 OSS community projects leads us to state some observations regarding characteristics that might affect the information rendering aspects of OSS communities and therefore their readiness for supporting selection. The most relevant ones suggest that two interrelated characteristics seem to affect the information rendering aspects of OSS projects: the involvement of commercial firms and the stratums that the projects belong to.

We observed substantial differences among the 3 Stratums. Surprisingly, all OSS projects from Stratum 1 have a close involvement of commercial firms in the community. This finding is in line with the results from the study reported in [5] that evidenced that firms coordinate, develop code for, or provide libraries to one third of the 300 most active OSS projects in SourceForge. Projects from Stratum 2 did not have commercial firms leading the projects, instead they referred to volunteer-based communities that fully adhered to the Free Software Foundation (FSF) and two of them (Gimp and Grub) were part of the GNU project that advocate for the "free software" philosophy. Projects from the Stratum 3 also show a high involvement of commercial firms. While MediaCoder is currently a purely commercial project, Fluent NHibernate and StatusNet are facing a transition stage for becoming business-oriented OSS communities. These facts are in line with the "commercialization" of OSS predicted by [12].

The involvement of firms in the OSS communities seems to influence their aesthetic appearance and information rendering aspects. For instance, Agilo for Scrum is an OSS project entirely governed and led by the company Agile42, and so its website is more oriented to business (i.e., selling services around the product) than to promote the involvement of potential contributors to the community. In the case of the Joomla!, the involvement of firms seems to be quite different as even if firms are quite involved in the project, the project is governed by the community. Thus, its website reflects a strong interest to promote resources for consolidating the community and attract contributors. It also offers several schemas for companies and organizations to participate in the community (i.e., donations, selling services around the product, merchandizing). The website of Subclipse is led by the company CollabNet and the provided resources are more oriented to final users (i.e., instructions on how to install the plugging) than to contribute to the community. Other examples are FluentNHibernate and StatusNet that are currently approaching business oriented models and are also improving the aesthetic appearance of their portals. Therefore, we suggest that: as higher the involvement of commercial firms is, the lesser seems to be the attention paid to promote the involvement of potential contributors in the community.

Communities without commercial firms involved shared several commonalities. It seems that they are mostly aimed to provide technical resources to strengthen the developers' community than aesthetic and attractive resources. This coincides with some studies that emphasize that some OSS projects mostly leaded by community programmers often value substance over form and some exhibit an antipathy for marketing and public relations work [13]. In all cases, these OSS projects provide mailing lists, forums and wikis aimed to enable the collaboration among the members.

Thus, our results might suggest that it would be useful to distinguish among: Commercial OSS, Foundation-based OSS, and Community-based OSS in order to better understand and assess the implications of selecting each kind of OSS product.

Table 3. Summary of results

Category/Subcategory/ Informational Need	Agilo for Scrum	Joomla	Subclipse	Gimp	GNU Grub	IpTables	Fluent NHibernate	Media Coder	StatusNet	Std Dev.
Compliance with client's functional requirements (21.18%)	21.18	21.18	21.18	21.18	21.18	21.18	21.18	21.18	21.18	0
List of main functional requirements of the OSS project.	✓	✓	✓	✓	✓	✓	✓	✓	✓	
To ensure technological stability and evolution of the OSS product and its provider (21.18%)	5.30	15.89	10.59	21.18	21.18	21.18	10.59	5.30	0	7.28
Is the project governed by a commercial firm or by the community?	✓ commercial	✓ community	✓ commercial	✓ GNU	✓ GNU	✓ community	Not clear	✓ commercial	Not clear	
List of companies/organizations collaborating in the community (others than the leader)	✗	✓	✗	✓	✓	✓	✗	✗	✗	
History of the product.										
Time of the product in the market.	✗	✓	✗	✓	✓	✓	✓	✗	✗	
Versions of the product available.	✗	✗	✓	✓	✓	✓	✓	✗	✗	
Evidence of successful OSS product usage (12.94%)	6.47	6.47	3.24	3.24	3.24	0	0	0	3.24	2.53
Number of registered users	✗	✓	✓	✗	✗	✗	✗	✗	✗	
List of companies using the product	✓	✓	✗	✓	✓	✗	✗	✗	✓	
Number of downloads	✗	✗	✗	✗	✗	✗	✗	✗	✗	
Ratings and comments from users	✓	✗	✗	✗	✗	✗	✗	✗	✗	
Ease of OSS product integration (12.94%)	9.71	9.71	12.94	12.94	9.71	9.71	6.47	9.71	9.71	1.94
Interoperability issues										
List of software system and subsystems required to ensure the correct functioning of the product.	✓	✓	✓	✓	✓	✓	✓	✓	✓	
Hardware requirements	✗	✗	✗	✓	✓	✗	✗	✓	✗	
Suitability of Code										
Well-commented code	✓	✓	✓	✓	✓	✓	✓	✗	✓	
Programming language	✓	✓	✓	✓	✓	✓	✓	✗	✓	
Availability of support (11.76%)	11.76	11.76	11.76	11.76	11.76	11.76	11.76	11.76	11.76	0
Free services	✓	✓	✓	✓	✓	✓	✓	✓	✓	
Non-Free services	✓	✓	✓	✗	✗	✗	✓	✓	✓	
Availability of test results (9.41%)	0	0	0	0	0	0	0	0	0	0
Tests done by the OSS community	✗	✗	✗	✗	✗	✗	✗	✗	✗	
Tests done by an external party	✗	✗	✗	✗	✗	✗	✗	✗	✗	
Licensing terms (7.06%)	7.06	7.06	7.06	7.06	7.06	7.06	7.06	0	0	2.77
Availability of detailed information about the licensing terms and explicitly state if they are listed by the OSS initiative	✓	✓	✓	✓	✓	✓	✓	✗	Not clear	
Availability of documentation (3.53%)	2.6	3.53	2.60	3.53	3.53	3.53	2.60	0.25	3.53	1.06
Documentation for final users	✓	✓	✓	✓	✓	✓	✓	✗	✓	
Documentation for integrators	✓	✓	✓	✓	✓	✓	✓	✗	✓	
Available languages of the documentation	Only English	Several	Only English	Several	Several	Several	Only English	Only English	Several	
Mean time spent by the two researchers for skimming the portal (hrs):	4:00	4:50	3:05	3:02	3:03	3.50	2:25	3:00	2:43	
Resulting percentage of coverage of the portal to the needs of selectors:	64.07	75.59	66.13	80.89	80.89	74.42	62.90	44.96	49.41	12.13

5   Discussion of Results

The previous section aimed to present a comprehensive view of the results. This section aims at emphasizing and discussing the most important findings and observations.

One of the main difficulties we faced in our OSS projects assessment was that the information—even if it was sometimes available—was not directly accessible. We had to browse the project website and explore among help files, manuals, or even demos. This fact increased the time spent on skimming the portal to find the information and definitely rules out any possibility of trying to automate the search for the information as previously stated by [3].

Even if we found that the involvement of commercial firms and the stratums seem to have a significant influence on information rendering aspects, the coverage of selectors' needs varies from project to project.

OSS communities are not aware of the importance of making some information available. We observed that some informational needs are actually known by OSS communities but are not explicitly provided by them. For instance, most of the analyzed projects did not explicitly offer information about the *Number or registered users* in the community, *Time of the product in the market*, or *List of companies using the product* (if any). So, we hope that the results provided here help to raise the awareness of the importance of providing such information. In addition, there are categories that seem to be almost always provided (i.e., *Compliance with client's functional requirements* and *Availability of support*) while there are others that are not fully covered (e.g., *To ensure technological stability and evolution of the OSS product and its provider*) or are not covered by any of the studied project as *Availability of test results*.

Providing such evidence is important to envisage the corresponding improvement strategies and increase the competitive advantage of the OSS products.

6   Threats to Validity

This section discusses the threats to validity of our study in terms of construct, internal, and external validity, as suggested by [24] and [27]. It furthermore emphasizes the corresponding strategies used to deal with these threats.

6.1   Construct Validity

Regarding construct validity, our study was supported by 2 main principles: rigorous planning of the study, and the establishment of protocols and instruments for data collection and data analysis. The data collection instrument was carefully designed taking into account the informational needs of selectors elicited from a semi-structured interview further reported in [2], as detailed in section 3.2. This allows us to focus the study on the information that is really needed by the industrial OSS selection practice. In addition, the data collection instrument was pre-tested and enhanced by creating categories and subcategories for grouping informational needs (as detailed in section 3.2). This allows us to improve its understandability and therefore to improve the data gathering process.

6.2 Internal Validity

Regarding internal validity, we tried hard to envisage and harmonize the data gathering and the subsequent data analysis strategies. With respect to the data gathering strategy, we took relevant decisions for approaching a better understanding of the availability of the information for covering the selectors' needs. One of the most relevant decisions was to avoid the non-deterministic factors inherent to the OSS selection processes. These non-deterministic factors refer to contextual issues that greatly affect the OSS selection decision. For instance, even if an OSS project provides a list of functional characteristics of the product, it might happen that such a list is not detailed enough for the context of the selection project and the selector have to face such a lack of detail by testing the product himself or by looking for further information in forums, email lists, etc. The strategies for facing (or not) such lack of information depend on the amount of time and resources that a company is willing and able to invest in the selection process [9]. Therefore, to avoid such potential issues we decided to focus our observations just on whether the informational need was covered or not and on how the information was provided.

In addition, we decided that two different researchers independently faced the assessment of each OSS community projects using the data collection instrument. Subsequently they discussed their results in order to agree and merge them. This helps us to deal with the potential subjectivity of the assessment of each researcher. Furthermore, it is important to mention that the researchers participating in the study are impartial parties and do not have any kind of involvement with any of the OSS communities analyzed. In this sense, we consider that there is no any intentional bias regarding the data gathered.

6.3 External Validity

Regarding external validity, it is important to highlight that the character of our study is exploratory, and hence we did not aim to make universal generalizations beyond the studied setting, but also provide some observations that might serve as a departing point for further investigations and improvements. Having this in mind, we discuss some mitigation strategies used in the study.

One of the main threats of external validity of the study is that we approached a small set of OSS community projects and these projects might not represent the whole variety of OSS projects. We tried to mitigate any possible bias related to this by having a stratified random sampling so that the studied OSS communities were diverse regarding size, application domain, popularity and success.

The informational needs used as a base to decide the informational coverage of OSS projects were elicited from industrial OSS selectors. While extracting such needs from the industrial practice is a good point to strengthen the external validity of our observations, we are aware that eliciting such needs from 23 selectors might not represent all real needs. However, we think that such results are useful to have a first approximation to the problem and might serve as a basis to envisage future studies.

Finally, other issues that might affect the presented results (especially the time spent skimming the OSS projects for finding and understanding the information) are:

a) As mentioned above, the assessment of each OSS project was performed using a strategy that avoids the non-deterministic nature of the OSS selection processes (i.e., just capturing whether the informational need was covered or not and further observations about how it was provided). At this respect, we are aware that in the industrial OSS selection practice the complexity and time for gathering the OSS products information is actually higher. Thus, we would like to stand out that the metrics (weights and time for assessing each OSS project) provided in Table 3 are just intending to offer insights of the coverage of each OSS community to the selectors' needs. In any case these metrics are aimed to be representative of the assessment of OSS projects for any specific OSS selection process.

b) The researchers in charge of analyzing the OSS communities were not experts in any of the approached domains. So, we may say that the performance of researchers that performed the data gathered process would be more similar to "novice selectors" than experienced selectors that might perform better in finding and analyzing the OSS communities.

7   Conclusions and Future Work

This study presents our results of exploring the current gap between the "required" information needed by 23 industrial OSS selectors for making informed decisions and the information "provided" by 9 OSS communities. The obtained results would contribute to research and practice: a) by informing OSS communities about information rendering aspects that could be improved to attract industrial users. b) by informing OSS selection researchers about informational limitations that might help them to calibrate their OSS selection proposals.

Our future work focus on complementing the results from the study reported here with further information that allows OSS communities to elaborate their tactics of improvement based on the selectors' feedback [29].

References

1. Ayala, C.P., Cruzes, D., Hauge, Ø., Conradi, R.: Five Facts on the Adoption of Open Source Software. IEEE Software, 95–99 (March-April 2011)
2. Ayala, C., Hauge, Ø., Conradi, R., Franch, X., Li, J.: Selection of Third Party Software in Off-The-Shelf-Based Software Development - An Interview Study with Industrial Practitioners. The Journal of Systems & Software 84, 620–637 (2011)
3. Bertoa, M., Troya, J.M., Vallecillo, A.: A Survey on the Quality Information Provided by Software Component Vendors. Journal of Systems and Software 79, 427–439 (2006)
4. Boeg, J.: Certifying Software Component Attributes. IEEE Software 23(3), 74–81 (2006)
5. Bonaccorsi, A., Lorenzi, D., Merito, M., Rossi, C.: Business firms' engagement in community projects. Empirical evidence and further developments of the research. In: Proceedings of the First International Workshop on Emerging Trends in FLOSS Research and Development (FLOSS 2007), pp. 1–5. IEEE Computer Society, Minneapolis (2007), doi:10.1109/floss.2007.3

6. Capiluppi, A., Lago, P., Morisio, M.: Evidences in the Evolution of OS Projects through Changelog Analy¬ses. In: Proc. 3rd IEEE Workshop Open Source Software Eng (WOSSE 2003), ICSE 2003, pp. 10–24 (2003)
7. Choi, N., Chengalur-Smith, I., Whitmore, A.: Managing First Impressions of New Open Source Software Projects. IEEE Software 73–77 (November-December 2010)
8. Ciolkowski, M., Soto, M.: Towards a Comprehensive Approach for Assessing Open Source Projects. In: Dumke, R.R., Braungarten, R., Büren, G., Abran, A., Cuadrado-Gallego, J.J. (eds.) IWSM 2008. LNCS, vol. 5338, pp. 316–330. Springer, Heidelberg (2008)
9. Cruz, D., Wieland, T., Ziegler, A.: Evaluation Criteria for Free/Open Source Software Products Based on Project Analysis. Software Process: Improvement and Practice 11(2), 107–122 (2006)
10. del Bianco, V., Lavazza, L., Morasca, S., Taibi, D.: Quality of Open Source Software: The QualiPSo Trustworthiness Model. In: Boldyreff, C., Crowston, K., Lundell, B., Wasserman, A.I. (eds.) OSS 2009. IFIP AICT, vol. 299, pp. 199–212. Springer, Heidelberg (2009)
11. Denner, C.S., Pearson, J., Kon, F.: Attractiveness of Free and Open Source Projects. In: The European Conference on Information Systems ECIS 2010 (2010) ISBN: 978-0-620-47172-5
12. Fitzgerald, B.: The Transformation of Open Source Software. MIS Quarterly 30(3) (2006)
13. Fogel, K.: Producing Open Source Software: How to Run a Successful Free Software Project. O'Reilly, Sebastopol (2006)
14. Golden, B.: Succeeding with Open Source. Addison-Wesley Professional, Reading (2004)
15. Hauge, Ø., Ayala, C.P., Conradi, R.: Adoption of open source software in software-intensive organizations - A systematic literature review. Information & Software Technology 52(11), 1133–1154 (2010)
16. Jadhav, A.S., Sonar, R.M.: Evaluating and Selecting Software Packages: A review. Information and Software Technology 51(3), 555–563 (2009)
17. Li, J., Conradi, R., Bunse, C., Torchiano, M., Slyngstad, O.P.N., Morisio, M.: Development with Off-The-Shelf Components: 10 Facts. IEEE Software 26(2), 80–87 (2009)
18. Li, J., Conradi, R., Slyngstad, O.P.N., Torchiano, M., Morisio, M., Bunse, C.: A State-of-the-Practice Survey of Risk Management in Development with Off-the-Shelf Software Components. IEEE Transactions on Software Engineering 34(2), 271–286 (2008)
19. Krippendorff, A.: Content Analysis. Sage Publications, London (1980)
20. Mahmood, S., Lai, R., Kim, Y.S.: Survey of Component-Based Software Development. IET Software 1(2), 57–66 (2007)
21. Merilinna, J., Matinlassi, M.: State of the Art and Practice of Open-Source Component Integration. In: Proceedings of the 32nd EUROMICRO Conference on Software Engineering and Advanced Applications, pp. 170–177. IEEE Computer Society, Los Alamitos (2006)
22. Openbrr, Business Readiness Rating for Open Source A Proposed Open Standard to Facilitate Assessment and Adoption of Open Source Software, Request For Comments, (2005),
    http://www.openbrr.org/wiki/images/d/da/BRR_whitepaper_2005R FC1.pdf
23. Open Source Initiative, http://www.opensource.org/
24. Robson, C.: Real World Research: A Resource for Social Scientists and Practitioner-researchers, 2nd edn. Blackwell Publishers Inc., Malden (2002)

25. Semeteys, R., Pilot, O., Baudrillard, L., Le Bouder, G., Pinkhardt, W.: Method for Qualification and Selection of Open Source software (QSOS) version 1.6, Technical report, Atos Origin (2006)
26. Simmons, G.L., Dillon, T.S.: Towards an Ontology for Open Source Software Development. In: Fitzgeralg, E., Scacchi, B., Scotto, W., Succi, M. (eds.) Open Source Systems. Damiani. IFIP, pp. 65–75. Springer, Boston (2006)
27. Wohlin, C., Runeson, P., Host, M., Ohlsson, M.C., Regnell, B., Wesslen, A.: Experimentation in Software Engineering - An Introduction. Kluwer Academic Publishers, Dordrecht (2000)
28. Winter, S.J., Saunders, C., Hart, P.: Electronic Window Dressing: Impression Management with Websites. European J. Information Systems 12, 309–322 (2003)
29. Wolfe, E., Bies, R.: Impression Management in the Feedback-Seeking Process: A Literature Review and Research Agenda. Academy of Management Review 15(3), 522–541 (1991)

To Fork or Not to Fork:
Fork Motivations in SourceForge Projects

Linus Nyman[1] and Tommi Mikkonen[2]

[1] Hanken School of Economics, Helsinki, Finland
linus.nyman@hanken.fi
[2] Tampere University of Technology, Tampere, Finland
tommi.mikkonen@tut.fi

Abstract. A project fork occurs when software developers take a copy of
source code from one software package and use it to begin an independent
development work that is maintained separately from its origin. Although
forking in open source software does not require the permission of the original
authors, the new version, nevertheless, competes for the attention of the same
developers that have worked on the original version. The motivations
developers have for performing forks are many, but in general they have
received little attention. In this paper, we present the results of a study of forks
performed in SourceForge (http://sourceforge.net/) and list the developers'
motivations for their actions. The main motivation, seen in close to half of the
cases of forking, was content modification; either adding content to the original
program or focusing the content to the needs of a specific segment of users. In a
quarter of the cases the motivation was technical modification; either porting
the program to new hardware or software, or improving the original.

1 Introduction

A project fork takes place when software developers take a copy of the source code
from one software package and use it to begin an independent development work. In
general, forking results in an independent version of the system that is maintained
separately from its origin. The beauty of open source software development is that no
permission from the original authors is needed to start a fork. Therefore, if some
developers are unhappy with the fashion in which the project is being managed, they
can start an independent project of their own. However, since other developers must
then decide which version of the project to support, forking may dilute the community
as the average number of developers per system under development decreases.

Despite some high-visibility forks, such as the forking of OpenOffice
(http://www.openoffice.org/) into LibreOffice (http://www.libreoffice.org/), the whole
concept of forking has seen little study. Furthermore, developers' motivations for
forking are understood even less, although at times it seems rational and
straightforward to identify frustration with the fashion in which the main project is
being managed as a core reason.

In this paper, we present the results of our investigation of SourceForge
(http://sourceforge.net/) for forked projects and the motivations the authors have

S.A. Hissam et al. (Eds.): OSS 2011, IFIP AICT 365, pp. 259–268, 2011.
© IFIP International Federation for Information Processing 2011

identified for performing a fork. Furthermore, we categorize the different motivations and identify some common misbeliefs regarding forking in general.

The rest of this paper is structured as follows: Section 2 discusses the necessary background for explaining some of the technical aspects associated with forking, Section 3 introduces the fashion in which the research was carried out, Section 4 offers insight into our most important findings, and Section 5 discusses them in more detail. Section 6 proposes some directions for future research, and Section 7 concludes the paper with some final remarks.

2 Background

When pushed to the extreme, forks can be considered an expression of the freedom made available through free and open source software. A commonly associated downside is that forking creates the need for duplicated development efforts. In addition, it can confuse users about which forked package to use. In other words, developers have the option to collaborate and pool resources with free and open source software, but this is enforced not by free software licenses, but only by the commitment of all parties to cooperate.

There are various ways to approach forking and its study. One is to categorize the different types to differentiate between, on the one hand, forks carried out due to amicable but irreconcilable disagreements and interpersonal conflicts about the direction of the project, and on the other, forks due to both technical disagreements and interpersonal conflicts [1]. Still, the most obvious form of forking occurs when, due to a disagreement among developers, a program splits into two versions with the original code serving as the basis for the new version of the program.

Raymond [2] considers the actions of the developer community as well as the compatibility of new code to be a central issue in differentiating code forking from code fragmentation. Different distributions of a program are considered 'pseudo-forks', because at first glance they appear to be forks, but in fact are not, since they can benefit enough from each others' development efforts not to be a waste, either technically or sociologically. Moody [3] reflects Raymond's sentiments, pointing out that code fragmentation does not traditionally lead to a split in the community and is thus considered less of a concern than a fork of the same program would be. These sentiments both echo a distinction made by Fogel [1]: it is not the existence of a fork which hurts a project, but rather the loss of developers and users. Here it is worth noting, however, that forking can potentially also increase the developer community. In cases in which developers are not interested in working on the original (for instance due to frustration with the project direction, disagreements with a lead developer, or not wanting to work on a corporate sponsored project), not forking would lead to fewer developers as the developers in question would likely simply quit the project rather than continue work on the original.

Both Weber [4] and Fogel [1] discuss the concept of forks as being healthy for the ecosystem in a 'survival of the fittest' sense; the best code will survive. However, they also note that while a fork may benefit the ecosystem, it is likely to harm the individual project.

Another dimension to forking lies in the intention of the fork. Again, several alternatives may exist. For instance, the goal of forking can be to create different branches for stable and development versions of the same system, in which case forking is commonly considered to serve the interests of the community. At the other extreme lies the hostile takeover, which means that a commercial vendor attempts to privatize the source code [5]. Perhaps somewhat paradoxically, however, the potential to fork any open source code also ensures the possibility of survival for any project. As Moody [6] points out, the open source community and open source companies differ substantially in that companies can be bought and sold, but the community cannot. If the community disapproves of the actions of an open source company, whether due to attempts to privatize the source code or for other reasons related to an open source program, the open source community can simply fork the software from the last open version and continue working in whichever direction it chooses.

3 Research Approach

In the study, we used SourceForge (http://sourceforge.net/) as the repository of open source programs from which we collected forks. SourceForge contains over 260,000 open source projects created by over 2.7 million developers. Creating new projects, participating in those that already exist, or downloading their contents is free, and developers exercise this freedom: programs are downloaded from SourceForge at a pace of more than 2,000,000 downloads daily.[1]

SourceForge offers programmers the opportunity to briefly describe their program, and these descriptions can be searched using keywords. Using this search function, we compiled a list of all of the programs with the word "fork" – as well as dozens of intentionally misspelled variations of the word fork, none of which turned up any hits – in their description. We then analyzed all the descriptions individually to differentiate between them and to sort out programs that the developers claimed had forked their code base from another program (which we call "self-proclaimed forks") from those which included the term 'fork' for some other reason, either to describe a specific functionality of the program or as part of its name (i.e. false positives). Consequently, a program that stated "This is a fork of ..." was considered a fork, while a program which noted that it "...can be used to avoid common security problems when a process forks or is forked" was not. If it was impossible to categorize a project based on the available data, it was discarded. Our data consisted of all programs registered on SourceForge from its founding in late 1999 through 31 December 2010, resulting in a time span of slightly more than 11 years. This search yielded a total of 566 programs that developers report to be forked.

We then analyzed the motivations stated in the descriptions of the forked programs. The coding process was done in three phases. First, we went through all of the descriptions and wrote a brief summary of the motivations, condensing the stated reasons to as few words as possible. Then, we went through all of the motivations and identified common themes, or subgroups of motivations, among them. In cases where the fork included elements from more than one theme, we placed it in the subgroup

---

[1] Source: http://sourceforge.net/about, accessed March 9, 2011

that seemed the most central to the motivation behind the fork. Finally, we examined the subgroups to identify overarching groups of themes.

To give some examples of the coding, one fork stated: "[Project name] is a fork of the [original project name] project. [The] purpose of [project name] is to add many new features like globule reproduction, text to speech, and much more." The motivation behind the fork was identified as belonging to the subgroup "add content", which in the final step was combined (with a subgroup of programs which sought to focus content) into a group called content modifications. A fork which sought to fix bugs, and a fork which was motivated by porting a program, were first put into separate subgroups, "technical: improvement" and "technical: porting", and then these subgroups were combined into the "technical modifications" group. Further examples from the data are presented in the next section.

Based on the descriptions entered by the developer, we were able to identify motivations for 381 of the forks. The group of forks which we were unable to categorize consisted of two main types of descriptions: firstly, descriptions which offered no insights as to underlying motivations, e.g. programs which simply stated which program they were forked from; secondly, cases in which it was unclear from the description if the elements described were added in the fork or if they existed in the original; in other words, one couldn't determine if the description of the program included the motivation behind the fork, for instance new technical features, or if they were describing pre-existing features common to both the original and the fork.

4 Reasons for Forking

Based on the data obtained, developers commonly attribute their reasons for forking the code to pragmatism. For a variety of reasons, some of which were well documented and some of which were unclear, the original version of the code failed to meet developers' needs. To expand the scope of the system, the developers then decided to fork the program to a version which serves their own needs. The descriptions of the forks include programs which note that certain changes have been made to the fork, as well as those programs which discuss which changes will or should be made to the forked version. In this paper, we have not distinguished between the two: both planned and already implemented changes are treated equally, since the goal was to study motivations rather than eventual implementations. In general, the forks appear to stem from new developers rather than the original developing team splitting into two camps. In fact, the data contain almost no references to disagreements among developers that might have led to the fork. However, this does not mean that such disagreements could not have existed.

In the following section, we provide a more detailed view of the different motivations we were able to find in the data (n = 381). The main motivations fall into two large groups (content and technical modifications) which comprise nearly three quarters (72%) of all forking motivations. Four smaller groups, all of similar size, comprise an additional 23% of the motivations. These four groups included the reviving of a project, license- or FOS-related motivations, language- or country-related reasons, and experimental forks. The remaining motivations, grouped simply as "other", consisted of diverse yet uncommon reasons. An overview reflecting the numbers of forks appears in Figure 1.

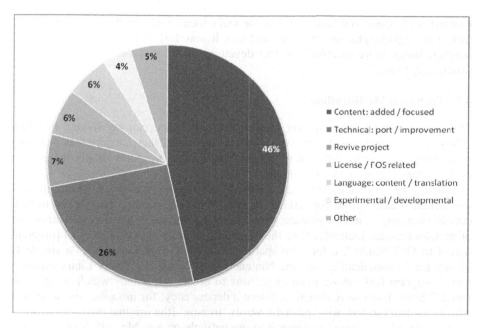

Fig. 1. Fork motivations in SourceForge projects

4.1 Content Modifications

Comprising almost half of all forks, content modifications is the largest group. The two main subgroups within the content modifications category, both of which are nearly equal in size, were the adding and the focusing of content; these are briefly discussed below.

Adding content is a self-explanatory reason for making a fork. The developers added new features or other content (e.g. adding better documentation, helper utilities, or larger maps to a game). Quite often, developers didn't describe additions in detail; one developer, for instance, simply noted that the program was a fork "that has the features I'm missing from [the original]." Another developer stated that the fork was "A [program name] fork with more features". In several cases, this group of forks also included bugfixes.

Focusing content implies focusing on the needs of a specific user segment. This category includes forks with both a technical and content-related focus, along with the addition of functionalities and features as well as the removal of elements or features unnecessary for a specific segment or purpose. Examples of content-related focus include programs forked in order to focus on serving the needs of dance studios, radio talk shows, catering companies, program developers, and astronomers, to name but a few. Examples of technical focus include forks "aimed at higher-resolution iOS devices", a fork which "features improvements and changes that make it more oriented for use in a Plone intranet site", and a fork intended "to run on machines that have 800x600 screen resolution". In a minority of the cases in the focusing content category, the original program was forked mainly to remove elements from the original. The main goal in this group was to create a lighter or simpler version of the original, with speed and ease of use as the main focus. One developer stated that the fork was "lightweight, less bloated" and that it was forked to "make [the original] simpler, faster, more useable." Another developer noted that the fork was "Smaller, faster, easy to use."

4.2 Technical Modifications

This group, comprising just over a quarter of all forks, can be divided into two subcategories: porting and improving. A characteristic of this category was that little if anything was visibly different to the user; the forked programs simply focused on either porting or improving the original.

Porting the original code to new hardware or software was the more common of the technical motivations for forking, usually involving porting the original to fit a certain operating system, hardware, game, plug-in, migrate to a different protocol, or other such reasons. Examples from the data for this group include a "fork of [program name] to GNU/Linux", a fork "compatible with the NT architecture", "a simple C library for communicating with the Nintendo Wii Remote [...] on a Linux system", and a program fork whose main target was to create a version "which works with ispCP." Some forks were ported to reduce a dependency; for instance, one developer who noted that the fork was "geared towards 'freeing' [the original program] from its system dependence, [thus] enabling it to run natively on e.g. Mac OS X or Cygwin." Another developer noted that the program was forked because the developer could not find a "good and recent [program type] without KDE dependency."

Improving the original program was slightly less common in the technical motivations category than porting, which focuses on improving already existing features and contains mostly bugfixes, code improvement and optimization, and security improvements. Some cases were very general in their descriptions, noting only that it was an "upgraded" or "improved" version of the original, or that the code was forked "to fix numerous problems in the code" or to "improve the quality of emulation". Others were more specific, as with the developer of one fork, who notes that "The main goal is to build a new codebase which handles bandwidth restrictions as well as upcoming security issues and other hassles which showed up [during] the last 6 months."

4.3 Reviving an Abandoned Project

The third common motivation for forking was to continue development of a project considered abandoned, deceased, stalled, retired, stagnant, inactive, or unmaintained. In several of these cases, the developers of the fork note who the original developers are and credit them. In a few cases, the developers of the fork note that they attempted (unsuccessfully) to reach the original developers; in other words, forking the code was the last available option for these developers, as the original developers could no longer be reached. One such example is a fork which the developer notes was "due to long-time inactivity" and then goes on to state "We want to thank the project founder [name] for starting this project and we intend to continue the work". In another case, also due to the inactivity of the original developer, the developer of the fork acknowledges the original author and notes that the fork "includes changes from comments made on his forum." Other examples from the data are: "This project is a fork of the excellent but dead [project name] project", "This project is a fork of the stalled [project name]", "a code fork from the (deceased) [project name] source", and, finally, "The previous maintainer is unresponsive since 2008 and the library [has] some deficiencies that need to [be] fixed. Anyway, thanks for creating this great library [name of original developer]!"

4.4 License/FOS-Related Issues

This group consists of forks which were motivated by license-related issues or a concern for the freedom of the code. Some of the forks appear to be simply a form of backup copies: stored open source versions of well-known programs. The motivation for this subgroup was a concern that the original version might become closed source. In one case, the developer stated that the fork was due to concern about the future openness of the code. In a similar case, a developer noted about the fork that "This is a still-GPL version of [program name,] just in case." One fork simply identifies the motivation as a "license problem". In five cases, the program was forked because the original was deemed to have become either closed source or commercial, and in one case, developers noted that the fork occurred because certain bits of the original code were closed source. One fork notes that the new version removes proprietary (boot) code from the program, but that "there is no need to use this version unless you are concerned about the copyright status of the embedded boot code."

4.5 Language- and/or Country-Specific Modifications

A small group of the forks were motivated by language and country. This group could well be considered a subcategory of the "focusing content" group, but was considered separate due to its clear language-related focus. The simplest, though not most common, form of forks included programs which were merely translated into one or more languages; in most cases, however, new content was also added to customize the fork for a specific country and/or group. Some examples are forks created for elections in New Zealand, the right-to-left reading of Hebrew texts, and a program "customized to meet German requirements regarding accounting and financial reporting."

4.6 Experimentation

This group consisted mostly of forks which declared that they existed for experimental purposes, with a handful citing development reasons. A feature common among many of these forks is that the developers state that the fork is temporary and that successful new features or improvements will be incorporated into the original program. Some describe the fork as simply "for testing", while others go into greater detail, noting for instance that the fork is "aimed at experimenting with a number of features turned up to maximum." One developer notes that the fork is simply "for fun", and then goes on to tell readers where they can find the original project.

4.7 Other Reasons

Of the remaining forks, a handful described it as a "community fork." In some of these cases, it was possible to identify an overarching motivation behind the community fork; in others it was not, the implications of the term in those cases remaining unclear. Two cases cite a reprogramming in a different programming language as the reason for the fork. The remaining reasons for the forks defied categorization, and included such motivations as a desire to create a study tool for the developer, as well as to test SourceForge for a different project.

Finally, the most surprising of the remaining groups was the group motivated by disagreement or breach of trust. In the beginning of the study, we assumed that a significant number of forks would stem from disagreement between developers. In reality, we were able to identify such forks, but their proportion is quite small: we identified only four cases, three of which stated that the users sought something the original developers did not intend to implement and one which noted that the fork was a reaction to a breach of trust. Furthermore, even some of these cases may be attributed to the original developers' loss of interest in the project.

5 Discussion

The data in this paper are based on information provided by developers themselves. Many of the cases of self-proclaimed forking – such as when a developer continues an abandoned project – could arguably be defined as something other than a true fork. However, determining forks any other way (other than through the self-proclaimed approach used here) would require a technical definition of a fork that would have to be mined from the project data. At present, no such mechanism seems to exist, and in general, differentiating between forked and fragmented code is an ambiguous practice, unless defined by elements outside of the code itself. Consequently, we have identified the developers as the most reliable source of information, at least at present.

Beyond the challenge of defining a fork, one here also needs to note two issues: how the choice of SourceForge as a sampling frame might affect the data, as well as how accurate, or complete, the descriptions offered there are. The choice of SourceForge could affect the data in several ways. The main question would seem to be whether the characteristics of the average program – or program fork – on SourceForge differ from those of programs hosted on other sites, or from independently hosted programs. For example, given that larger projects often have their own hosting, it is possible that we are seeing only a small number of forks in some categories because projects that would face such issues are not using SourceForge. As to the completeness of the motivations offered by developers, there could be a number of reasons why the information offered is incomplete. For instance, the low frequency of disagreements as a motivational factor in forking may perhaps in part be explained by either a reluctance to mention such disagreements or the limited space offered by SourceForge in which to describe the program. It is also possible that such information, while not stated on SourceForge, would be available on project homepages. Indeed, we came across a project which noted elsewhere that a disagreement among the developers of the original was a factor in the fork; however, the same project did not mention this disagreement in their description on SourceForge.

In general, the results of our study suggest that forking is not a particularly extreme situation in real-life projects. For the most part, developers' motivations are easily understandable, and forking can be considered a reasonable action. However, this does not mean that hostile takeovers are absent from high-profile projects, but simply that in the vast majority of cases, developers appear simply to seek to satisfy their own needs and to develop interesting systems. Such motivations were evident in the documentation in many ways. Some of the forks note that the changes or improvements have already been made, whereas others announce the intended direction of the fork and mention features to be added to it. Furthermore, crediting the original developers was a rather common practice among those who forked a program, which further emphasizes the fact that forks sought to achieve certain goals, not to compete with existing communities. Perhaps more telling still is that a number of forks noted that they hoped to be temporary, and clearly stated their desire that the bugfixes and improvements introduced in their fork be incorporated into the original program.

6 Future Work

Future work regarding issues associated with forking could take numerous directions. Below we list some of the most promising directions that merit further investigation.

Defining a fork. All of the programs in the data for this article define themselves as forks. In practice, upon more careful review, many of them could perhaps more accurately be categorized as pseudo-forks, code fragmentation, or simply different distributions of a code. The creation of a commonly agreed-upon view of forking vs. fragmentation (or distributions) vs. code reuse would be a very practical step that could benefit both researchers as well as the entire open source community. It could also be possible to define a fork based on technical details, rather than depend on information provided solely by the developers.

Licenses before and after forking. Future researchers could conduct a survey of developers who have forked a program in which they explain their choice of license in comparison to the license of the original program from which they forked.

Perception of forking. Another practical aspect related to forking is how programmers view it; in other words, when is it acceptable to fork, and when is it not? Furthermore, discovering whether certain behaviors make forking more acceptable among developers would be an important direction for such work.

Expanding the data set. Performing a similar study for other sites that host open source projects would contribute to a deeper understanding of forking. Because all the data come from only one source, certain aspects may skew the results. Furthermore, it would be interesting to test if one can tie the observed categories to antecedents or consequences, e.g., are particular kinds of software more likely to fork in particular ways or are particular kinds of forks more successful?

Forking in relation to business. A number of forks we have identified occurred because the original project became closed source. Examining what happened to these projects would deepen our understanding and view of forking in relation to business.

7 Conclusions

Forking is one of the least understood topics in open source development. While often perceived initially as something malicious, the developers who perform the actual forking cite rather straightforward reasons for their actions.

In this paper, we addressed the motivations of developers for performing a fork. The data used in the project originate from SourceForge (http://sourceforge.net/), one of the best-known hosts of open source projects, and focus on "self-proclaimed forks", or programs that the program developers themselves consider to be forks. The motivations behind forking are based on developer input, not on mining technical qualities of the project. However, using only the latter to determine forking would be difficult, as separating forking from other open source-related phenomena is problematic and inconclusive. At the very least, additional data from developers are needed to define forking.

In conclusion, while hostile takeovers and the hijacking of a project as well as a loss of developers after a fork are often associated with forking, the reality is that forks seem to be a lot less dramatic. In fact, forking appears to be more or less business as usual, and developers fork because doing so provides certain benefits for their own goals. While we were able to find forks where the rationale for forking lay in disagreement or trust issues, such cases were few in comparison to the total number of projects we studied.

References

[1] Fogel: Producing Open Source Software. O'Reilly, Sebastopol (2006)
[2] Raymond: The Cathedral & the Bazaar: Musings on Linux and Open Source by an Accidental Revolutionary. O'Reilly, Sebastopol (2001)
[3] Moody: The Deeper Significance of LibreOffice 3.3. ComputerWorld UK (January 28, 2011)
[4] Weber: The Success of Open Source. Harvard University Press, Cambridge (2004)
[5] Lerner, Tirole: Some Simple Economics of Open Source. The Journal of Industrial Economics 50(2), 197–234 (2002)
[6] Moody: Who owns commercial open source and can forks work? Linux Journal (April 23, 2009)

An Analysis of Author Contribution Patterns in Eclipse Foundation Project Source Code

Quinn C. Taylor, Jonathan L. Krein,
Alexander C. MacLean, and Charles D. Knutson

SEQuOIA Lab — Brigham Young University — Provo, Utah, USA
{quinntaylor,jonathankrein,amaclean}@byu.net, knutson@cs.byu.edu

Abstract. Collaborative development is a key tenet of open source software, but if not properly understood and managed, it can become a liability. We examine author contribution data for the newest revision of 251,633 Java source files in 592 Eclipse projects. We use this observational data to analyze collaboration patterns within files, and to explore relationships between file size, author count, and code authorship. We calculate author entropy to characterize the contributions of multiple authors to a given file, with an eye toward understanding the degree of collaboration and the most common interaction patterns.

1 Introduction

Software development is an inherently complex activity, often involving a high degree of collaboration between multiple individuals and teams, particularly when creating large software systems. Interactions between individual contributors can affect virtually all aspects of software development, including design, implementation, testing, maintenance, complexity, and quality.

Collaboration involves cooperation, communication, and coordination, and generally implies some governing organizational structure. The organization has an effect on the structure of the software being developed, as per "Conway's Law" [4]; presumably applying equally to proprietary and open source software. Brooks noted that potential communication channels increase as the square of the number of contributors [2]. Thus, there is benefit to understanding and managing collaboration so it does not become a liability.

Analyzing collaboration data can help explain how people work together to develop software. Studies by Bird [1], Ducheneaut [6], Gilbert [7], Mockus [12], Dinh-Trong [5], and others have examined interactions between open source developers by correlating communication records (such as email) with source code changes. Such approaches can expose patterns which reinforce contributor roles and module boundaries, but may not be feasible for all projects (particularly if email archives are unavailable) and can be difficult to compare or aggregate across disparate projects.

In addition to examining collaboration across projects and modules, there is value in understanding how contributors collaborate *within* files. Having a sense

S.A. Hissam et al. (Eds.): OSS 2011, IFIP AICT 365, pp. 269–281, 2011.

of what constitutes "typical" collaboration for a project can provide valuable context. For example, if most files in a project have one or two authors, a file with 10 authors may merit additional scrutiny. In open source projects, unorganized and organic contributions may be evidence of the bazaar rather than the cathedral [13]. In any case, simply knowing can help set expectations.

This paper both replicates and extends earlier results [15]. Our research goals center around detecting, characterizing, and understanding patterns of collaboration within source code files. Our primary research questions are:

1. *How often do* n *authors contribute to a given file?*
   We anticipate that most files have a single author, and as the number of authors increases, the count of files with that many authors will decrease.
2. *Is there a higher degree of collaboration in small or large files?*
   We anticipate that there will be a positive correlation between file size and author count, partially because larger files have more code, and the potential for more distinct functionalities and individual responsibilities.
3. *Do files contain similar proportions of contributions from each author, or is there a dominant author who is the clear "owner" of a given file, and if so, how dominant is that author?*
   We anticipate that most source files will have one author who contributes significantly more code than any other single author, and that this author's dominance will be inversely related to the number of contributing authors.
4. *Is there a uniform or uneven distribution of collaboration across projects?*
   We anticipate that there will be a few "core" projects which are highly collaborative, and many ancillary projects which are less collaborative.

2 Methodology

We conducted an observational study on existing Eclipse projects by extracting author attribution data for Java source code files from git repositories. In this section we describe the process we used to select and obtain the data.

2.1 Project and File Selection

We chose to analyze Eclipse Foundation projects for several reasons, including:

– the number and variety of Eclipse-related projects,
– use of easily-recognizable programming languages,
– the ability to locally clone remote git repositories,
– a track record of sustained development activity,
– the existence of corporate-sponsored open source development projects.

We selected Java source files for our analysis, since over 92% of the source files in the repositories are Java, and Eclipse is so closely aligned with Java. We mined data from 251,633 files in 592 projects. We included auto-generated code in our analysis, since the inclusion of such files allows us to accurately characterize the state of the project to which they belong.

2.2   Extraction and Calculation

The first step in calculating author collaboration is to count how many authors have contributed to a file and the number of lines contributed by each one. Summarizing raw line counts with a single representative statistic per file allows for detailed statistical analysis of collaboration trends. In this paper, we use: (1) the percentage of lines attributed to the most dominant author in each file, and (2) author entropy (see Section 3 for details). These numbers can help characterize some aspects of author contribution patterns.

We created a bash script to locally clone each remote git repository and use 'git blame' to count the number of lines attributed to each author for each matching file. For each file in a project, the file path and line counts attributed to each author were recorded.

We then wrote a simple CLI tool to process this data and calculate the percentage of lines written by each author. Author entropy for each file was calculated using Equation 1. We also normalized entropy by dividing by the maximum possible entropy for each file, shown in Equation 2.

2.3   Limitations of the Data

We draw data only from git, a source control management (SCM) system that preserves snapshots of file state over time. We do not consider other collaboration mechanisms, such as email archives, forums, etc., although this could be a very interesting extension of this work.

It it important to note that the SCM record of who "owns" a line of code only identifies the individual who committed the most recent change affecting that line. It does not guarantee that the contributor actually conceived of, wrote, or even understands the code. By itself, it also does not tell us the genesis of a line; it could be new, a minor tweak, or a formatting change.

Because we consider only the latest revision of each file, this data cannot be used to make any inferences about collaboration over time. Without historical data, we can see the result of collaboration, but not the nature of the evolution of such collaboration.

Lastly, because we record author counts but not relative ordering of contributions from various authors, this data does not fully capture or express the amount of disorder. Because only percentages by each author are considered, the data makes no distinction between files with orderly, segregated blocks of contributions and files in which authors' contributions are all mixed together.

3   Author Entropy

Author entropy is a summary statistic that quantifies the mixture of authors' contributions to a file. Contribution percentages are weighted using logarithms and summed; the resulting value conveys more information about the distribution than a simple average, and can expose interesting authorship patterns more readily than raw line counts. Taylor et al [15] introduced author entropy and examined distributions in a proof-of-concept study with SourceForge data. A follow-on paper [3] examined author entropy in GNOME application source.

Entropy originated in the field of thermodynamics, which defines it as the disorder or randomness of molecules in a system. Entropy has also been defined in terms of probability theory and used in the fields of information theory [14] and machine learning [11].

We apply entropy as a measure of collaboration between individual contributors. Specifically, we consider entropy of source code by counting the number of lines attributed to each author. This definition of entropy allows us to quantify the mixture of author contributions to a file.

3.1   Calculating Entropy

Entropy formulae are nearly identical across domains, and generally vary only in symbolic representation and constant multipliers. We use a formulation very similar to that used in machine learning.

If F is a file, A is the number of authors, and p_i is the proportion of the text attributed to author i, then the entropy of the file is defined as:

$$E(F) \equiv - \sum_{i=1}^{A} (p_i \cdot log_2 \ p_i) \qquad (1)$$

$E(F)$ is maximized when all authors contributed equal proportions of text in a file ($\forall \ i, p_i = \frac{1}{A}$). The upper limit of $E(F)$ is a function of A:

$$E_{max}(F) \equiv log_2 \ A \qquad (2)$$

We use log_2 for historical reasons tied to information theory (essentially, calculating the number of bits required to encode information). Although any logarithmic base would suffice, it is convenient that using log_2 results in entropy values in the range (0,1] for a binary classification.

3.2   Normalizing Entropy

Because the maximum possible entropy for a file is a function of the number of authors, intuitive understanding of entropy can be difficult. For example, an entropy value of 1.0 is the maximum possible for a file with 2 authors, but comparatively low for a file with 10 authors. Dividing E by E_{max} produces a normalized value in the range (0,1] which represents the percentage of maximum entropy. Normalized entropy can be easier to understand, and in some cases more appropriate for comparisons between disparate files.

4   Interpreting Collaboration

A high degree of collaboration within a given source file is not inherently good or bad; as with any metric, context is key. Without knowledge about additional factors such as a project's state, organization, and development conditions, interpreting collaboration is purely speculative. To illustrate this point, we list below a variety of factors that could influence author entropy.

Low entropy could result from factors as varied as:

- Well-architected and modular software.
- Excellent communication and coordination.
- Lack of involvement from potential contributors.
- A disciplined team in which each person "owns" a module.
- A gatekeeper who gets credit for code written by others.
- Code that few people understand.
- Code that was reformatted and old attributions lost.
- Code with exhaustive unit tests, known to be correct.
- Code with negligible unit tests and unknown defects.
- Auto-generated code that no human actually "wrote."
- Critical code that few people are allowed to modify.
- Mature code with little or no need for maintenance.
- Stale code that isn't touched, even if it needs fixing.
- Dead code which is no longer used or modified.

High entropy could result from factors as varied as:

- Code with high coupling or many inter-dependencies.
- Unrelated code entities being stored in a single file.
- Adding manpower to a late project (Brooks' law).
- Extremely buggy code that is constantly patched.
- Extremely stable code that is well-maintained.
- Enhancements or fixes that touch several files.
- Contributors joining or leaving a project team.
- Actively evolving code or refactoring activity.
- Miscommunication or lack of clear direction.
- Healthy collaboration between contributors.
- Overlapping responsibilities of contributors.
- Agile development or team programming.
- Potential for integration-stage problems.
- Continuous integration testing and fixes.

Such a menagerie of disparate factors is not a flaw in the metric itself, but rather suggests that any metric can easily be misinterpreted without proper context. For example, a file with high entropy written by several experts is likely of higher quality than a file written by one novice author. Two files may have similar entropies despite a large size difference. A recent contributor may understand a file better than the original author who wrote it years ago. Correlating author entropy with other metrics and observations can help distinguish between "good" and "bad" entropy and provide valuable new insights.

Author entropy cannot directly indicate other attributes of the source code. For example, file length is obscured since files of different size but equal proportions of contribution have the same entropy. Entropy also does not reflect quality or the relative importance of contributions, such as new functionality, bug fixes, comments, whitespace, or formatting. Although different entropy calculation techniques could opt to account for such factors, there is no way to deduce the weighting of such factors from a single number.

5   Results

The line count of the source files we examined ranged from 1 to 228,089, with a median of 89. The extreme right-tail skew (97.5% have 1,000 lines or fewer, 92.5% have 500 or fewer) suggests that the data may have an exponential distribution. Plotting the data with a log_{10} transformation produces a histogram (Figure 1) that closely resembles a normal distribution. A Q-Q plot showed that the population fits a log-normal distribution quite well, although the long tail caused higher residual deviations in the upper range. We also examined the files with 10 lines or fewer and found that nearly all of them were related to unit tests; several projects have extensive tests with Java source files that specify in and out conditions, but have little or no code. Excluding these left-tail outliers greatly improved the fit of the Q-Q plot in the low range.

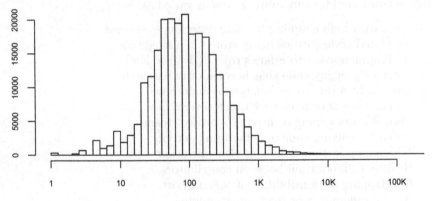

Fig. 1. Frequency of file sizes (in number of lines)

To answer our first research question, we plotted the frequencies of files with n authors. The resulting histogram was an exponential decay curve, and when plotted with a logarithmic scale, a near-perfect log-linear decay is evident (see Figure 2). This confirms our hypothesis that most files have a single author, and that the number of files with n authors decreases as n increases. It is also strikingly similar to Lotka's Law [10], which states that the number of authors who contribute n scientific publications is about $1/n^a$ of those with one publication, where a is nearly always 2. Lotka's law predicts about 60% of authors publish only once; in our data, 58.22% of the files have one author.

To answer our second research question, we plotted file size distributions grouped by author count (see Figure 3). The log-linear increase in average file size as the number of authors increases confirms our hypothesis that, on average,

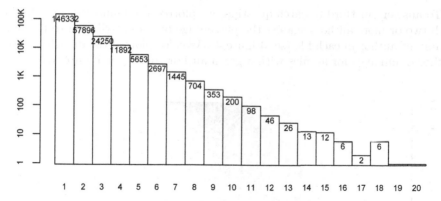

Fig. 2. Frequency of number of authors contributing to a given file

there is more collaboration (i.e., more authors) in large files. However, we must note that there is a degree of uncertainty due to the decreasing sample sizes for higher author counts and the extreme outliers for lower author counts.

We augmented Figure 3 with two additional data series: (1) the average number of lines in a file, and (2) the average number of lines contributed *per author* to a file. Note that there is a pronounced dip between 1 and 10 authors, but a fairly consistent average throughout. Although evaluating the causes and ramifications of this trend are beyond the scope of this paper, we find this to be an interesting topic for future work.

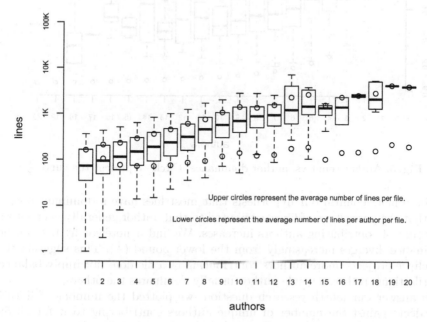

Fig. 3. Author count vs. file size (in number of lines)

To answer our third research question, we plotted the number of lines in files with two or more authors against the percentage of lines attributed to the most dominant author in each file (see Figure 4). We also plotted the distributions of author dominance for all files with a given author count (see Figure 5).

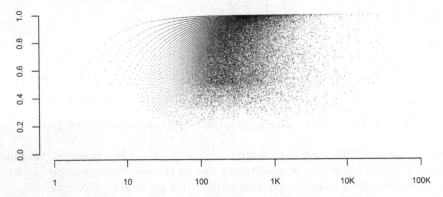

Fig. 4. Line count vs. percent written by dominant author for files with 2+ authors

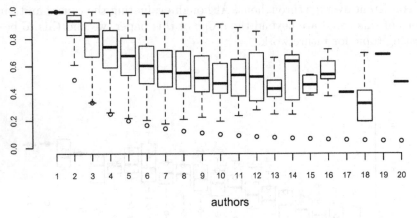

Fig. 5. Author count vs. author dominance. Circles represent the curve $\frac{1}{x}$.

These plots confirm our hypothesis that most files have a dominant author, and that the percentage of lines attributed to that author generally decreases as the count of contributing authors increases. We find it noteworthy that author dominance diverges increasingly from the lower bound ($\frac{1}{x}$). This suggests that in Eclipse projects, more authors contributing to a file does not imply balanced contributions; rather, a single author usually dominates the others.

To answer our fourth research question, we plotted the number of lines in a project against the number of unique authors contributing to it for all 592 projects (see Figure 6). Over 83% of the projects have 10 or fewer unique authors, and some significant outliers have much larger numbers of authors.

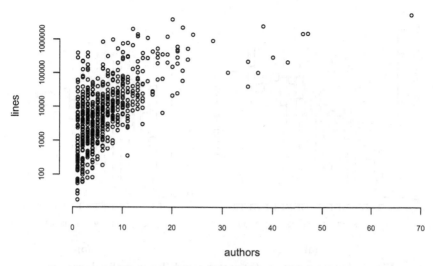

Fig. 6. Author count vs. total number of lines for all 592 projects

We also manually examined the 211 files with 11 or more authors. Nearly all of these files came from a handful of projects, all of which were among the top 25 projects with the most authors. These projects include:

- org.eclipse.ui.workbench (Eclipse IDE interface)
- org.eclipse.jdt.core (Java Development Tools)
- org.eclipse.cdt (C/C++ Development Tooling)
- org.eclipse.pdt (PHP Development Tools)
- org.eclipse.birt.report (Business Intelligence and Reporting Tools)
- org.eclipse.jface (UI application framework/toolkit based on SWT)

The nature and role of these highly-collaborative projects confirms our hypothesis that collaboration is not distributed uniformly, but is concentrated in a few core projects. This phenomenon is also related to our second research question, about the relationship between collaboration and file size.

5.1 Additional Questions

In addition to our primary research questions, we also replicated some results from prior related work to verify whether the assertions made therein still hold for broader data. These results are related to distributions of author entropy (see Section 3) over varying file sizes and author counts.

In [15] we found a positive relationship between author count and entropy (entropy rises as the number of authors increases). We found the same trend in Eclipse source code, although it breaks down somewhat for 11 or more authors due to sparseness of data (see Figure 7).

Although the entropy metric is inherently biased toward higher values (particularly when there are more authors), any file can have low entropy when one

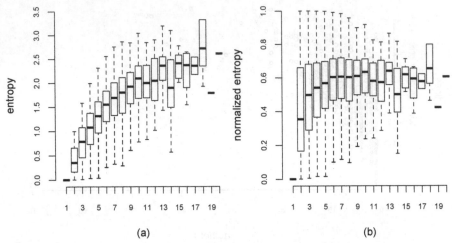

Fig. 7. Author count vs. (a) entropy and (b) normalized entropy

of the authors is extremely dominant. However, because the maximum possible entropy for a given file is a function of the number of authors, it can be difficult to compare entropies for files with different number of authors. For this reason, we use normalized entropy, which always falls in the range [0,1] regardless of author count, and thus represents the percentage of maximum possible entropy.

Interestingly, the data exhibits a trend previously observed [15] in a very constrained set of SourceForge data: distributions of normalized entropy tend to center around 0.6 (or 60% of maximum possible entropy) as the author count increases. Even as the data becomes more sparse for higher author counts, the distributions tend to converge on the same range.

Casebolt [3] examined two-author source code files and observed an inverse relationship between file size and entropy (small files have high entropy and vice versa). A similar pattern occurs in our data, as shown in Figure 8(a). Unfortunately, it is impossible to discern how many data points share the same location. The task is even more hopeless when all files (not just those with two authors) are included in the same plot, as in Figure 8(b). To better understand the distribution and density of these data, we borrow a tool used by Krein [9] to visualize language entropy: 3D height maps. This technique generates an image in which densely-populated regions appear as elevated terrain (see Figure 9).

Figure 9 is an adaptation of both Figure 8(b) (adding height) and Figure 1 (adding depth). Starting from the back/left, the curves represent files in which one author "owns" all lines but one, two, etc. The disproportionate distribution of files on the furthest curves suggests that one- and two-line edits are probably extremely common occurrences in the projects we examined. This may be a manifestation of many small bug fixes, refactorings, interface changes, etc.

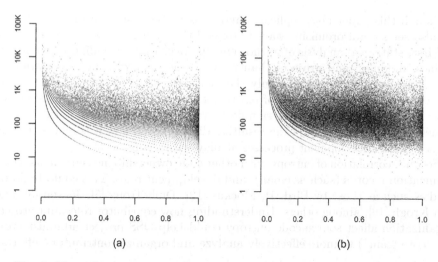

Fig. 8. Normalized entropy vs. line count for (a) two authors and (b) all files

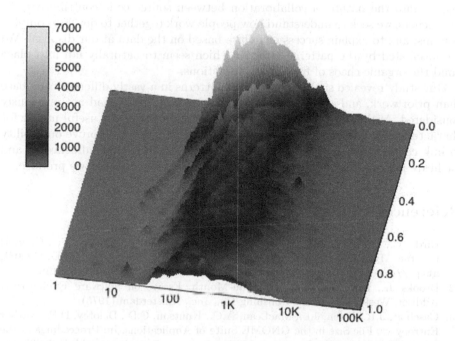

Fig. 9. Height map of line count vs. normalized entropy (same data as Figure 8b)

6   Future Work

Although this paper both replicates and adds to the results of prior work [15], we also see several promising ways to extend this research.

First, statistical analysis of author entropy over time, including how entropy of files, modules, and projects change over time, and why. One limitation of this paper is that we examine only the most recent version of each file; we do not consider previous versions of existing files, or files which once existed but have since been deleted. We see significant value in understanding not only code ownership, but the degree of the resulting disorder, and how it is related to and caused by the development processes at play within a project.

Second, correlation of entropy with other code ownership measurements, communication records (such as email), and development roles. This could build on studies such as those by Bird [1], Mockus [12], Dinh-Trong [5], Jensen [8], and von Krogh [16], among others. Understanding how conributor roles and project organization affect source code entropy could help OSS project administrators (or "core team") to more effectively analyze and organize contributors' efforts.

7   Conclusion

We discovered that author attribution data for source code files can provide insight into the nature of collaboration between source code contributors. As researchers, we seek to understand how people work together to develop complex systems, and to explain success or failure based on the data at our disposal. We are fascinated by the patterns of order which seem to naturally fall into place amid the organic chaos of free-form interactions.

Our study revealed similar authorship patterns in a vastly different code base than prior work, and suggested interesting new patterns we had not previously considered. Author entropy continues to be an interesting and useful metric for characterizing contributor interactions. Future research will improve our ability to link collaborative activity with the underlying factors that influence it, and facilitate improvements that enhance the quality of the software we produce.

References

1. Bird, C., Pattison, D., D'Souza, R., Filkov, V., Devanbu, P.: Chapels in the Bazaar? Latent Social Structure in OSS. In: FSE (2008), http://citeseerx.ist.psu.edu/viewdoc/summary?doi=10.1.1.138.2158
2. Brooks Jr., F.P.: The Mythical Man-Month: Essays on Software Engineering. Addison-Wesley Longman Publishing Co., Inc., Amsterdam (1975)
3. Casebolt, J.R., Krein, J.L., MacLean, A.C., Knutson, C.D., Delorey, D.P.: Author Entropy vs. File Size in the GNOME Suite of Applications. In: Proceedings of the 6th IEEE Working Conference on Mining Software Repositories (MSR 2009), pp. 91–94 (May 2009)
4. Conway, M.E.: Do Committees Invent? Datamation 14(4), 28–31 (1968)

5. Dinh-Trong, T.T., Bieman, J.M.: The FreeBSD Project: A Replication Case Stdy of Open Source Development. IEEE Transactions of Software Engineering 31(6), 481–494 (2005)
6. Ducheneaut, N.: Socialization in an Open Source Software Community: A Socio-Technical Analysis. In: Computer Supported Cooperative Work, vol. 14, pp. 323–368 (2005)
7. Gilbert, E., Karahalios, K.: CodeSaw: A social visualization of distributed software development. In: Baranauskas, C., Abascal, J., Barbosa, S.D.J. (eds.) INTERACT 2007. LNCS, vol. 4663, pp. 303–316. Springer, Heidelberg (2007)
8. Jensen, C., Scacchi, W.: Role Migration and Advancement Processes in OSSD Projects: A Comparative Case Study. In: 29th International Conference on Software Engineering, ICSE 2007 Minneapolis, MN, pp. 364–374 (May 2007)
9. Krein, J.L., MacLean, A.C., Knutson, C.D., Delorey, D.P., Eggett, D.L.: Impact of Programming Language Fragmentation on Developer Productivity: a SourceForge Empirical Study. International Journal of Open Source Software and Processes (IJOSSP) 2(2), 41–61 (2010)
10. Lotka, A.J.: The frequency distribution of scientific productivity. Journal of the Washington Academy of Sciences 16(12), 317–324 (1926)
11. Mitchell, T.M.: Machine Learning, pp. 55–57. McGraw-Hill, New York (1997)
12. Mockus, A., Fielding, R.T., Herbsleb, J.D.: Two Case Studies of Open Source Software Development: Apache and Mozilla. ACM Transactions on Software Engineering and Methodology 11(3), 309–346 (2002)
13. Raymond, E.S.: The Cathedral and the Bazaar: Musings on Linux and Open Source by an Accidental Revolutionary. O'Reilly and Associates, Inc., Sebastopol (2001)
14. Shannon, C.E.: A Mathematical Theory of Communication. The Bell System Technical Journal 27, 379–423 (1948)
15. Taylor, Q.C., Stevenson, J.E., Delorey, D.P., Knutson, C.D.: Author Entropy: A Metric for Characterization of Software Authorship Patterns. In: Proceedings of the 3rd International Workshop on Public Data about Software Development (WoPDaSD 2008), Milan, Italy (September 2008)
16. von Krogh, G., Spacth, S., Lakhani, K.R.: Community, Joining, and Specialization in Open Source Software Innovation: A Case Study (Open Source Software Development). Research Policy 32(7), 1217–1241 (2003)

Cliff Walls: An Analysis of Monolithic Commits Using Latent Dirichlet Allocation

Landon J. Pratt, Alexander C. MacLean, Charles D. Knutson,
and Eric K. Ringger

Computer Science Department, Brigham Young University, Provo, Utah
{landonjpratt,amaclean,knutson,ringger}@byu.edu

Abstract. Artifact-based research provides a mechanism whereby researchers may study the creation of software yet avoid many of the difficulties of direct observation and experimentation. However, there are still many challenges that can affect the quality of artifact-based studies, especially those studies examining software evolution. Large commits, which we refer to as "Cliff Walls," are one significant threat to studies of software evolution because they do not appear to represent incremental development. We used Latent Dirichlet Allocation to extract topics from over 2 million commit log messages, taken from 10,000 SourceForge projects. The topics generated through this method were then analyzed to determine the causes of over 9,000 of the largest commits. We found that branch merges, code imports, and auto-generated documentation were significant causes of large commits. We also found that corrective maintenance tasks, such as bug fixes, did not play a significant role in the creation of large commits.

1 Introduction

Artifact-based software engineering research may in some respects be compared to archaeology, a field that has been defined as "the study of the human past, through the material traces of it that have survived"[2]. Much like archaeologists, empirical software engineering researchers often seek to understand people. The software engineering researcher, while not isolated from a target population by eons, faces other obstacles that often make direct observation impossible. Many organizations are loath to allow researchers through their gates, in an effort to protect trade secrets or merely to hide shortcomings. Even in cases where researchers are allowed to directly observe engineers building software, the Hawthorne effect threatens the validity of such observations. Time investment and organizational complexity are also issues that pose problems in software engineering research. Direct observation requires significant time investment, making it impossible for a single researcher to observe everything that takes place within a given software organization.

As a result of these barriers, software researchers, like their archaeologist counterparts, take advantage of artifacts—work products left behind in software

S.A. Hissam et al. (Eds.): OSS 2011, IFIP AICT 365, pp. 282–298, 2011.

project burial grounds. Artifacts are collected after the fact, minimizing the confounding influence of the presence of a researcher. Artifacts also help researchers deal with the requirements of studying complex organizations. By leveraging artifacts of the software process, researchers are able to study thousands of pieces of software in a relatively short period of time, an otherwise impossible task.

The open source movement is particularly important to software engineering research, since project artifacts, such as source code, revision control histories, and message boards, are openly available to software archaeologists. This makes open source software an ideal target for researchers with a desire to understand how software is built.

1.1 Threats to Artifact-Based Research

Unfortunately, the study of artifacts in software engineering is not all sunshine and double rainbows; serious challenges threaten the results of artifact-based research involving open source software projects (such as those hosted on SourceForge). Since artifact data is examined separate from its original development context, identifying the development artifacts actually recorded in the data can be difficult. It is challenging enough to ensure that measurements taken for a specific purpose actually measure what they claim to measure [5]. It is all the more difficult (and necessary), therefore, to validate artifact data, which is generally collected without a targeted purpose.

Many phenomena can easily be overlooked by software archaeologists, such as auto-generated code, the presence of non-source code files in version control, and the side effects of branching and merging. Understanding the limitations of artifact data represents an important step toward validating the results of numerous studies (for example, [3, 6, 9, 13, 18, 20, 22]). Despite on-going efforts to identify and mitigate the limitations of artifact-based research, new threats are constantly emerging.

The focus of this paper is one such threat—the presence of monolithic commits in open source repositories. A close look at the version control history of many projects in SourceForge reveals some worrisome anomalies. Massive commits, which we refer to as "Cliff Walls," appear with alarming frequency. One investigation of Cliff Walls in SourceForge found that out of almost 10,000 projects studied, half contained a single commit that represented 30% or more of the entire project size [16]. These Cliff Walls indicate periods of unexplained acceleration in development, threatening some common assumptions of artifact-based research, especially for studies of project evolution. Cliff Walls thwart attempts to tie authors to contributions, mask true development history, and taint studies of project evolution. We must better understand Cliff Walls if we are to paint an accurate picture of software evolution through the use of artifacts.

2 Cliff Walls

In [16] we introduced the concept of "Cliff Walls" as unexpected increases in version control activity over short periods of time (see Fig. 1). In the most extreme cases, Cliff Walls represent millions of lines of source code contributed to the repository in less than a day. Such activity is problematic for researchers, especially those investigating the evolution of software, because such sudden surges of source code commits cannot possibly be the result of common incremental development activities. Even a team of "supercoders" would be hard-pressed to produce such massive quantities of code in a short period of time, all the more impossible for the single "author" to which a version control commit is attributed. We are forced to ask what these "Cliff Walls" truly represent.

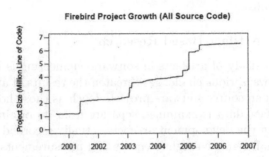

Fig. 1. Cliff Walls in the Firebird Project

2.1 Definitions

When examining contributions to version control systems, it is logical to discuss them in the context of "commits." However, some researchers have defined "commit size" in terms of the number of files changed in a commit [12, 11, 10], while others treat the size of a commit as a function of the number of LOC added, removed or modified [1].

Within the context of this study, a *commit* refers to the set of files and/or changes to files submitted simultaneously to a version control system by an individual developer. Since we are primarily concerned with the growth of projects over time, *Commit size* is defined as the total number of lines added or removed from all source code files in a given commit, which allows us to perceive code growth (or shrinkage) within a project. For example, if an existing line in one file is modified, and no other changes are made, the size of the resulting commit would be zero LOC. In this study, we include code comments in the measurement of commit size, allowing us to detect code growth attributable to the insertion of comments within source files.

We use the term *Cliff Wall* to describe any single commit with size greater than 10,000 lines of code (LOC). We believe this threshold to be sufficiently high that it avoids capturing most incremental development contributions for individual authors. While it is possible that an extremely productive developer could produce 10,000 lines of code in a period of days or weeks, it is unlikely. Additionally, the methods employed in this study should allow us to identify instances where the threshold fails.

2.2   Commit Taxonomies

The ability to distinguish between different types of cliff walls is critical for many artifact-based studies. For example, a researcher attempting to measure developer productivity on a project would likely want to be able to distinguish between large commits caused by auto-generated code and branch merges, as they must be handled differently when calculating code contributions.

In [12], the authors present a taxonomy used to manually categorize large commits on a number of open source software projects. Hattori and Lanza also present a method for the automatic classification of commits of all sizes in open source software projects [10]. The taxonomies developed in these studies focus on classifying the type of change made (for example, development vs. maintenance). Both these studies concluded that corrective maintenance activities, such as bug fixes, are rarely the topic of large commits. These studies also found that a significant portion of large commits were dedicated to "implementation activities." The authors consider all source code files added to the repository to be the result of implementation activities.

In [1], the authors categorize commits from open source projects into three groups, using a rough heuristic based on commit size. Their study makes no attempt to analyze commit log messages. Rather, they divide each commit into one of three groups, based on the commit size: 1) single individual developer contributions, 2) aggregate developer contributions, and 3) component or repository refactoring or consolidations.

In our study, we seek to identify specific behaviors that play a significant role in the creation of monolithic commits. In the next section, we discuss the methods that we will apply in our search for the causes of Cliff Walls.

3   Latent Dirichlet Allocation

Latent Dirichlet Allocation (LDA) is an unsupervised, hierarchical Bayesian topic model for analyzing natural language document content that clusters the terms in a corpus into topics [4]. In LDA, the model represents the assumption that documents are a mixture of these topics and that each topic is a probability distribution over words. In short, LDA allows us, without any knowledge engineering, to discover the topical content of a large collection of documents in an unsupervised fashion. Given a corpus of documents such as our commit log messages, inference techniques such as Gibbs sampling or Variational Bayes can be used to find a set of topics that occur in that corpus, along with a topic assignment to each word token in the corpus.

A common technique for visualizing the topics found using LDA is to list the top n most probable terms in each topic. For example, one of the topics we found using LDA that concerns bug correction, consisted of the following terms:

> fix bug fixes crash sf problems small quick
> closes blah deleting weapon decoder lost
> hang weapons delphi noted led

When brevity is required, it is also common to refer to a topic by just the first two or three most probable terms, sometimes in hyphenated form. Using this notation, the above topic could be referred to as the "fix-bug" topic. Both of these methods for indicating a particular topic (word list and first words) are used throughout this paper.

Because LDA also assigns a topic to each individual token within a commit message, a single commit can, and usually does, contain many topics. This is one benefit of using LDA: each "document" may belong to multiple classes. Such is not the case with many of the supervised classification methods, which typically assign each document to a single class. In the case of commit log messages, allowing for multiple topics in a single message allows us to conduct a more fined-grained analysis.

The approach taken by LDA is an attractive alternative to other classification and clustering methods. Supervised classification methods require a training set of "tagged" data, for which the appropriate class for each instance has previously been specified. In many situations this tagged data does not exist, requiring the researcher to manually assign tags to some portion of the data. This is a tedious, time-consuming, and potentially biased process which can be avoided through the use of unsupervised methods such as LDA. Unsupervised methods may also compensate for a measure of "short-sightedness" by the researcher. In supervised methods, since the classes must be predefined by the researcher, it is possible to degrade the usefulness of the model through the omission of important classes, or the inclusion of irrelevant classes. Unsupervised methods avoid some of these errors since no predefined clusters are specified, allowing the data to better "speak for itself." For our analysis, we used the open source tool "MALLET" which includes an implementation of LDA [17].

4   Methods

Our data set consists of the version control logs of almost 10,000 projects from SourceForge, acquired in late 2006. This data set has been previously used in a number of studies [6, 7, 8, 13, 14, 15, 16]. The logs for all projects in our data set were extracted from the CVS version control system. In creating the data set, the original authors filtered projects based on development stage; only projects labeled as Production/Stable or Maintenance were included in the data set. For further description of the data, see [6]. In calculating commit size, we excluded non-source code files, based on file extension. Over 30 "languages" were represented in the final data set, including Java, C, C++, PHP, HTML, SQL, Perl and Python.

Because CVS logs do not maintain any concept of an atomic multi-file "commit" it was necessary to infer individual commits. We utilized the "Sliding Time Window" method introduced by Zimmerman and Weißgerber [23]. This resulted in a set of almost 2.5 million individual commits, extracted from over 26 million file revisions. Applying our pre-defined threshold of 10,000 LOC yielded over 10,000 Cliff Walls. We also found that a number of the commits contained log messages that were uninformative. Commits with empty log messages or with **"empty log message"** were removed from the data to prevent degradation in the quality of topics identified. The resulting set contained 2,333,675 commits, with 9,615 Cliff Walls. We later removed other uninformative commits (see discussion in Sec. 6), ultimately resulting in the exclusion of 6.6% of commits in our data set due to log messages that conveyed no information about the development activities they represent. A disproportionate number of the commits removed were Cliff Walls (an ultimate exclusion of 14.8% of all Cliff Walls). Additionally, very common English adverbs, conjunctions, pronouns and prepositions belonging to our "stop-word" list were removed from the commit messages in order to ensure the identification of meaningful topics.

The LDA algorithm, as implemented in MALLET, requires three input parameters: the number of topics to produce in its analysis, the number of iterations, and the length of the burn-in period. In our study, we elected to identify 150 topics with MALLET. The authors Hindle and German identified 28 "types of change" for the commits classified as a part of their taxonomy [12]. Hattori and Lanza, in their study of commit messages, identified 64 "keywords" that were used to classify commits [10]. These prior results gave us reason to believe that 150 topics would be a sufficient number to capture the motivations behind the commits in our data set, with an appropriate level of detail.[1]

As one of the steps to understanding the Cliff Wall phenomenon, we compare the most prevalent topics found in Cliff Wall commits to those found in the entire body of commits. Instead of running LDA separately on the two subsets of our data, we run LDA once on all of the data and then filter the results to gain a picture of Cliff Walls in contrast to All Commits. This approach ensures that the topics found are consistent across both groups, which helps yield a meaningful comparison.

5 Analysis and Discussion

Figure 2 provides a first glance at some of the variation exhibited by Cliff Walls. In these graphs, each horizontal bar represents one of the 150 topics generated. The thickness of each bar represents the proportion of tokens in the entire corpus of commits that were assigned to that topic. Commit log messages are fairly evenly distributed over the topics for the general population of commits. However, a small number of topics are considerably more prevalent in the large commits. Tables 1 and 2 list the 15 most prevalent topics for all commits and Cliff Wall commits.

---

[1] The other two parameters, number of iterations and length of burn-in period, are required by the Gibbs Sampling inference method employed by MALLET. We refer the reader to [21] for a description of LDA as it is implemented within MALLET, including a description of Gibbs Sampling. For these two parameters we used the default values provided by MALLET; 1,000 iterations with 200 dedicated to burn-in. Further work should investigate the possibility of more appropriate values for all three parameters, as discussed in Sec. 6.

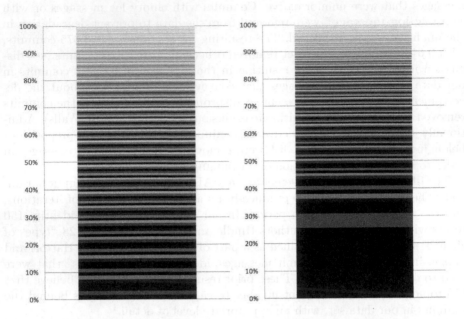

Fig. 2. Topic distribution for All Commits (left) and Cliff Wall Commits (right)

Similarly, the tag clouds in Figs. 3, 4, and 5 begin to give us an idea of the most common topics for our two groups of interest.[2] Each tag in the cloud represents a topic that has been summarized using the "first words" method described in Sec. 3. Like a stratum from Fig. 2, each tag is sized based on the proportion of tokens belonging to that topic. Thus, the largest tag in Fig. 4, "initial-import," is also the largest stratum in the bar chart for Cliff Walls. Tag position and color do not convey any additional information. Figure 5 is an alternate view of Fig. 4, with the dominant "initial-import" topic removed to improve readability. These images provide an overall view of the topics and their proportions for the two groups of interest. We next discuss some of the most prevalent topics and their interpretations.

One of the goals of this paper is to compare the topics produced by LDA with previously hypothesized causes of Cliff Walls. We examine each of these causes to see if LDA is able to identify them as prominent features of large commits. We also examine some of the causes for Cliff Walls that were previously over-looked, but were consequently suggested by LDA. In pursuit of these goals, we have found two views of the data that provide insight into the causes of Cliff Walls: Overall Topic Proportion and Topic Relative Rank. We discuss these in the next two subsections.

---

[2] All tag cloud images were generated using Wordle (www.wordle.net).

Fig. 3. Topic Tag Cloud: All Commits

Fig. 4. Topic Tag Cloud: Cliff Walls

Fig. 5. Topic Tag Cloud: Cliff Walls ("initial-import" excluded)

5.1   Overall Topic Proportion

Tables 1 and 2 display the 15 most prevalent topics in the Cliff Wall and All Commits groups, as determined by proportion of tokens belonging to that topic. For example, the topic "version-release" in Tbl. 1 has a proportion of 1.75%, suggesting that 1.75% of all the words in all of the commit messages in our data set were assigned to this topic by the LDA algorithm. In other words, these tables list the topics most frequently discussed in commit log messages belonging to our two groups of interest. We refer back to these tables frequently throughout this section as we discuss the various causes of Cliff Walls.

5.2   Topic Relative Rank

Additional insight into Cliff Walls can be gained through the use of another simple metric. Within each of our two groups, "All Commits" and "Cliff Walls," each topic can be ranked based on its proportion within the group. The difference between a topic's ranking in the two groups is a good indicator of the prevalence of a given topic relative to other topics.

For example, in Tbl. 1 we see that "initial-import" is the 5th ranked topic. In Table 2, the same topic is ranked 1st, for a rank difference of +4. In contrast, as we see from Tbl. 3, the topic "cvs-sync" holds ranks of 101 and 4, resulting in a rank difference of +97. In essence this means that, relative to other topics, "cvs-sync" is discussed more frequently within Cliff Wall commits than it is for the general population of commits.

It is important to note that the difference between topic ranks is not synonymous with a similar difference in proportion. The difference between proportions for the "initial-import" topic is an approximately 25% increase for the Cliff Walls group. This is a very large change in proportion which results in a relatively small difference in rank. It is even possible, given the distinct distributions of our two groups (see Fig. 2) that a *negative* change in proportion could still result in a *positive* rank difference.

Table 1. Top 15 Topics for All Commits

#	Proportion	Key Terms				
1	1.75%	version	release	updated	update	final
2	1.43%	file	branch	initially	added	java
3	1.42%	fixes	minor	small	cleanups	updates
4	1.38%	data	minor	updates	fixes	bugfixes
5	1.38%	initial	import	commit	checkin	revision
6	1.36%	added comments	comment	documentation		javadoc
7	1.29%	fixed	bug	bugs	incorrect	couple
8	1.28%	added	support	info	extended	basic
9	1.28%	error	message	errors	handling	checking
10	1.26%	removed	code	template	commented	unnecessary
11	1.25%	fix	bug	fixing	small	bugs
12	1.18%	fix	typo	corrected	correct	errors
13	1.17%	fixed	bug	wrong	crash	introduced
14	1.17%	page	link	updated	links	url
15	1.14%	code	cleanup	source	clean	cleaned

Table 2. Top 15 Topics for Cliff Walls

#	Proportion	Key Terms				
1	25.74%	initial	import	commit	checkin	revision
2	5.11%	version	release	updated	update	final
3	2.30%	removed	deprecated	sources	constant	imported
4	1.63%	cvs	sync	tabs	real	converted
5	1.60%	error	message	errors	handling	checking
6	1.58%	message	project	messages	error	testing
7	1.51%	merged	merge	head	merging	main
8	1.47%	code	cleanup	source	clean	cleaned
9	1.29%	files	added	dialogs	library	directories
10	1.21%	directory	moved	common	dir	structure
11	1.17%	xml	updated	api	latest	version
12	0.98%	update	added	format	updating	creation
13	0.93%	script	makefile	install	configure	sh
14	0.89%	added	comments	comment	documentation	javadoc
15	0.89%	double	beta	v1	v2	values

5.3   Code Imports

A "Code Import" occurs when a significant amount of source code is added to the repository. Code Imports differ from "Off-line Development" in that the code added was not developed as part of the project of interest, but instead originated from some other project. The most common example of a code import is probably the addition of the source code for an externally developed library.

We found a good deal of evidence that Off-line Development is a significant cause of Cliff Walls. As shown in Tbl. 2, a few prominent topics (particularly 1,

Table 3. Largest "Positive" Rank Differences

	All Rank	Cliff Wall Rank	Key Terms			
+97	101	4	cvs	sync	update	repository
+97	112	15	beta	v1	v2	v3
+93	131	38	org	cvs	synchronized	packages
+91	98	7	merged	merged	trunk	stable
+91	100	9	files	added	directories	library
+89	99	10	directory	moved	structure	location
+88	142	54	cc	net	users	sourceforge
+80	128	48	module	python	py	libsrc
+73	119	46	system	specific	platform	devel
+71	133	62	http	www	urls	net
+71	105	34	web	site	resource	component
+67	109	42	php	index	http	forum

Table 4. Largest "Negative" Rank Differences

	All Rank	Cliff Wall Rank	Key Terms			
-120	13	133	fixed	bug	wrong	crash
-118	12	130	fix	typo	corrected	correct
-116	7	123	fixed	bug	bugs	incorrect
-105	44	149	button	selection	tab	dialog
-90	23	113	fixed	problems	issue	bad
-89	30	119	string	return	null	true
-80	51	131	variable	global	unused	define
-77	26	103	output	debug	print	messages
-76	49	125	problem	fixed	patch	solution
-75	36	111	size	buffer	limit	bytes
-75	54	129	function	calls	static	inline
-74	3	77	fixes	minor	cleanups	cosmetic

4 and 9) deal with the first-time addition of files to the repository. The addition of files alone, however, does not indicate a Code Import. Table 1 indicates that topic 1 is quite prominent for all commits, because files are constantly being added to version control systems. In the case of Cliff Walls, however, the size of the commit gives us good reason to believe that the files added contain a great deal of code, and therefore do not represent files added as part of incremental development.

Also, a few of the topics do provide additional evidence of Code Imports. Topic 9 contains the term "library" which indicates that this topic relates to the addition of library files to version control. Similarly, the topic "module-python" appears in Tbl. 3 as a topic with a much higher relative rank for Cliff Wall commits. Examination of log messages for which this topic had high proportion yielded messages such as "bundle all of jython 2.1 with marathon so all python standard library is available" and "Add the Sandia RTSCTS module to the code base." These messages are indicative of Code Imports.

5.4 Off-Line Development

In this paper, we use the term "Off-line Development" to refer to large quantities of code that were developed as part of a project, but for which we have no record. This may be code that was developed without the benefit of version control, or that was developed in a separate repository and then added to the SourceForge CVS repository in a monolithic chunk.

Much of the evidence for Off-line Development is similar to that of Code Imports. Many of the same topics that may refer to code imports ("initial-import," "cvs-sync," and "files-added") could equally be attributed to Off-line Development. Thus it is difficult to distinguish between the true source of many large commits, because it is hard to tell if the files added were developed in conjunction with the current project or separately. Further investigation is required to elucidate the differences between Cliff Walls attributable to Code Import and those due to Off-line Development.

5.5 Branching and Merging

Merging is a major factor in the creation of Cliff Walls. The 7th ranked topic for Cliff Walls is a topic dealing with the merging of a branch in the repository. This same topic also appears as one of the largest positive rank differences in Tbl. 3. This indicates that not only are merges a significant factor behind the creation of Cliff Walls but also that the merge topic is significantly more prevalent within Cliff Walls than it is for All Commits. In contrast, the "initial-import" topic is one of the highest ranked topics in both groups.

5.6 Auto-Generated Code

Topics pertaining to Auto-generated Code are a bit more difficult to identify. The topic "target-generated" appears to capture auto-generated code quite well:

> **target generated rules rule generate mark
> reports make generation targets automati-
> cally linked policy libraries based generator
> jam dependencies building**

Surprisingly, this topic is of relatively little importance, with rank 67 (Cliff Walls) and 113 (All Commits). Such low ranks would seem to indicate that Auto-generated code does not play a significant role in the explanation of Cliff Walls. We did, however, find another example that suggests that Auto-generated Code may be a more significant factor. We were surprised to find that the topic "added-comments" was the 14th ranked topic for Cliff Walls (see Tbl. 2). Non-source code files had been excluded from our study, and code commenting seemed an unlikely cause for commits on the magnitude of 10,000 lines or greater. Upon appeal to the commit log messages, we found a large number of messages containing text such as "Add generated documentation files," "Documentation generated from javadoc," and "Updated documentation using new doxygen configuration."

Further examination revealed that, at least in the cases mentioned above, these commits consist almost entirely of HTML files. The above comments contain 81, 120, and 448 HTML files, respectively. This suggests that large, comment or documentation related commits may be the result of auto-generated HTML files from documentation systems such as javadoc and doxygen.

It is possible that there may be other significant sources of Auto-generated Code expressed in the topics obtained from LDA. In the above case, the tools that generated the "code" were more effective identifiers of Auto-generated code than were the terms "automatically" and "generated." Further investigation is required to determine whether other such cases exist.

5.7   Other Findings

As we hoped, the application of LDA to this problem suggested some potential Cliff Wall causes we had not forseen. Additionally, a few interesting observations served to confirm some of our suspicions about Cliff Walls. In this section we discuss some of these findings.

One discovery of note was the importance of activities related to project releases and new versions. The 2nd most prevalent topic discussed in Cliff Wall log messages is "version-release." Topics 11 and 15 in Tbl. 2 also deal primarily with project releases and versioning, with prevalent terms such as "latest," "version," "beta," "v1" and "v2." It is difficult to tell exactly what is occurring with these commits; most provide little information other than a version number. We suspect that many of these may be the result of merges, and further investigation may determine the true cause.

We were able to gain some understanding of topics which were infrequently discussed in the log messages of Cliff Wall commits. Table 4 shows some of the topics for which the topic rank dropped significantly for Cliff Wall commits. This drop would indicate topics that were discussed much less frequently for Cliff Walls than All Commits, when compared to all other topics. Some of the trends in the table include topics discussing corrective maintenance ("fixed-bug," "fix-typo," "fixes-minor," "output-debug"), gui tweaks ("button-selection"), and minor implementation details ("string-return," "variable-global," "size-buffer"). It is not surprising that these topics do not significantly occur in the log messages of large commits, but these trends lend credibility to our results.

Table 3 provides another interesting insight. We observe that two topics appear to deal with web technologies: "http-www" and "php-index." In many cases, we found that these topics indicated the presence of a URL in the log message. It is intriguing that this topic surpassed so many others in the Cliff Walls category. We believe that these URLs could convey valuable information about the commit, and may help to identify library code that is being imported, or the location of an external version control repository utilized by the project.

6   Threats

Some of the most significant threats to our results arise from the data set employed. As previously stated, the data was gathered in late 2006, and is now relatively old. It is possible that the results that we have found do not correspond to the current state of projects in SourceForge. This study is also limited to projects using the CVS version control system. According to our estimates, almost all new projects in SourceForge are now using Subversion instead of CVS. While the two technologies are similar in many ways, it is possible that our analysis would produce different results if conducted using data from Subversion logs.

It should be noted that when the original data was gathered, projects were filtered to include only those projects listed as "Production," "Stable," or "Maintenance," in an effort to limit the data set to include only "successful" projects [6]. As a result, when we talk about topics across "All Commits," we are actually unable to generalize to the entire population of projects in SourceForge. This is significant, because as one estimate found, only about 12% of projects in Source-Forge were being actively developed [19]. It is possible, even likely, that a similar analysis, not limited to "Production/Stable/Maintenance" projects would produce different results. However, we do not feel that the depiction of Cliff Walls would change dramatically, as we presume they are rare in defunct projects.

Another significant threat to the validity of our results is the presence of "low quality" topics. We found two types of low quality topics in our results: topics with contradictory terms and topics generated from dirty data. One example of a topic with contradictory terms is the "removed-deprecated" topic ranked 3rd in Tbl. 2. This topic contains the contradictory terms "removed," and "imported" as important terms in the topic. This leads to a topic that is difficult to interpret, as the "meaning" of the topic can vary based on the document in which it is present. To better understand this topic we examined the log messages of 233 Cliff Walls containing the topic. Of those 233, the term "removed" occurred in only 1 message, while "imported" occurred in 154. Obviously, for large commits, "imported" is a much more appropriate description of log messages with this topic.

We found two low quality topics resulting from dirty data in our results. The topics "error-message" and "message-project," the 5th and 6th most prevalent topics for Cliff Walls, are also misleading. We looked at 286 Cliff Wall log messages containing at least one of these two topics, and found that 211 (74%) of them contained only the message "no message." These commits should have been removed from the data set prior to the analysis. Exclusion of these commits would result in a data set containing 2,301,620 commits, with 9,199 Cliff Walls, a minor decrease in size. We do not feel that this issue greatly affected the outcomes of this study. However, these topics possibly prevented more appropriate topics from being considered.

In order to improve the results of future studies applying LDA to the Cliff Walls problem, greater effort should be made in LDA model selection. The three input parameters that we were required to specify (number of topics, number of iterations, and burn-in period) could likely be tuned to produce higher quality topics. In particular, this may help us to avoid the issue of topics with contradictory terms.

7 Conclusions

We are excited by the promise that LDA shows for automated analysis of large commits. Through the use of tag clouds and other views of the data, we have been able to gain an insightful picture into the causes behind Cliff Walls. We found that in most cases, our suspicions of Cliff Walls were confirmed. We found significant evidence that library imports, externally developed code, and merges were the subjects of topics frequently discussed in log messages of large commits. We also found evidence that auto-generated code can, in some cases, result in the creation of cliff walls. LDA also helped us to confirm that maintenance tasks, such as bug fixes, do not occur in large commits with much frequency. These conclusions agree with previous studies on the causes of large commits [12, 10].

We found that it was difficult to use commit log messages to distinguish library code imports from imports of large amounts of project code. However, in some cases we are able to identify library imports. We also hope that, in the future, the URLs included in some Cliff Wall commit messages may be used to identify other instances of library code imports.

We believe that LDA is a welcome alternative to many of the methods that have previously been used for classification of commit log messages. While we invested a great deal of time manually interpreting the results produced by LDA, we were able to avoid the tedium of data tagging required by most supervised classification tasks.

8 Future Work

The role of large commits in software evolution is still largely unclear. In this study, we have examined the causes of Cliff Walls for a particular subset of all software projects—those that are relatively successful, are hosted on Source-Forge, and that use CVS for version control. In order to better understand Cliff Walls, we need to build upon this subset. First, research should consider investigating Cliff Walls as they occur in other version control systems. CVS is no longer the most significant version control system, since others, such as Subversion and GIT have risen to take its place.

SourceForge is only one of many environments in which open source software is developed. There are various open source forges and foundations, each with its own tools, communities, policies, and practices that influence the software development that occurs therein. It is possible that some of these other environments may prove more welcoming to those interested in studying the evolution of software. An effort should be made to characterize and compare the Cliff Walls that exist in other open source development communities, such as the Apache and Eclipse Foundations, RubyForge, and GITHub, to name a few. Of course the study of large commits should not be limited to only open source organizations, but should be investigated wherever possible.

More information may also be gained through a more in-depth analysis of the Cliff Walls themselves. The largest commit in our data set was over 13 million lines of code. In contrast, the smallest Cliff Wall contained 10,001 LOC. In this study, both of these commits, as well as everything in between, were lumped into the same class: Cliff Walls. It is likely that such a large level of granularity hides much that can be learned about the causes of Cliff Walls. We believe that there are opportunities to better understand this phenomenon by examining more closely the causes behind Cliff Walls of differing magnitudes.

Acknowledgements. The authors would like to thank Dan Walker of the BYU Natural Language Processing Lab for his willingness to provide insight and guidance on the methods used in this paper.

References

[1] Arafat, O., Riehle, D.: The Commit Size Distribution of Open Source Software. In: 42nd Hawaii International Conference on System Sciences, HICSS 2009, pp. 1–8. IEEE, Los Alamitos (2009)

[2] Bahn, P., Bahn, P.G., Tidy, B.: Archaeology: a very short introduction. Oxford University Press, USA (2000)

[3] Bird, C., Gourley, A., Devanbu, P., Swaminathan, A., Hsu, G.: Open borders? immigration in open source projects. In: International Workshop on Mining Software Repositories, p. 6 (2007)

[4] Blei, D.M., Ng, A.Y., Jordan, M.I.: Latent dirichlet allocation. The Journal of Machine Learning Research 3, 993–1022 (2003)

[5] Briand, L.C., Morasca, S., Basili, V.R.: Defining and validating measures for object-based high-level design. IEEE Transactions on Software Engineering 25(5), 722–743 (1999)

[6] Delorey, D.P., Knutson, C.D., Chun, S.: Do programming languages affect productivity? a case study using data from open source projects. In: 1st International Workshop on Emerging Trends in FLOSS Research and Development (FLOSS 2007) (May 2007)

[7] Delorey, D.P., Knutson, C.D., Giraud-Carrier, C.: Programming language trends in open source development: An evaluation using data from all production phase sourceforge projects. In: 2nd International Workshop on Public Data about Software Development (WoPDaSD 2007) (June 2007)

[8] Delorey, D.P., Knutson, C.D., MacLean, A.: Studying production phase sourceforge projects: A case study using cvs2mysql and sfra+. In: Second International Workshop on Public Data about Software Development (WoPDaSD 2007) (June 2007)

[9] Hassan, A.E.: Predicting faults using the complexity of code changes. In: Proceedings of the 31st International Conference on Software Engineering (ICSE 2009), pp. 78–88. ACM, New York (2009)

[10] Hattori, L.P., Lanza, M.: On the nature of commits. In: 23rd IEEE/ACM International Conference on Automated Software Engineering-Workshops, ASE Workshops 2008, pp. 63–71. IEEE, Los Alamitos (2008)

[11] Hindle, A., German, D.M., Godfrey, M.W., Holt, R.C.: Automatic classication of large changes into maintenance categories. In: IEEE 17th International Conference on Program Comprehension, ICPC 2009, pp. 30–39. IEEE, Los Alamitos (2009)

[12] Hindle, A., German, D.M., Holt, R.: What do large commits tell us?: a taxonomical study of large commits. In: Proceedings of the 2008 International Working Conference on Mining Software Repositories, pp. 99–108. ACM, New York (2008)

[13] Krein, J.L., MacLean, A.C., Delorey, D.P., Knutson, C.D., Eggett, D.L.: Language entropy: A metric for characterization of author programming language distribution. In: 4th Workshop on Public Data about Software Development (2009)

[14] Krein, J.L., MacLean, A.C., Delorey, D.P., Knutson, C.D., Eggett, D.L.: Impact of programming language fragmentation on developer productivity: a sourceforge empirical study. In: International Journal of Open Source Software and Processes (IJOSSP); Publication Pending

[15] MacLean, A.C., Pratt, L.J., Krein, J.L., Knutson, C.D.: Threats to validity in analysis of language fragmentationon sourceforge data. In: Proceedings of the1st International Workshopon Replicationin Empirical Software Engineering Research (RESER 2010), p. 6 (May 2010)

[16] MacLean, A.C., Pratt, L.J., Krein, J.L., Knutson, C.D.: Trends that affect temporal analysis using sourceforge data. In: Proceedings of the 5th International Workshop on Public Data about Software Development (WoPDaSD 2010), p. 6 (June 2010)

[17] McCallum, A.K.: MALLET: A machine learning for language toolkit (2002), http://mallet.cs.umass.edu

[18] Mockus, A., Fielding, R.T., Herbsleb, J.D.: Two case studies of open source software development: Apache and mozilla. ACM Transactions on Software Engineering and Methodology 11(3), 309–346 (2002)

[19] Rainer, A., Gale, S.: Evaluating the Quality and Quantity of Data on Open Source Software Projects (2005)

[20] Tarvo, A.: Mining software history to improve software maintenance qual- ity: A case study. IEEE Software 26(1), 34–40 (2009)

[21] Wallach, H., Mimno, D., McCallum, A.: Rethinking LDA: Why priors matter. Advances in Neural Information Processing Systems 22, 1973–1981 (2009)

[22] Xu, J., Gao, Y., Christley, S., Madey, G.: A topological analysis of the open souce software development community. In: HICSS 2005: Proceedings of the 38th Annual Hawaii International Conference on System Sciences, vol. 7 (2005)

[23] Zimmermann, T., Weißgerber, P.: Preprocessing CVS data for fine-grained analysis. In: Proceedings 1st International Workshop on Mining Software Repositories (MSR 2004), Citeseer, pp. 2–6 (2004)

Package Upgrade Robustness: An Analysis for GNU/Linux® Package Management Systems

John Thomson[1], Andre Guerreiro[1], Paulo Trezentos[1], and Jeff Johnson[2]

[1] Caixa Mágica Software
Edificio Espanha - Rua Soeiro Pereira Gomes
Lote 1 - 8 F, 1600-196 Lisboa
{first.surname}@caixamagica.pt
[2] rpm5.org
jbj@rpm5.org

Abstract. GNU/Linux systems are today used in servers, desktops, mobile and embedded devices. One of the critical operations is the installation and maintenance of software packages in the system. Currently there are no frameworks or tools for evaluating Package Management Systems (PMSs), such as RPM, in Linux and for measuring their reliability. The authors perform an analysis of the robustness of the RPM engine and discuss some of the current limitations. This article contributes to the enhancement of Software Reliability in Linux by providing a framework and testing tools under an open source license. These tools can easily be extended to other PMSs such as DEB packages or Gentoo Portage.

1 Introduction

Installation of software in Linux systems is mostly performed by installing pre-compiled binary code using a Package Management System (PMS). The most frequently used package installers are RPM Package Manager (RPM) and dpkg (Debian format). We identify methods in which package upgrades can be analysed for their reliability and to ascertain how often failures occur. By formalising the failures we hope to provide the basis for future work where failures can be classified and detected that provides a method to quantitatively assess package installers.

1.1 Background

Source code compilation on a user machine has largely been superseded by having dedicated build machines. GNU/Linux distributions then make these 'packages'[1, Sec. 3, p. 9] available as pre-compiled binary packages (Fig. 2). The MANCOOSI project[1], is dedicated to solving problems associated with various package installers and provides the most recent research in this area.

---

[1] http://www.mancoosi.org

S.A. Hissam et al. (Eds.): OSS 2011, IFIP AICT 365, pp. 299–306, 2011.

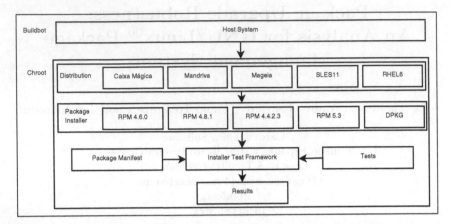

Fig. 1. An indication of the proposed architecture of the test framework, with an internal python testing system and external buildbot automation suite

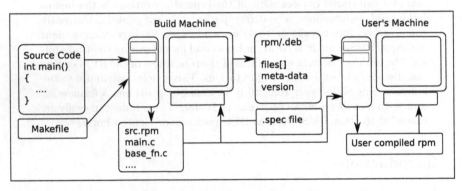

Fig. 2. A simplified topology demonstrating how sources and packages combine on a host and user machine

There have been few investigations into package installer reliability and robustness[11,12,2]. Applications on GNU/Linux systems are distributed and installed through PMSs, therefore it makes sense to systematically test and analyse their reliability. RPM is the baseline PMS chosen by the Linux Standard Base (LSB) as the definitive installation system for GNU/Linux OS's[2].

There is a distinct fork of RPM Package Manager (RPM), '@rpm5', referred to as RPM 5. One of the main development activities for RPM 5 is that of creating a fully Atomicity, Consistency, Isolation, Durability (ACID) compliant transactional system for installation of packages.

One other main alternative to RPM is that of .deb packages[3], used in Debian systems and there are some subtle differences between them[1, Sec. 3.1-3.2].

---

[2] http://refspecs.freestandards.org/LSB_3.1.0/LSB-Core-generic/LSB-Core-generic/swinstall.html#SWINSTALL-INTRO

[3] http://www.debian.org/doc/FAQ/ch-pkg_basics.en.html

Failures relating to dependency resolution which recently have been better defined [5,7] are a matter for solvers[9][8].

1.2 The Aim

To investigate the extent that RPM systems ensure that in event of a failure that the package transaction leaves the system in a consistent state.

2 Concepts

Packages are collections of self-sufficient software that can direct PMSs. Our framework considers pre-compiled binary packages for testing.

Installers create a database of the files associated with packages and the status on the system.

An 'upgrade' is defined as when a package is modified from one version to another.

2.1 Package Upgrade Failure

Used to describe when a package upgrade operation has resulted in an erroneous installation and can occur due to a lack of physical resources and scriptlet failure (amongst others [3, Ch. 2]). Maintainer script failures remain the single biggest cause of failures.

Database Consistency Failure. For RPM, there is a centralised database where package upgrades are recorded. When upgrading a package the database can fail to commit the transaction and can be left in an inconsistent state (see Sect. 3).

3 Test Framework

Software fault-injection and 'black-box' approaches are recognised methodologies for testing the robustness of systems[6,4,10]. The aim is to have a cross-distribution, test framework that can identify common faults and also identify unique failures, specific to architectures.

3.1 Analysable Elements

A package installer like RPM performs operations both on the file-system and on its internal database of package meta-data.

The database side of the problem is analysed since the new developments in PMSs are mainly improvements in that area.

Multiple package transactions differ from single because they have more requirements to be fulfilled: either they install all of the packages or none of them.

Within each test suite the input parameters are the Error Injection Time and the package(s) composition.

Upgrade transactions enclose two different sub-atomic operations: installation and removal. The final expected result is that only one version of a package is installed.

Failed upgrade for Individual package tests case (Sect. 4.2):

- Database consistency test: There are zero, two or more different versions present in the RPM database after an upgrade transaction;

Failed upgrade for Groups package tests case (Sect. 4.3):

- Database consistency test: Number of failures in an upgrade due to invalid database entries.
- Group Atomicity test: More than one package matches the failure case of an individual transaction.

3.2 Injecting Faults

We introduce to the normal upgrade procedure an external interruption in the form of a SIGKILL, signal #9 on most POSIX compliant systems, forcing a termination.

SIGKILL is useful for testing the robustness of RPM in a worst-case situation.

4 Test Results

Two versions of RPM are being tested with this version of the framework as can be seen in Table 1. Future versions of RPM 5 will likely be fully ACID compliant, therefore group tests should show fewer transaction failures.

4.1 Test Environment

Tests were performed on Linux Caixa Mágica (CM) 14 virtual machines with Gnome. BerkeleyDB error meant that RPM 4.6 needed a rebuild of the database after a set of interrupted package upgrades (unnecessary for RPM 5.3).

Table 1. Details of the versions of RPM being examined

Version	Type	Release	Build Date
4.6.0	Package	2.3xcm14	2009/08/03
5.3.1	Built from Source	N/A	2010/05/24

RPM 4.6.0 being tested is from a Caixa Mágica build [4].
RPM 5 version used is from the RPM 5.3.1 tarball[5].

---

[4] http://contribsoft.caixamagica.pt/trac/browser/packages/cm14/rpm
[5] http://rpm5.org/files/rpm/rpm-5.3/rpm-5.3.1.tar.gz

4.2   Individual Package Tests

The results shown in Table 2 are from running 100 iterations of upgrading packages with a random kill-time. Although the file-system failure rate is higher for RPM 5 the number of database failures are lower than for RPM 4.

Table 2. Individual packages transaction: File-system and database consistency failure rate after error injection

		File-System		Database	
Pkg No.	Pkg size MB	RPM 4.6 Failure Rate	RPM 5.3 Failure Rate	RPM 4.6 Failure Rate	RPM 5.3 Failure Rate
1	0.01	0/100	0/100	7/100	1/100
2	0.06	1/100	4/100	13/100	4/100
3	0.20	46/100	63/100	19/100	19/100
4	0.30	41/100	65/100	11/100	0/100
5	0.46	10/100	15/100	22/100	14/100
6	0.50	7/100	13/100	17/100	5/100
7	1.80	39/100	60/100	0/100	0/100
8	2.30	50/100	56/100	0/100	0/100
9	6.10	68/100	56/100	0/100	0/100
10	21.00	49/100	45/100	0/100	0/100
Avg.	3.27	31/100	38/100	9/100	4/100

4.3   Group Packages Tests

If a package upgrade fails in a transaction it increments the number of database failures. A single package failure in a group package upgrade transaction indicates a group transaction failure (Txn. Failure Rate). Table 3 shows the results of database consistency whereas Table 4 presents file-system consistency.

4.4   Individual Packages Against Time

If a PMS has no time to perform an upgrade there is no chance for a failure to be introduced. Figs. 3 & 4 indicate such behaviour and possibly provide an explanation for why in Table 2 the larger packages do not exhibit any database failures.

Table 3. Group transactions: Database consistency for individual packages in transaction and group atomicity failure rate after error injection

| Group (No. Pkgs) | Database Consistency | | Group Atomicity | |
	RPM 4.6 Package Failure rate	RPM 5.3 Package Failure rate	RPM 4.6 Txn. Failure Rate	RPM 5.3 Txn. Failure Rate
1 (4)	263/400	19/400	92/100	17/100
2 (4)	189/400	45/400	83/100	36/100
3 (4)	59/400	76/400	30/100	65/100
4 (3)	11/300	31/300	10/300	31/100
5 (3)	29/300	9/300	16/100	7/100
6 (3)	30/300	39/300	18/100	53/100
7 (3)	121/300	60/300	77/100	55/100
8 (3)	73/300	65/300	73/100	65/100
9 (3)	89/300	39/300	47/100	31/100
10 (4)	55/400	1/400	55/100	1/100
Avg	92/340	38/340	50/100	36/100

Table 4. Group transactions: File-system consistency for individual packages in transaction and group atomicity failure rate after error injection

| Group (No. Pkgs) | File-system Consistency | | Group Atomicity | |
	RPM 4.6 Package Failure rate	RPM 5.3 Package Failure rate	RPM 4.6 Txn. Failure Rate	RPM 5.3 Txn. Failure Rate
1 (4)	0/400	83/400	0/400	82/100
2 (4)	206/400	33/400	87/100	26/100
3 (4)	66/400	50/400	66/100	50/100
4 (3)	40/300	123/300	30/100	74/100
5 (3)	0/300	3/300	0/100	3/100
6 (3)	238/300	0/300	89/100	0/100
7 (3)	184/300	151/300	100/100	84/100
8 (3)	80/300	75/300	80/100	75/100
9 (3)	235/300	201/300	93/100	86/100
10 (4)	118/400	83/400	100/100	83/100
Avg	117/340	80/340	65/100	56/100

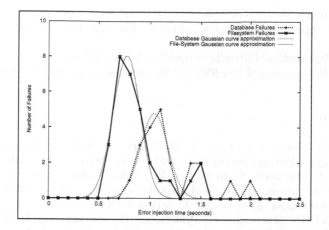

Fig. 3. RPM 4.6 - Package Failures vs. Error Injection Time for a package of 300KB over ten iterations

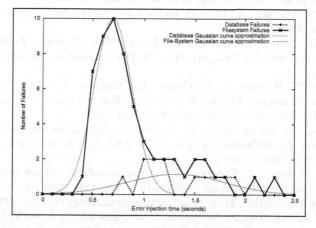

Fig. 4. RPM 5.3 - Package Failures vs. Error Injection Time for a package of 300KB over ten iterations

5   Conclusions

Linux systems are today deployed worldwide in different environments. The majority of such systems rely on a functioning kernel and a reliable Package Management System.

The SIGKILL test is one of the most extreme types of test likely indicating worst case behaviour. Although generally RPM 5.3 performed better in terms of database consistency, it still doesn't support atomicity in group transactions. There is the possibility to extend this study to Debian packages. Other types and permutations of errors can be injected to explore the recovery mechanisms of different PMSs.

The outcomes of this analysis, can be used to resolve problems and then suggest novel approaches for more robust package upgrade transactions.

Acknowledgements. Partially supported by the European Community's 7th Framework Programme (FP7/2007-2013), grant agreement n°214898.

References

1. Barata, P., Trezentos, P., Lynce, I., di Ruscio, D.: Survey of the state of the art technologies. Mancoosi project deliverable D3.1, Mancoosi (June 2009)
2. Crameri, O., Knezevic, N., Kostic, D., Bianchini, R., Zwaenepoel, W.: Staged deployment in mirage, an integrated software upgrade testing and distribution system. SIGOPS Oper. Syst. Rev. 41(6), 221–236 (2007)
3. Di Ruscio, D., Thomson, J., Pelliccione, P., Pierantonio, A.: First version of the DSL. Mancoosi Project deliverable D3.2, Mancoosi (November 2009), http://www.mancoosi.org/reports/d3.2.pdf
4. Duraes, J.A., Madeira, H.S.: Emulation of software faults: A field data study and a practical approach. IEEE Transactions on Software Engineering 32, 849–867 (2006)
5. Le Berre, D., Parrain., A.: On SAT technologies for dependency management and beyond. In: ASPL (2008)
6. Madeira, H., Costa, D., Vieira, M.: On the emulation of software faults by software fault injection. In: DSN 2000, pp. 417–426. IEEE Computer Society, Washington, DC (2000)
7. Mancinelli, F., Boender, J., Di Cosmo, R., Vouillon, J., Durak, B., Leroy, X., Treinen, R.: Managing the complexity of large free and open source package-based software distributions. In: ASE, pp. 199–208 (2006)
8. Manquinho, V., Marques-Silva, J., Planes, J.: Algorithms for weighted boolean optimization. In: Kullmann, O. (ed.) SAT 2009. LNCS, vol. 5584, pp. 495–508. Springer, Heidelberg (2009)
9. Trezentos, P., Lynce, I., Oliveira, A.L.: Apt-pbo: solving the software dependency problem using pseudo-boolean optimization. In: ASE 2010, pp. 427–436. ACM Press, New York (2010), http://doi.acm.org/10.1145/1858996.1859087
10. Voas, J.: Fault injection for the masses. Computer 30(12), 129–130 (1997)
11. Yoon, I.C., Sussman, A., Memon, A., Porter, A.: Effective and scalable software compatibility testing. In: ISSTA 2008, pp. 63–74. ACM, New York (2008)
12. Zacchiroli, S., Cosmo, R.D., Trezentos, P.: Package upgrades in foss distributions: Details and challenges. In: First ACM Workshop on HotSWUp (October 2008)

Applying Open Source Practices and Principles in Open Innovation: The Case of the Demola Platform

Terhi Kilamo[1], Imed Hammouda[1], Ville Kairamo[2], Petri Räsänen[2], and Jukka P. Saarinen[3]

[1] Tampere University of Technology
firstname.lastname@tut.fi
[2] Uusi Tehdas/New Factory
firstname.lastname@hermia.fi
[3] Nokia Research Center
jukka.p.saarinen@nokia.com

Abstract. In numerous fields, businesses have to rely on rapid development and release cycles. Variant new ideas and concepts can emerge through open innovation as the participants are not limited to the company scope. This makes open innovation an increasingly appealing option for the industry. One such open innovation platform, Demola, allows university students to work on real life industrial cases of their own interest. We have identified similarities with its way of operation to open source software development and find that it offers a viable motivational, organizational and collaborative solution to open innovation.

1 Introduction

Constant, lightning-fast innovation is becoming an essential element to companies in software business. Innovation can lie in any commodity it being something novel that can be put to actual use. Many companies rely on innovation on a daily basis to create better products and to improve their internal processes [2]. Traditionally such advantages have been kept within the company.

Opening up the option to innovate to a wider group of partners can enforce and expand the scope of the innovation process, which becomes free of the boundaries of the company and what knowledge is available within. Open innovation helps in identifying the best ideas by combining internal and external ideas into architectures and systems [11,2]. The open innovation process typically involves proof of concepts, trials, and, perhaps most importantly, the right people to identify what should (or must) be focused on.

Open innovation, however, comes with a number of challenges such as motivation, integration and exploitation of innovation [5]. It needs a governance framework [4] that enables organizational alignment of the different partners, proper handling of intellectual property rights issues, and the emergence of new kinds of business opportunities. These challenges have to be taken into account

S.A. Hissam et al. (Eds.): OSS 2011, IFIP AICT 365, pp. 307–311, 2011.

when building any open innovation platform with the goal of driving future development and solutions.

In this paper, we argue that the open source model of development and knowledge creation brings a set of principles of practices that could be adapted to the context of open innovation, in the same way as observed in [10]. We focus on open innovation in the context of academia-industry co-operation. In order to support our arguments, we have analyzed an open innovation platform for students called Demola [12].

We identify characteristics of open source software development in the motivational, organizational, and collaboration aspects of open innovation. The main research question answered is: How Demola's approach shares similarities with other community driven development methods, mainly open source? In Section 2 we give background on the Demola organization and discuss the practices of open source within the open innovation context of Demola in Section 3. Section 4 then concludes the paper with some discussion and final remarks.

2   Platform for Open Innovation and Learning

There is a real need for increased opportunities for innovation projects that can lead to new business ideas. Open innovation environments allow businesses to reach beyond the company scope in the search for new concepts and ideas. A governance framework is needed with practices and working principles to bring innovation partners together and to ensure ongoing innovation work.

Demola is one such open innovation platform intended for students. It aims to multidiciplinary and agile development of innovative products and product demos. The project ideas come from the industry and public organisations and thus concepts that have practical business importance are developed. The student work is supported by both the industrial and the academia partners that provide guidance throughout the project. Demola offers a governance framework

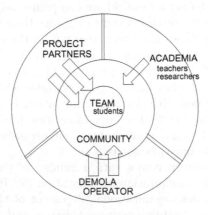

Fig. 1. Demola Partners

that facilitates team building and supports emerging business ideas. It also incorporates a model for managing immaterial rights that supports startups and respects the authors. On a practical level, Demola provides workspaces that support team work and co-creation. Demola is a modern and actual learning environment to students from different universities.

Figure 1 shows the partners in Demola innovation and the flow of communication and support for the project work. The team is at the heart of development while others direct, aid and facilitate the work. In terms of numbers, there are currently 35 companies involved in Demola as project partners. During 2011 the aim is to reach a yearly level of around 100 projects running. The Demola operator itself employs three people: one manager and two assistants.

3   Adopting Open Source

Demola was built on the basis of openness. There are different aspects and challenges that need to be addressed in making the platform open and functional. A set of principles and practices of free/libre open source (FLOSS) can been identified in Demola.

Motivation. A famous quote from Raymond [1] claims that "Every good work of software starts by scratching a developer's personal itch." This is commonly seen as one of the driving forces behind open source software quality and success. The participant's motivation is also one of the main characteristics of Demola team building and work. Similarly Raymond's restatement of the itch: "To solve an interesting problem, start by finding a problem that is interesting to you" is a major driving force in the Demola way of doing.

Internal motivation as driver: The participant's internal motivation is the main driving factor for the Demola team work. Participation is fueled by their own background, motivation and goals that range beyond normal school work.

Participant chooses the project: The way teams are formed in Demola is similar to how open source communities come into being. Students search for project topics that are meaningful and interesting to them and apply for participation in it. The reasons for choosing a project are personal to the applicant with widely varied factors behind the selection. The applicants have no knowledge of the possible other team members in advance and it is not possible to choose the people you form the team with.

Collaboration. Jukka Saarinen, one of the key people behind the the foundation of Demola, has said about the platform: *"What is special about Demola is the way of doing things· anyone and everyone can contribute ideas to a demo which is then built together. The let's do it attitude without bureaucracy and formal processes makes the atmosphere fruitful"* [8]. This reflects the philosophical standpoint of open source software development. Those with the interest and skill can contribute their work to the community.

Co-creation: Demola has been built on the notion of bringing the right people together and to enabling collaboration between participants. Demola itself is a developer community where anyone can contribute their work based on their own interest and skill. Similarly the development of the project concepts and demos in Demola is done through collaborative teams. Each team member brings his or her own knowledge and expertise into the team and each team is different.

Community spirit: The student teams, active academia members and the project partners form an innovation ecosystem where all participants benefit from the Demola platform. Demola acts like a community of developers where the teams share ideas and work and where the project partners benefit from the work done in teams for other partners.

Legal Concerns. What is special about open source is its philosophy on intellectual property rights (IPR). The approach chosen for managing the IPR of the project teams in Demola is akin to the idea of licensing in open source. The open innovation approach in Demola respects the IPR of the teams: the students own the rights to the project results. The originator of the project idea can buy wide and parallel usage rights to the results by paying the project team an agreed reward, i.e. the team licences their work to the industrial partner.

4 Discussion and Conclusions

We have identified the best practices and principles of FLOSS development within an open innovation platform, Demola. When its way of doing is juxtaposed with the FLOSS principles and practices the common factors are identifiable. FLOSS also enables a better and wider exploitation of the results as the teams hold the rights to their work. Demola not only provides support for the project partners to buy rights to the work but also for the students themselves to start new businesses on top of the results.

Traditional open source principles and practices, however, may fall short in other aspects such as timely delivery, communication, and quality. Such challenges in the daily workflow of the project development need futher management methods on top of FLOSS. How these challenges are met is a focus of future research. Our findings suggest that the open source model offers a viable solution to open innovation in terms of motivational, organizational, and collaborational aspects.

References

1. Raymond, E.S.: The Cathedral and the Bazaar. O'Reilly Media, Sebastopol (1999)
2. Chesbrough, H.: Open Innovation: Researching a New Paradigm, chapter Open Innovation: A New Paradigm for Understanding Industrial Innovation. Oxford University Press, Oxford (2006)
3. Takeuchi, H., Nonaka, I.: The New New Product Development Game. Harvard Business Review, 137–146 (January-February 1986)

4. Feller, J., Finnegan, P., Hayes, J., O'Reilly, P.: Institutionalising information asymmetry: governance structures for open innovation. Information Technology & People 22(4), 297–316 (2009)
5. West, J., Gallagher, S.: Challenges of Open Innovation: The Paradox of Firm Investment in Open-Source Software. R&D Management 36(3), 319–331 (2006)
6. Beck, K.: Embracing Change With Extreme Programming. Computer 32(10), 70–77 (1999)
7. Beck, K., Beedle, M., van Bennekum, A., Cockburn, A., Cunningham, W., Fowler, M., Grenning, J., Highsmith, J., Hunt, A., Jeffries, R., Kern, J., Marick, B., Martin, R.C., Mellor, S., Schwaber, K., Sutherland, J., Thomas, D.: Manifesto for Agile Software Development (March 2002), http://agilemanifesto.org/ (last visited March 2011)
8. Facilitating Innovation at Demola. Open Threads: Open Innovation Newsletter (April 2009)
9. Abrahamsson, P., Salo, O., Ronkainen, J., Warsta, J.: Agile Software Development Methods Review and Analysis. VTT Publications 478 (2002)
10. Goldman, R., Gabriel, R.P.: Innovation Happens Elsewhere: open source as business strategy. Morgan Kaufmann, San Francisco (2005)
11. Davis, S.: How to Make Open Innovation Work in Your Company. Visions Magazine (December 2006)
12. Demola Innovation Platform, http://www.demola.fi (last visited March 2011)

KommGame: A Reputation Environment for Teaching Open Source Software

Veerakishore Goduguluri, Terhi Kilamo, and Imed Hammouda

Department of Software Systems, Tampere University of Technology
Korkeakoulunkatu 1, FI-33720 Tampere, Finland
firstname.lastname@tut.fi

Abstract. The importance of teaching open source software in universities is increasing with the advent of open source as a development and business model. A novel, student centric approach of teaching open source was tried out at Tampere University of Technology where a new environment called KommGame was introduced to assist in teaching open source development. This environment includes a reputation system to motivate learners to participate. In this paper, we present our approach of teaching open source and how the KommGame environment was employed to teach open source software.

1 Introduction

With the advent of open source software (OSS) as a development and business model, the number of job vacancies valuing open source knowledge and experience has been rising on a regular basis. This in turn has motivated many universities and professional schools to introduce new courses and programmes related to teaching OSS principles and practices (e.g. [1], [3]). So far OSS teaching has mostly been organized in a traditional lecture course format, for example taking the form of a seminar where students present specific OSS related topics. Other attempts rely on sending students out into real open source projects and communities (e.g. [2]).

Such approaches to teaching open source software face two major challenges. First, classical teaching methods may not fully convey all the special aspects involved in OSS development such as community collaboration, peer review, and co-creation. Second, students may find it hard to participate in real OSS project as a first experience. This is because OSS projects typically have own principles, practices, processes, and tools.

A more attractive approach is to provide a learning environment for OSS where students could collaborate collectively to achieve a common goal. Such constructivist approach [4] to learning allows students to generate new knowledge through the interaction of the group's past experience and new ideas. A constructivist learning method however needs individual's active participation, which from the OSS perspective means student contribution to the community. An important question is, therefore, how to keep students' motivation high for the purpose of learning OSS concepts through active contribution. It has been argued that reputation systems could play an important role in maintaining student motivation [5].

S.A. Hissam et al. (Eds.): OSS 2011, IFIP AICT 365, pp. 312–315, 2011.
© IFIP International Federation for Information Processing 2011

In this paper, we argue that reputation systems can be applied in a learning environment for open source software. Our approach is also inspired by the experiences of using reputation systems to reward and recognize developers in OSS communities such as Qt [6]. Towards this aim, we present an example reputation model and a concrete reputation environment known as KommGame that mimics real open source projects. The environment has successfully been tested at Tampere University of Technology (TUT) to introduce OSS concepts and practices to software engineering students.

The remaining of this paper is structured as follows: Section 2 reputation systems for teaching open source. Section 3 presents the KommGame environment. Finally we conclude in Section 4.

2 Reputation Model for Teaching Open Source Software

Reputation systems are used to measure the contribution of individuals in an online community; they are also applied in different fields such as e-commerce, search engines, and social news. As reputation systems are applied for measuring online activities one can see that reputation systems can be applied for e-learning in the educational context where most of the activities happen online. In [3] Farmer has explained about different reputation models. It is discussed in [7] that reputation systems suites a small group of young participants; they have high competitive sprit which makes learning more active and motivated.

In OSS development all kinds of contribution are treated as equally important and there is no good metric with which to compare or quantify different types of contribution with each other. This is the reason why most of the open source communities have not adopted a reputation system. In an educational context, however, the course moderator may decide which types of contribution should be emphasized. A reputation model can be designed accordingly.

We argue that the karma reputation model fits well the activities and the nature of OSS communities, where the object subjected to reputation is human. The final karma value of the participants is the sum of weight times of each contribution. The universal karma model can be written as

$$Karma = \sum_{k=1}^{n}(f_k(contribution_k)) + f(Favorites) + g(Weekly\,Quality\,Tokens) \qquad (3.1)$$

Here n corresponds to the total number of contributions. f_k is the weight function corresponding to contribution type. "Favorites" is the number of like bookmarks a content author gets. "Weekly Quality Tokens" corresponds to the number of time the particular participant was selected as the best quality contributor of the week by the rest of the members of community.

For example, a sample karma model which covers activities related to bugs, features, improvements and wiki is given below. In the formula each activities is multiplied with its associated weight. Total karma is sum of all karmas from each activity.

Karma= 6*√(number of bugs reported) + 3*√(number of bug comments) + 2*(number of bugs closed) + 4*√(number of feature requests) + 3*√(number of bug comments) + 2* (number of closed new features) + 4*√(number of request) + 3*√(number of improvement comments) + 2*√(number of closed improvements) + 4*√(number of edits) + 4*√(number of likes) + 4* √(number of weekly quality tokens)

3 KommGame Environment

We have developed an OSS learning environment based on the reputation model presented earlier. The learning environment, called KommGame [8], maintains karma values as a motivational factor for a community of learners. The KommGame environment forms an infrastructure required for collaborative and student centric learning.

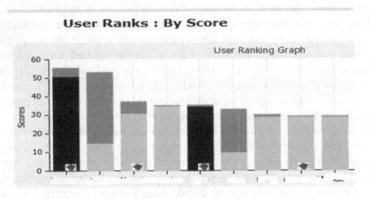

Fig. 1. Karma reporting interface

The KommGame infrastructure has been developed to mimic the infrastructure of a real open source community. The environment has features to add and edit open content, a user management system to manage users of the community, a system to track user activities, a communications channel, a bug management system, a source code base to maintain source code of the project, a reputation system to calculate the karma of each community member and an user interface to publish karma values.

Figure 1 shows the KommGame interface for karma value reports. The graph shows different categories of users (i.e. committers and reporters), illustrated using different colors. Each vertical bar in the graph represents the score of each user of the system. Each vertical bar has two parts with different colors, the bottom part indicates the score of the previous weeks. The upper part indicates the score of the current week. The hat icons, shown in some bars, indicate the best contributors of the week.

4 Conclusions

The approach of KommGame for OSS education allows students to practice OSS project in safe and realistic OSS environment. The KommGame motivates the students to make more contribution to the OSS project and thus give them a valuable OSS project experience. This kind of realistic setting gives the students a good starting point to work in real OSS development.

The future plans for KommGame are to research how this can be applied in traditional programming courses where, students have to collaborate and participate in programming exercises. Future work includes applying the karma model to other courses and using KommGame as a standard system to issue certificates for OSS learners.

References

1. Megías, D., Serra, J., Macau, R.: An International Master Programme in Free Software in the European Higher Education Space. In: Proceedings of the First International Conference on Open Source Systems, pp. 349–352 (July 2005)
2. Lundell, B., Persson, A., Lings, B.: Learning Through Practical Involvement in the OSS Ecosystem: Experiences from a Masters Assignment. In: Open Source Development, Adoption and Innovation. IFIP, vol. 234, pp. 289–294. Springer, Boston (2007)
3. German, M.D.: Experience teaching a graduate course in Open Source Software Engineering. In: Proceedings of the First International Conference on Open Source Systems (OSS 2005), Genova, Italy, July 11-15, pp. 326–328 (2005)
4. Piaget, J.: The Child's Conception of the World. Rowman and Allenheld, New York (1960)
5. Temperini, M., Sterbini, A.: Learning from Peers: Motivating Students through Reputation Systems. In: International Symposium on Applications and the Internet, pp. 305–308 (2008)
6. Qt developers network reputation system, http://developer.qt.nokia.com/ranks (last visited on March 2011)
7. Temperini, M., Sterbini, A.: Learning from Peers: Motivating Students through Reputation Systems. In: International Symposium on Applications and the Internet, pp. 305–308 (2008)
8. OSS Learning environment at TUT, http://osscourse.cs.tut.fi/mantis/login_page.php (last visited on March 2011)

Virtual Health Information Infrastructures: A Scalable Regional Model

Ann Séror

eResearch Collaboratory, 352 Rue Lavigueur, Quebec City, Canada G1R 1B4
annseror@eresearchcollaboratory.com

Abstract. Integrating research, education and evidence-based medical practice requires complex infrastructures and network linkages among these critical activities. This research examines communities of practice and open source software tools in development of scalable virtual infrastructures for the regional Virtual Health Library of the Latin American and Caribbean Health Sciences System (Bireme) and embedded national cases. Virtual infrastructures refer to an environment characterized by overlapping distribution networks accessible through Internet portals and websites designed to facilitate integrated use of available resources. Case analysis shows engagement of interdisciplinary communities of practice for scalable virtual infrastructure design. This research program considers theory and methods for study of transferability of the Latin American model to large health care systems in other cultures.

Keywords: virtual infrastructures, open source systems, communities of practice, Bireme, culture.

Information technologies and telecommunication infrastructures are transforming institutions at the foundation of evidence-based research, education and practice in medicine and the health sciences. Emerging health information networks offer integrated systems to link these complex activities for innovation in health care service delivery. There is growing recognition in scientific communities of the critical importance of networks in social systems in general as well as in health care. [1,2] The OSS community offers substantial resources to build infrastructures at the foundation of health information networks. [3] Recent literature recognizes that access to health information for health care providers, policy makers, researchers, publishers, and systematic reviewers is essential to global development of equitable health care systems as well as achievement of the Millennium Development Goals. [4] In this fast developing field of inquiry, only fragmented research has focused on the design and scaling of collaborative health information systems taking into account the institutional roles of virtual infrastructures.

Studies of the Latin American and Caribbean Health Sciences System (Bireme) suggest the importance of local research and publication as well as regional leadership to integrate medical education and practice. [5-8] Emerging open access publishing systems reveal new dynamics between international and national as well as academic and practitioner communities, while the social medicine tradition of the Latin American region illustrates the power of ideology as a factor shaping integrated

S.A. Hissam et al. (Eds.): OSS 2011, IFIP AICT 365, pp. 316–319, 2011.

regional and local health information systems. Virtual infrastructures supporting these systems refer to an environment characterized by overlapping distribution networks, systems brokerage functions, and the adoption of a software perspective emphasizing the devices and channels through which information is processed and distributed. [9] The diversity of such systems means that technology varies, in particular as a function of technological choices based on local and regional ideologies and traditions. The health care system is defined here as a dynamic set of interconnected individuals, institutions, organizations, and projects offering products and services in health care markets. [10] Ruef [11] suggests that inclusion in an ecological field – such as a complex national health care system - should be determined in the context of broad system functions and their linkages. These linkages may be social, functional, geographical, and temporal as well as virtual.

Nonaka has described the complex Japanese concept of "ba" as "a shared space for emerging relationships" that may be physical, virtual or mental in nature reflecting the individual in the collective "all". [12,13] "Ba" poses the foundation for translation of information into knowledge, and the temporal and spatial frame for its use. In this study, "ba" represents the health care system knowledge ecology linking communities of practice and virtual infrastructures contributing to translation of information into knowledge and care delivery. Communities of practice are defined here as groups of people sharing a focus on a common interest or task and interacting regularly to improve their knowledge or performance. [14-17] While research, education and practice in health care may form independent professional communities; researchers, learners and practitioners may also create both social and intellectual capital through mutual and reciprocal engagement. [18]

Effective system integration at the country and regional levels of analysis is particularly evident in the Latin American social medicine model where communities of practice share common ideologies [19] associated with universal health care, public education in medicine and the health sciences, open access publishing of health information and research, and infrastructure creation through open source software development. Some experience suggests that the open source model for software development may be extended to research in the health sciences, pointing to the ideological coherence between infrastructure development and productive activities conducted within such infrastructure. [20-22] Virtual infrastructures, accessible on the Internet offer visible evidence of ecological domains for mapping and analysis of their configurations at micro-, meso- and meta-levels of structure. The research questions considered here are:

- What is the emergent configuration of virtual infrastructures integrating regional and national health information systems and knowledge ecologies?
- What are the roles of human resources and open source communities of practice in these ecological systems?
- What factors determine the transferability of this scalable model to other large system contexts?

The methodology for this qualitative research program is embedded case analysis suggested by Yin. [23] The Latin American and Caribbean Health Sciences System (Bireme) forms the context in which national information systems have developed.

This choice of case is revelatory of the reciprocal roles of regional and country level leadership in this process. While qualitative methods may not yield generalizable conclusions, they contribute to rich description of regional and national knowledge ecologies and the ideological role of the social medicine model shared among countries of the Latin American region. [24] Data are drawn from published accounts of system development [25] and the websites of the constituent organizations, networks and services to describe the configuration of virtual infrastructures. [26,27] E-mapping software is used to visualize maps of the linkages among resources identified in the knowledge ecology of the regional Virtual Health Library of the Latin American and Caribbean Health Sciences System (Bireme). [28,29] The maps generated using this methodology show how global, regional and national open source resources are shared and integrated by users in the virtual infrastructure. [30,31] Open source ideology and culture are considered in development of a model of transferability of the Bireme system to other large regional systems such as those in China and India.

References

1. Barabási, A.-L.: Linked: How Everything is Connected to Everything Else and What It Means for Business, Science, and Everyday Life. Basic Books, New York (2003)
2. PloS Medicine (eds) It's the Network, Stupid: Why Everything in Medicine is Connected. PLoS Med. 5, e71 (2008)
3. Reynolds, C., Wyatt, J.: Open Source, Open Standards, and Health Care Information Systems. J. Med. Internet Res. 13, e24 (2011)
4. Godlee, F., Pakenham-Walsh, N., Ncayiyana, D., Cohen, B., Packer, A.: Can We Achieve Health Information for All by 2015? Lancet 364, 295–300 (2004)
5. Meneghini, R., Mugnaini, R., Packer, A.: International versus National Oriented Brazilian Scientific Journals: A Scientometric Analysis Based on SciELO and JCR-ISI Databases. Scientometrics 69, 529–538 (2006)
6. Meneghini, R., Packer, A.: Is There Science Beyond English? EMBO Reports 8, 112–116 (2007)
7. Packer, A.: SciELO as a Model for Scientific Communication in Developing Countries: Origins, Evolution, Current Status, Management and Perspectives of the SciELO Network of Open Access Collections of Ibero-America Journals. CODATA 2007 - Strategies for Open and Permanent Access to Scientific Information in Latin America: Focus on Health and Environment Information for Sustainable Development. Sao Paulo, Brazil, May 8-10 (2007)
8. Packer, A.: The SciELO Model for Electronic Publishing and Measuring of Usage and Impact of Latin American and Caribbean Scientific Journals. In: Second UCSU/UNESCO International Conference on Electronic Publishing in Science, February 20-23 (2001)
9. Séror, A.: A Case Analysis of INFOMED: The Cuban National Health Care Telecommunications Network and Portal. J. Med. Internet Res. 8, e1 (2006)
10. Alliance for Health Policy and Systems Research: Strengthening Health Systems: The Role and Promise of Policy and Systems Research. Global Forum for Health Research, Geneva (2004)
11. Ruef, M.: The Emergence of Organizational Forms: A Community Ecology Approach. Am. J. Sociol. 106, 658–714 (2000)

12. Nonaka, I.: The Concept of "Ba": Building a Foundation for Knowledge Creation. Calif. Manage. Rev. 40, 40–54 (1998)
13. Nonaka, I.: A Dynamic Theory of Organizational Knowledge Creation. Organ. Sci. 5, 14–37 (1994)
14. Wenger, E.: Communities of Practice: Learning as a Social System. Systems Thinker 9, 1–8 (1998)
15. Wenger, E.: Communities of Practice: Learning, Meaning, and Identity. Cambridge University Press, Cambridge (1998)
16. Edwards, K.: Epistemic Communities, Situated Learning and Open Source Software Development. MIT Working Paper (2001)
17. Haas, P.: Introduction: Epistemic Communities and International Policy Coordination. Int. Organ. 46, 1–35 (1992)
18. McDonald, P.W., Viehbeck, S.: From Evidence-Based Practice Making to Practice-Based Evidence Making: Creating Communities of (Research) and Practice. Health Promot. Pract. 8, 140–144 (2007)
19. Committee for Economic Development (CED): Open Standards, Open Source, and Open Innovation: Harnessing the Benefits of Openness. Washington, D.C. (2008)
20. Anonymous: An Open-source Shot in the Arm? The Economist (June 10, 2004)
21. Maurer, S., Rai, A., Sali, A.: Finding Cures for Tropical Diseases: Is Open Source an Answer? PLoS Med. 1, 180–183 (2004)
22. Mueller, M.: Info-Communism? Ownership and Freedom in the Digital Economy. First Monday 13 (2008)
23. Yin, R.: Case Study Research: Design and Methods. Sage, London (2002)
24. Collier, D., Mahoney, J.: Insights and Pitfalls: Selection Bias in Qualitative Research. World Polit. 49, 56–91 (1996)
25. VHL Guide (2011),
    http://guiabvs2011.bvsalud.org/en/presentation/
26. Vidal Ledo, M.C., Febles Rodríguez, P., Estrada Sentí, C.V.: Mapas Conceptuales. Educación Médica Superior 21 (2007)
27. Rodriguez Pina, R.A., Guerra Avila, E.: Mapas Conceptuales y Geo-referencias en Productos y Servicios de Inteligencia Empresarial. In: ACIMED, vol. 17 (2008)
28. Egbu, C., et al.: Knowledge Mapping and Bringing about Change for the Sustainable Urban Environment. In: Engineering and Physical Sciences Research Council (EPSRC), Glasgow (2006)
29. Ruffini, M.: Using E-Maps to Organize and Navigate Online Content. EDUCAUSE Quarterly 31, 56–61 (2008)
30. Ebener, S., Khan, A., Shademani, R., Compernolle, L., Beltran, M., Lansang, M.A., et al.: Knowledge Mapping as a Technique to Support Knowledge Translation. B. World Health Organ. 84, 636–642 (2006)
31. Lavis, J., Lomas, J., Hamid, M., Sewankambo, N.: Assessing Country-Level Efforts to Link Research to Action. B. World Health Organ. 84, 620–628 (2006)

Something of a Potemkin Village? Acid2 and Mozilla's Efforts to Comply with HTML4

Matthijs den Besten[1] and Jean-Michel Dalle[2]

[1] Chaire Innovation et Régulation, Ecole Polytechnique, Paris, France
`matthijs.den-besten@polytechnique.edu`
[2] Université Pierre et Marie Curie, Paris, France
`jean-michel.dalle@upmc.fr`

The real point here is that the Acid3 test isn't a broad-spectrum standards-support test. It's a showpiece, and something of a Potemkin village at that. Which is a shame, because what's really needed right now is exhaustive test suites for specifications— XHTML, CSS, DOM, SVG.[2]

Acid3 is the third of three benchmark tests that have been devised to challenge browsers to comply with Internet standards [6]. While Firefox developers at Mozilla had fully embraced the predecessor to Acid3, Acid2, they showed themselves much more reticent this time around. As the quote above indicates they had come to feel that Acid3 would divert attention from the real issues and might actually make it more difficult to achieve "deep compliance" as developers would scramble to come up with quick fixes just to pass the benchmark test. But were these fears justified? To find out, we retrieved the bug reports for bugs in Mozilla's Bugzilla bug tracker concerning compliance with the HTML4 standard and tried to analyze the differences in the process of bug resolution between bugs that were linked to Acid2 and bugs that were not. In Bugzilla, the bug resolution process passes a number of well-defined stages. Based on the transition rates that we observe we conclude that the process of bug resolution is markedly different for bugs associated with Acid2. In particular, bug resolution appears to be much more chaotic in case of Acid2. This might be symptomatic for "scrambling", which would explain why developers were not so keen to repeat the experience when Acid3 came around. Further investigations, however, are needed to corroborate this hypothesis.

Bugs reports in Bugzilla are often part of Bug Report Networks [3]. That is, they are part of a network of dependencies as bugs can be declared to depend on, block, or duplicate other bugs. Note that the dependencies between bugs are not always purely technical. In fact, an important type of bugs in Bugzilla is the "meta-bug", also known as the "tracker bug", which is a bug at the root of a dependency tree whose leafs are bugs that are related to the issue that the meta-bug is trying to address. For instance, meta-bug 7954 is the bug that tracks issues related to the implementation of the HTML4 standard and the meta-bug 289480 tracks the issues related to Acid2. For our investigation we took advantage of the efforts of the administrators of these meta-bugs to list all bugs related to HTML4 and Acid2 respectively. Yet, as these meta-bugs facilitate coordination among a group of people with a particular interest, much like project-pages in case of

S.A. Hissam et al. (Eds.): OSS 2011, IFIP AICT 365, pp. 320–324, 2011.
© IFIP International Federation for Information Processing 2011

Wikipedia [5], it might be that the differences in bug resolution behavior, which we observe, are a reflection of internal project-culture rather than the effect of external pressure from a public challenge.

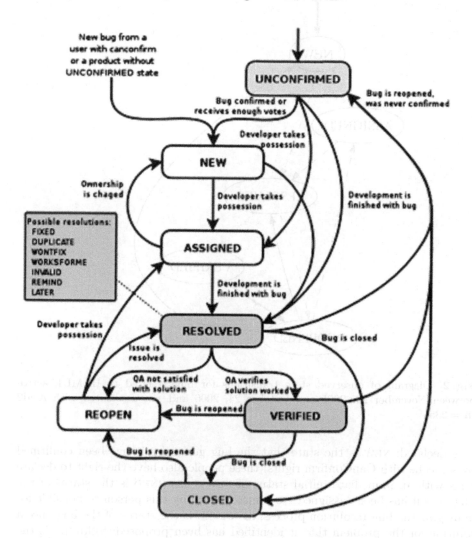

Fig. 1. Bug resolution process according to Bugzilla (source: [4])

As of March 2011, there are 2904 bugs in the dependency tree of the HTML4 meta-bug and 2195 in the tree of Acid2 (and except for 51 bugs, these bugs also appear in the HTML4 tree). In order to inspect the process of bug resolution, we code each bug report as a sequence of states (cf. [1]), where the duration of states is defined by the number of messages posted on the bug's discussion forum and a state by the bug's Bugzilla bug status. Bugzilla distinguishes 7 forms of "status" [4]: UNCONFIRMED is the default initial state assigned to a bug when

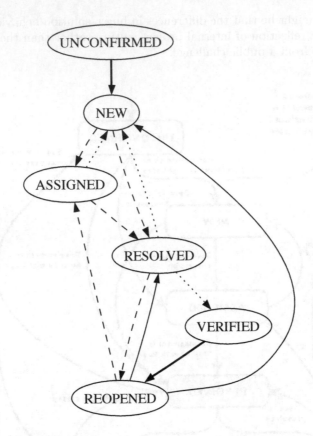

Fig. 2. Diagram of observed state transitions for bugs related to HTML4, active between November 29, 2005 and October 24, 2006 and *not* associated with Acid2 ($n = 235$)

it is declared; NEW is the state that the bug gets once it has been confirmed by someone with CanConfirm rights (these people also have the right to declare bugs with an immediate initial status of NEW); ASSIGNED is the status of the bug once it has been "assigned" to someone, making this person responsible for managing the bug resolution process; RESOLVED is the status of the bug once a solution for the problem that it identified has been proposed; VERIFIED is the status of the bug once the solution has gone through a review; REOPENED is the status of the bug signalling that it has been decided that the proposed solution is not valid or not sufficient; and finally a bug can have status CLOSED to indicate that comments to the bug are no longer welcome.[1] There is a canonical path of bug treatment from UNCONFIRMED to NEW, from NEW to ASSIGNED, etcetera, ending with VERIFIED and/or CLOSED that is proposed in the Bugzilla manual

---

[1] Note, however, that status CLOSED is no longer used (see bug 169885).

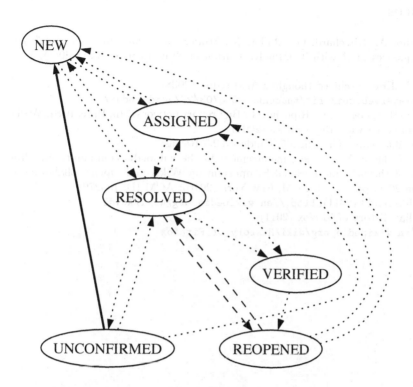

Fig. 3. Diagram of observed state transitions for bugs related to HTML4, active between November 29, 2005 and October 24, 2006 and associated with Acid2 ($n = 274$)

(see Figure **??**. Deviations from this path are allowed, but that are supposed to be exceptions rather than the rule.

Figure 2 is a state diagram based on the transition rates for bugs related to HTML4, but not to Acid2, on which there was activity between the releases of Firefox 1.5 (November 29, 2005) and 2.0 (October 24, 2006).[2] The shape of the edges indicates the likelihood of a transition.[3] The pathways depicted in the diagram are very close to the canonical path proposed in the Bugzilla manual. In contrast, the state diagram in Figure 3 for bugs related to Acid2 during the same period, which falls just after the launch of Acid2 and includes the passing of the test by a development version of Firefox 3.0 a year later, shows a very different picture full of loops and shortcuts. Further investigations will help us determine whether this is a sign of chaos or due to greater efficacy in solving bugs since the availability of a public test suite makes it easy to verify the resolution of a bug and since people with shared interest come to know each others' competences.

---

[2] Release dates according to [7].

[3] *Bold* for a rate higher than 0.2; *solid* if > 0.1; *dashed* > 0.05; and *dotted* > 0.

References

1. Gabadinho, A., Ritschard, G., Müller, N., Studer, M.: Analyzing and visualizing state sequences in R with TraMineR. Journal of Statistical Software 40(4), 1–37 (2011)
2. Meyer, E.: Eric's archived thoughts: Acid redux (2008), http://meyerweb.com/eric/thoughts/2008/03/27/acid-redux/
3. Sandusky, R.J., Gasser, L., Ripoche, G.: Bug report networks. In: Proc. ICSE Workshop Mining Software Repositories (2004)
4. The Bugzilla Team. The Bugzilla Guide - 3.0.5 Release
5. Ung, H., Dalle, J.-M.: Project management in the wikipedia community. In: Proceedings of the 6th International Symposium on Wikis and Open Collaboration, WikiSym 2010, pp. 13–14. ACM, New York (2010); ACM ID: 1832790
6. Wikipedia. Acid3 (2011), http://en.wikipedia.org/wiki/Acid3
7. Wikipedia. History of Firefox (2011), http://en.wikipedia.org/wiki/History_of_Firefox

Aspects of an Open Source Software Sustainable Life Cycle

Flávia Linhalis Arantes and Fernanda Maria Pereira Freire

Nucleus of Informatics Applied to Education (NIED)
State University of Campinas (UNICAMP)
Campinas, SP, Brazil
{farantes,fernanda}@unicamp.br

Abstract. In this paper we present a literature overview about OSS sustainability, considering not only financial resources, but also community growth, source code and tools management. Based on these aspects, we define an OSS life cycle that may contribute to OSS projects sustainability.

Keywords: OSS Sustainability, OSS Communities, Financial Resources, Software Maintenance.

1   Introduction

With the popularity of OSS (Open Source Software), governments, universities and other institutions around the world are adopting free platforms aiming to save millions [1,12].

A natural question that arises when one looks at the increasing use of such software for which you do not pay any fee or license is: how open source projects are supported?

Researches related to OSS sustainability [6,9,14] show that the profits come from offering services such as support, consulting and training. In this work, we mean by sustainable the management model that can support the community through various resources. It is not, however, financially maintained by a single company or institution.

We present a study on other aspects, besides financial ones, that are also important for OSS projects sustainability. In the next sections we treat OSS projects sustainability broadly, considering the dynamics involved in a sustainable life cycle.

2   An OSS Sustainable Life Cycle

Based on a literature review, we can say that the sustainability of OSS is closely related to three factors – community growth, financial resources and software management, as illustrated in Figure 1.

S.A. Hissam et al. (Eds.): OSS 2011, IFIP AICT 365, pp. 325–329, 2011.

According to Figure 1, growth and continuity of the community can result more naturally in financial resources that, if well managed, can be reversed in benefits to the community and encourage its growth, feeding the OSS sustainable life cycle.

Fig. 1. Sustainable life cycle model for OSS projects.

The proper management of community, software and tools can decrease the financial resources necessary to maintain the project. With management improvements in that subjects, it is possible to minimize tasks and human factors, and to promote a vision of a robust computational development for the community. As a result, it is possible to reduce costs and, consequently, to contribute to the sustainability of OSS communities.

2.1 Community Growth

When we talk about OSS communities management, it is natural to think of reasons why communities continue to exist and to grow. Some important factors for the existence and continuity of an OSS community are: interactivity, variety of participants, large number of members, web space that allows interaction, matters of common interest, and cooperation among participants. A basic premise is that people should bring information and share it openly with the group. Another important attribute of a community is the population size. A community becomes more valuable as more members join it [18].

Researches also point out that community growth and development process openness can contribute to software quality and reliability [16,8,7]. This statement assumes that the more people are paying attention to the software and interacting with each other, the easier it is to find and to fix its bugs.

Researches also show that the continuity of an OSS community depends on the adoption of the software that it produces [18]. There are several factors for companies or individuals to take the decision to adopt certain free software. These factors include availability of support, size of community, technical attributes of software such as reliability, safety, quality and performance [11,5]. Software adoption can increase the visibility of the community and generate employment opportunities in bundled services (support, enhancements, upgrades) that can contribute to the community sustainability.

2.2   Financial Resources

In most cases, OSS projects profits come from offering services such as support, consulting, training, product maintenance, and development of new software customizations [6,9,14]. Free software can still be funded by various segments of society such as government, academia and corporations [3,4]. In order to receive more direct support from government, the groups associated with OSS often turn to NGOs, foundations and even micro companies [18,14].

Many OSS economic aspects have been investigated in the literature. Riehle [13] tried to answer one of the main questions in the economic field of OSS projects: How is developers payment determined? Lerner and Tirole [10] discuss questions about economic justification of OSS projects. Shirali-Shahreza [15] discuss various OSS aspects, including economic ones.

2.3   Software Management

Factors related to source code structure can greatly contribute to OSS projects sustainability – it can lead to faster changes with less bugs. The code with a modular structure is an incentive for developers to enter and to continue in the project development [2]. Terceiro and colleagues [17] claim that structural complexity increases the cost of maintaining a software project, because the code becomes harder to understand, and consequently more difficult to modify. In OSS projects, this increased effort may represent an additional difficulty for the entry of new developers, and a sustainability problem.

The discontinuity of the development team and the geographically distributed nature of OSS projects make them even more difficult to manage. For issues like these, the need to communicate, interact and socialize using communication tools with computational support is even greater.

In order to coordinate their work, OSS communities members use the Internet with simple and widely available tools. There are two categories of tools used in OSS projects. The first one is related to communication between community members and the second concerns the management of source code.

A challenge to be exploited in this context is to investigate solutions on how to improve integration and communication between OSS tools, such as discussion forum, issue tracking, wiki, among others. The possibilities of using the communication that is made through these tools can reduce, for example, support costs and other activities that require time from the development team.

3   Conclusion

This study provided a brief overview, bringing together different aspects which contribute to OSS projects sustainability. With such a study we defined a sustainable life cycle aiming to provide guidance on the creation and maintenance of sustainable OSS development and communities.

References

1. Allen, J.P.: Open source deployment at the city and county of san francisco: From cost reduction to rapid innovation. In: Hawaii International Conference on System Sciences, pp. 1–10 (2010)
2. Baldwin, C.Y., Clark, K.B.: The architecture of participation: Does code architecture mitigate free riding in the open source development model? Manage. Sci. 52, 1116–1127 (2006), http://portal.acm.org/citation.cfm?id=1246148.1246160
3. Capek, P.G., Frank, S.P., Gerdt, S., Shields, D.: A history of ibm's open-source involvement and strategy. IBM Syst. J. 44, 249–257 (2005), http://dx.doi.org/10.1147/sj.442.0249
4. Capra, E., Francalanci, C., Merlo, F., Rossi Lamastra, C.: A survey on firms participation in open source community projects. In: Boldyreff, C., Crowston, K., Lundell, B., Wasserman, A.I. (eds.) OSS 2009. IFIP AICT, vol. 299, pp. 225–236. Springer, Heidelberg (2009)
5. Dedrick, J., West, J.: An exploratory study into open source platform adoption. In: Proceedings of the Proceedings of the 37th Annual Hawaii International Conference on System Sciences (HICSS 2004) - Track 8, vol. 8, pp. 1–10. IEEE Computer Society, Washington, DC (2004), http://portal.acm.org/citation.cfm?id=962756.963252
6. Hecker, F.: Setting up shop: The business of open-source software. IEEE Softw. 16, 45–51 (1999), http://dx.doi.org/10.1109/52.744568
7. Joode, R.W., Bruijne, M.: The organization of open source communities: Towards a framework to analyze the relationship between openness and reliability. In: Hawaii International Conference on System Sciences, vol. 6, p. 118b (2006)
8. Lakhani, K., Wolf, R.G.: Why hackers do what they do: Understanding motivation and effort in free/open source software projects. Social Science Research Network, 1–27 (2003), http://www.ssrn.com/abstract=443040
9. Lawton, G.: The changing face of open source. Computer 42, 14–17 (2009)
10. Lerner, J., Tirole, J.: Some simple economics of open source. The Journal of Industrial Economics 50(2), 197–234 (2002), http://dx.doi.org/10.1111/1467-6451.00174
11. Morgan, L., Finnegan, P.: Benefits and drawbacks of open source software: An exploratory study of secondary software firms. In: Proceedings of the 3rd IFIP International Conference on Open Source Systems (OSS), pp. 307–312. Springer, Heidelberg (2007)
12. Richter, D., Zo, H., Maruschke, M.: A comparative analysis of open source software usage in germany, brazil, and india. In: Proceedings of the 2009 Fourth International Conference on Computer Sciences and Convergence Information Technology, ICCIT 2009, pp. 1403–1410. IEEE Computer Society, Washington, DC (2009), http://dx.doi.org/10.1109/ICCIT.2009.169
13. Riehle, D.: The economic motivation of open source software: Stakeholder perspectives. Computer 40, 25–32 (2007), http://portal.acm.org/citation.cfm?id=1251559.1251741
14. Riehle, D.: The economic case for open source foundations. Computer 43, 86–90 (2010)
15. Shirali-Shahreza, S., Shirali-Shahreza, M.: Various aspects of open source software development. In: International Symposium on Information Technology, ITSim 2008, vol. 4, pp. 1–7 (2008)

16. Stamelos, I., Angelis, L., Oikonomou, A., Bleris, G.L.: Code quality analysis in open source software development. Inf. Syst. J. 12(1), 43–60 (2002)
17. Terceiro, A., Rios, L.R., Chavez, C.: An empirical study on the structural complexity introduced by core and peripheral developers in free software projects. In: Proceedings of the 2010 Brazilian Symposium on Software Engineering, SBES 2010, pp. 21–29. IEEE Computer Society, Washington, DC (2010), http://dx.doi.org/10.1109/SBES.2010.26
18. Vincentin, I.C.: Desenvolvimento de Software Livre no Brasil: estudo sobre a percepção dos envolvidos em relação às motivações ideológicas e de negócios. Ph.D. thesis, Faculdade de Economia, Administração e Contabilidade da Universidade de São Paulo (2007)

Open Source and Open Data: Business Perspectives from the Frontline

Juho Lindman and Yulia Tammisto

Aalto University School of Economics, Information Systems Science,
PL 21210, 00076 Aalto, Helsinki, Finland
{Yulia.Tammisto,Juho.Lindman}@aalto.fi

Abstract. Open data initiatives on governmental data seem often to be linked to small software companies, which also use and release software under OSS licenses. This paper calls for more research to understand the similarities between open data and open source software vendors. We build a theoretical linkage between the more established OSS research and emerging research on open data in the context of small software companies.

1 Introduction

Governmental Open Data projects in different countries have created new opportunities for small software companies [11], but the possibilities of Open Data (OD)[1] are not limited to governmental data. A better understanding of the changes in the ecosystems where these small software companies operate helps to better understand the transformation of the software marketplace driven by OD and Open Source Software (OSS)[2].

There is a gap in research traditions between research on OSS and OD. This is surprising at the outset, as most OD advocates have invested heavily in OSS; many of the tools used in OD publication are licensed under OSS licenses; and often the actual companies are similar or even operate in both OSS and OD. OD also enjoys a wide popularity in OSS communities. We propose that this gap should be bridged and theoretical linkages built between OSS and OD research.

2 OSS and OD

Voluntary collective action systems often include a public or semipublic good [5]. These public goods can be for example OSS or OD. Mixing open and proprietary product strategies offers potential to many software companies [3]. Another way to

---

[1] OD refers to "information that has been made technically and legally available for reuse" [22]. In addition to the technological details our definition stresses the legal and organizational aspects of open data that are similar in OSS research.

[2] In this paper we rely on the following OSS definition: "Open Source is a development method for software that harnesses the power of distributed peer review and transparency of process" (http://www.opensource.org).

S.A. Hissam et al. (Eds.): OSS 2011, IFIP AICT 365, pp. 330–333, 2011.
© IFIP International Federation for Information Processing 2011

benefit from more open development is to change internal software production based on the lessons from the OSS world [4].   Concepts used to describe OSS inspired practices within an organization include: Corporate Source [2] and Inner Source [9]. Open Source can also be considered as a sourcing strategy and defined as a governance model, where software development tasks are opensourced to an unknown workforce [13].

Open government data has been claimed to offer possibilities for economic growth by providing data sets which can be used in the provisioning of new services [6]. Tim Berners-Lee [1] has provided a categorization of five levels of open data for linked open data. The process of data transformation and publication can be theorized in several ways. Latif et al. [8] offered a model to describe the roles of entities in OD business: 1) raw data provider, 2) linked data developer and 3) applications developer. Elsewhere [12], we have developed a conceptualization, building on Latif [8] and Rajala's [10] classification, which focuses on the different business models of the actors. Based on our findings, it seems that value capturing (of the small software companies of open data) may follow three different paths: 1) consultancy, 2) conversion, and 3) application development.

3 Findings

We conducted a small round of interviews about OD using interpretive interview approach [7] and compared the results with the earlier collected data on OSS. Through the course of the analysis we detected a certain similarities between OSS and OD companies that are reported in Table 1. All the respondents are from Finland, their profiles are listed in Table 2.

Table 1. Similarities between OD and OSS business

	Open Data	Open Source	Similarity
Competition environment	Market is divided between small software companies and large software companies	Market is divided between small, medium- and large software companies	Most of the large competitors are the same in both OD and OSS. Some companies are the same and they use and develop the same software.
Customers	So far emphasis on public organizations (cultural institutions, municipalities), potential in the media-industry	Emphasis on public organizations (schools) and private actors	Public sector as a large customer
Revenue sources	Consultancy, conversions, application development, maintenance	Consultancy, application development, maintenance	Not based on traditional software sales, develop services on top of public goods

Table 1. (*Continued*)

Communities	Often enjoy popularity and community support	Often enjoy popularity and community support	Developer-communities are the same and have "activist" components
Openness of activities	*"I think the added value [of OD] comes from having more clever people to look at it."*	*"More eyeballs make bugs shallow"*	Favor openness in the innovation activity

Table 2. Informants of the interviews

OD	*Company*	*Position*
1	Small (5 persons) web technology and application development company	Project manager / Consultant
2	Small (5 persons) web technology and application development company (same as above)	CEO / Consultant / Developer
3	Small (10 persons) software development company	Project manager / Developer
4	Small (2 persons) consultancy and software development company	CEO / Consultant

OS	*Company*	*Position*
1	Small (3 persons) OSS company developing collaborative learning tools	CEO
2	Small (3 persons) OSS company developing collaborative learning tools (same as above)	Developer
3	Small (1 person) OSS company developing relational database tools	Entrepreneur
4	Small (10 persons) OSS company developing web services	Developer

4   Conclusion

The aim of this paper was to look for some similarities between OD and OSS in the context of small software companies engaged in OD and OSS. We speculate that there are interesting lessons to be learned to the OD research from OSS business model research related to service design and delivery relying on public goods. Research on OSS communities can in some cases be applicable also to the emerging OD communities. By this paper we only scratched the surface of the potential contribution for the research. We call for a further research on comparison of OD and OSS to realize all the benefits of the combination of these two phenomena.

References

1. Berners-Lee, T.: Linked Data Design Issues (July 2006),
   http://www.w3.org/DesignIssues/LinkedData.html
2. Dinkelacker, J., Garg, P., Miller, R., Nelson, D.: Progressive Open Source. In: The Proceedings of ICSE 2002, May 19-25, pp. 174–184 (2002)
3. Fosfuri, A., Giarratana, M., Luzzi, A.: The Penguin Has Entered the Building: The Commercialization of Open Source Software Products. Organization Science 19(2), 292–305 (2008)
4. Gurbani, V., Garvert, A., Hersleb, J.: Managing a Corporate Open Source Asset. Communications of the ACM 53(2), 155–159 (2010)
5. Heckathorn, D.: The Dynamics and Dilemmas of Collective Action. American Sociological Review 61(2), 250–277 (1996)
6. Huijboom, N., Van den Broek, T.: Open Data: an International Comparison of Strategies. European Journal of ePractice 12 (March/April 2011),
   http://www.epractice.eu/files/European%20Journal%20epractice%20Volume%2012_1.pdf
7. Klein, H., Myers, M.: A set of Principles for Conducting and Evaluating Interpretative Field Studies in Information Systems. MIS Quarterly 23(1), 67–94 (1999)
8. Latif, A., Saeed, A.U., Hoefler, P., Stocker, A., Wagner, C.: The Linked Data Value Chain: A Light Weight Model for Business Engeneers. In: Proceedings of I-SEMANTICS 2009 International Conference on Semantic Systems, Graz, Austria, pp. 568–575 (2009)
9. Linden, F., Lundell, B., Marttiin, P.: Commodification of Industrial Software – A Case for Open Source. IEEE Software (July-August 2009)
10. Rajala, R.: Determinants of Business Model Performance in Software firms. Doctoral Dissertation, Aalto University School of Economics, Helsinki, Finland (2010)
11. SOMUS. Social media for citizens and public sector collaboration) project – final report (January 2011),
    http://www.vtt.fi/inf/pdf/publications/2011/P755.pdf
12. Tammisto, Y., Lindman, J.: Open Data Business Models. In: The Proceedings of the 34th IRIS Seminar, Turku, Finland, August 16-19, pp. 16–19 (accepted, 2011)
13. Ågerfalk, P., Fitzgerald, B.: Outsourcing to an Unknown Workforce: Exploring Opensourcing as a Global Sourcing Strategy. MIS Quarterly 32(2), 385–409 (2008)

Forge.mil: A Case Study for Utilizing Open Source Methodologies Inside of Government

Guy Martin and Aaron Lippold

Forge.mil Community Management Team
gmartin@collab.net, aaron.lippold@disa.mil
http://www.forge.mil

Abstract. In late 2008, DISA (Defense Information Systems Agency), the global IT arm of the US Department of Defense, embarked upon a project to create an internal collaboration and software application lifecycle management system. Beyond simply fielding yet another tool, the Forge.mil effort was designed to fundamentally change the way the DoD developed and acquired software technology and systems. The method of this change was the application of Open Source principles **inside** of the larger DoD community, including ideas such as meritocracy and code sharing, as well as Agile and collaborative software development. This lightning talk will explain the rationale behind Forge.mil, how it was developed using Open Source principles, and how it continues to influence technology acquisition within the DoD in both practice and policy changes.

1 Introduction

The US Department of Defense has had a long (and at times challenging) relationship with the Open Source world. While there have been successes along the way (including work on projects such as SE Linux), by and large, the use of Open Source within the DoD has been limited to certain server side applications.

There have been groups within the DoD community who recognized that there was value not only in working within the Open Source world where appropriate, but also in applying the same principles to internal development that made projects such as Linux, Apache, and Subversion successful.

This goes beyond simple 'code sharing' or 'reuse repositories,' which, while a laudable goal, are rarely successful without a large upfront effort at cultural upheaval. The DISA team (civilian, military, and contractors) that came together in 2008 to start Forge.mil realized early on that they were not going to change years of entrenched cultural resistance to 'the Open Source way.' What they hoped to provide, however, was not only a simple set of integrated tools to make life easier for all project stakeholders (developers, program managers, testers, senior executives), but a 'safe place' to begin the cultural change process necessary to make technology acquisition in the DoD much more streamlined.

Forge.mil is the third incarnation of this effort and arguably, the most successful, with 10,000 registered users and a new 'Community layer' built upon one of the most successful Open Source projects to date – the Drupal content management engine.

S.A. Hissam et al. (Eds.): OSS 2011, IFIP AICT 365, pp. 334–337, 2011.

However, there are still entrenched cultural issues that need to be addressed to make the Forge.mil effort more successful and pervasive.

1.1 Vision

Forge.mil's official mission statement is: *"Improve DoD's ability to rapidly deliver dependable software, services and systems"*

Unofficially, the team works very hard at breaking down cultural barriers and 'silos of excellence' through the application of Open Source principles such as: meritocracy, transparency, reuse, trust, and community. While there has to be a base of operations, per se, with a unified tool set, a lot of the work done by the Forge.mil community management team revolves around coaching projects on collaboration techniques and software reuse strategies. Additionally, the team has shown leadership by helping lawmakers craft new guidelines on DoD acquisition strategy[1]. These new guidelines direct the US Secretary of Defense to establish *"a new acquisition process for information technology systems, designed to include:*

- *Early & continual involvement of users*
- *Multiple, rapidly executed increments of capability*
- *Early, successive prototyping to support an evolutionary approach*
- *A modular, open-systems approach"*

Forge.mil helps provide a place to continue defining this improved style of 'innersourcing' for DoD technology acquisition.

1.2 Implementation Specifics

Forge.mil serves two specific types of audiences:

- Projects and developers looking to share and collaborate on internal code/projects
- Teams looking to collaborate with a limited set of government/contract workers

Because of these diverse audiences, and to make administration simpler, Forge.mil is primary composed of two main capabilities:

- **Software.forge.mil**
  o   Freely available for project hosting for 'internal open' projects
- **Project.forge.mil**
  o   Fee-for-service offering allowing for limited participation

Additionally, there is a Content Management System (utilizing Drupal) called **Community.forge.mil** that provides 'social development' tools such as voting, reputation management, and project activity streams to provide visibility across multiple projects on each project site that may be part of a larger community of interest.

---

[1] US Congress: HR 2647, National Defense Authorization Act (2010), Section 804.

It is important to note again at this point that these systems are not freely available to the outside Open Source community – they are behind DoD firewalls and protected by Public Key Infrastructure systems. These systems are designed to replicate the Open Source 'dynamic' present in successful projects such as Linux, Apache, Drupal, etc.

The primary human interface between the community of users and the system is the Forge.mil Community Management team, composed of two half-time community managers (contractors), and one quarter-time government employee. This team's roles include:

- Project onboarding and adjudication
- Coaching for Open Source and Agile best practices
- Detailed consulting/support to remove barriers to entry

One of the primary roles of this team is to help in determining what projects should be allowed on the 'Internal Open Source' system (Software.forge.mil). Projects that are forks of existing Open Source projects are usually rejected and their requesters sent to the proper external Open Source community, so that unmaintainable DoD-specific forks do not occur.

1.3 Challenges

Attempting culture change of this magnitude inside of an organization like the US Department of Defense is not without significant challenges. Among those are:

- Contractor resistance to 'loss of intellectual property'
- Fear of sharing and/or showing substandard code/systems
- Perceived increases in cost to develop systems (easily countered – see Metrics/Outcomes below)
- Cultural resistance to change

There is no single 'silver bullet' to address all of these issues, but increasing economic pressures have provided a lever to help most teams 'cross the chasm.' Additionally, the government can invoke 'Government Purpose Rights' on software they have paid a contractor to develop, though there are sometimes additional costs for that.

1.4 Metrics/Outcomes

Forge.mil closely tracks several key metrics and ROI figures. In three years of operation, the program has had the following successes:

- 10,000 registered users
- 500 projects
- 57,000 software commits
- 51,000 software downloads
- 4,000 discussion posts
- 15,000 shared documents
- 1,000 software repositories created
- $175M in ROI savings (cost avoidance and software asset reuse)

1.5 Conclusion

The implementation of Forge.mil provides a useful case study in how to apply Open Source methodologies to internal development problems within corporations or governments for maximum benefit. There are many challenges along the way, but a collaborative approach can be utilized to overcome most stakeholder concerns and issues.

Health Informatics: The Relevance of Open Source and Multilevel Modeling

Luciana T. Cavalini[1] and Timothy W. Cook[2]

[1] Institute of Community Health, Fluminense Federal University (UFF)
Rua Marquês de Paraná, 303 - 3rd Floor Annex HUAP
Niterói, RJ – Brazil 24220-331
lutricav@vm.uff.br
[2] National Institute of Science and Tecnology – Medicine Assisted by Scientific Computing
(INCT-MACC)
timothywayne.cook@gmail.com
http://www.mlhim.org

Abstract. Health information features significant spatial-temporal and domain complexities, which brings challenges to the implementation of patient-centered, interoperable and semantically coherent healthcare information systems. This position paper supports the idea that the multilevel modeling approach is essential to ensure interoperability at the semantic level, but true interoperability is only achieved by the adoption of open standards, and open source implementations are needed for promote competition based on software quality. The Multilevel Healthcare Information Modelling (MLHIM) specifications are presented as the fully open source multilevel modeling reference implementation, and best practices for the development of multilevel-based open source healthcare applications are suggested.

Keywords: Health informatics, open source software, multilevel modeling.

1 Introduction

The information related to human health is inherently complex at the intersections of space, time and knowledge [17]. This complexity precludes the feasibility of one, single, all encompassing electronic health record for individuals in a population, since various pieces of information from a number of different applications may be needed at any point in time [21]. Semantic interoperability is crucial in recording information in purpose specific applications that need to synchronize to larger databases. This is the way to provide useful information from the point of care to the healthcare services in a timely manner [15].

Regarding public health, the typical turnaround time now for acquiring the healthcare status of a population can be weeks or even months while experts pour over the data and try to merge pieces from various applications and paper forms into something meaningful [14]. By basing applications on a common information model and using a constraint based approach to define the knowledge components, we can achieve semantic interoperability and near real-time information regarding the

S.A. Hissam et al. (Eds.): OSS 2011, IFIP AICT 365, pp. 338–347, 2011.

healthcare status of a population and its individuals so that faster action can be taken to meet the needs in that area. This approach also helps empowering application developers at the local level, so they can develop healthcare applications fitted to the very specific local needs without losing semantic coherence and interoperability with other local services and with the highest levels of the healthcare systems [11].

Healthcare inequalities will not be easily solved by adopting 21st Century healthcare based on 20th Century information systems [16]. There are no remaining obstacles related to hardware, including mobile computing and pervasive medicine, but software based on traditional data models are not fitted to deal with the significant spatial and temporal complexities of healthcare information [4].

That is the case because health information systems based on traditional data models are not interoperable and have high maintenance costs. These problems have a significant negative impact on the use of these systems to the emerging situations and dynamics currently found in healthcare [5,25]. In fact, the development of healthcare applications is a complex challenge, especially given the large number of concepts in constant evolution, which makes it difficult to reach a consensus on any concept [9].

Some solutions to these problems have been proposed over the past two decades, such as the work of Yoder et al. [35]. However, the solution most fitted to the specific features of healthcare information involves the separation between domain model and persistence of data. This multilevel modeling approach proposes the definition of at least two levels: the Reference Model, which defines generic types of data and data structures and a Domain Model, defined by restrictions on the Reference Model [19].

Health information systems based on multilevel modeling are more easily interoperable and can be implemented on any hardware. The adoption of a common Reference Model and a Domain Model for different systems allows a transparent and shareable interface with geographic information systems and statistical analysis tools that can analyze information collected from various remote systems [22].

Nowadays, multilevel modeling specifications for healthcare information systems are openly available and proven in software. Based on these specifications, it is possible to develop healthcare applications centered on the citizen, with the capability of recording longitudinal data [12].

Furthermore, decision support systems and standardized reports can be implemented in the systems and still ensuring semantic interoperability at any level, since the development of algorithms for decision support based on a common domain model allows the reuse of decision rules in different implementations. Thus, in the point and time of care, control measures can be implemented immediately, allowing for greater effectiveness of healthcare and, at the governance level, larger areas can be monitored and priority areas can be identified for intervention [2].

However, despite its technical advantages, multilevel modeling-based solutions have not been widely implemented in real healthcare settings, except for some few academic projects [6,20,24].

There is one aspect that is essential but seldom addressed regarding interoperability of healthcare information systems, which is related to the modality of software licensing. In fact, the general business model of proprietary software companies may be considered unfriendly to interoperability, since the competition between companies has the goal to establish hegemony or monopoly, in order to concentrate capital, and that is based on the secrecy of the software source code. Actually, one can state there is no proven interoperability without the development of, at least, open specifications, since it is necessary for one system to be compliant to the other system's features that are related to interoperability, and that can only be attained if the systems were developed based on a common set of specifications. Thus, expanding this argument, it is possible to deduce that a complete condition of interoperability between all systems is only possible if they share a common set of specifications, at least at the level of data extract exchange. Therefore, full interoperability requires open standards and open source software [1,8].

This business model has not shown any differences when applied to healthcare. Actually, it is stated that open standards facilitate competition between open source software and proprietary software, since it allows the competition between different implementations of the same specification [30]. This is a key issue related to software quality, which is crucial in healthcare, since the quality of the software is directly related to the quality of care [2].

Taking into account the centrality of open source and multilevel modeling to ensure the development of high quality, citizen-centered, interoperable, semantically coherent health information systems, our objective is to describe the essential features of a open specification for multilevel healthcare information modeling, and to propose a set of best practices for the development of healthcare applications based on those specifications.

2   Method

2.1   Summary of the Specifications

The "Multilevel Healthcare Information Modelling" (MLHIM) specifications are a fully open set of specifications for the development of health information systems based on multilevel modeling. The MLHIM documentation is published under the Open Document Format (ODF) at http://www.mlhim.org.

The technological choices for the development of the MLHIM specifications were made because of the distributed and diverse nature of healthcare information systems; thus, its goal is interoperability and standardization is the path. The basis of MLHIM are the dual-level openEHR specifications [3] and the healthcare-specific data types as defined by the ISO 21090 standard. These specifications and standards are articulated in a single specification, with the specific purpose of creating a path for semantic interoperability among different health applications, including legacy systems.

In the MLHIM specifications, the classes of the Reference Model are persistent and should be kept as stable as possible over time. In the Domain Model, the Constraint Definitions on the Reference Model provide the semantic interpretation of the objects stored by the Reference Model.

The idea behind the multilevel modeling is that changes in structure and rules of inference are reflected on the Constraint Definitions and not on the Reference Model. Thus, change requests on the persistence mechanisms of the information systems are reduced. Furthermore, the Constraint Definitions are created and edited by domain experts and not by computer scientists, which avoids the need for interpretation of the knowledge extracted from an *ad hoc* interaction. Once the domain expert is responsible for modeling the knowledge, concepts are thoroughly and accurately expressed as Concept Constraint Definitions (CCDs).

This approach is compliant to the following standards developed by the International Organization for Standardization (ISO):

- ISO/TS 18308:2004 - Health informatics - Requirements for an electronic health record architecture;
- ISO/TR 20514:2005 - Electronic health record - Definition, scope and context;
- ISO 13606-1:2008 - Health informatics - Electronic health record communication - Part 1: Reference model;
- ISO 13606-2:2008 - Health informatics - Electronic health record communication - Part 2: Archetype interchange specification;
- ISO 13606-3:2009 - Health informatics - Electronic health record communication - Part 3: Reference archetypes and term lists;
- ISO/TS 13606-4:2009 - Health informatics - Electronic health record communication - Part 4: Security;
- ISO 13606-5:2010 - Health informatics - Electronic health record communication - Part 5: Interface specification;
- ISO/FDIS 21090:2011 - Health informatics - Harmonized data types for information interchange;

2.2 Knowledge Modeling

The MLHIM specifications adopt XML Schema Documents (XSDs) for the elaboration of the Concept Constraint Definitions (CCDs). A CCD is a XSD file that expresses a defined healthcare concept. This concept is expressed on the CCD as constraint definitions on the MLHIM Reference Model.

XML is regarded as the most widely adopted solution to system interoperability and semantic coherence; therefore, in order to fit its purposes, XML must be open source [26]. There are some recent publications (since 2005) about the use of XML Schema languages as an attempt to perform *a posteriori* standardization of data types and metadata [28,32], development of templates for structured documents [27,35], or any combination of the techniques cited above [10,13,29], what can be understood as solutions to promote data interchange between one-level based information systems.

Some of those studies adopt the concept of the domain expert as the author of the knowledge modeling [18]. Although that approach solves some of the semantic loss derived from *ad hoc* interactions between the domain expert and the system developer, it still does not ensure interoperability and semantic coherence for the attempts of data interchange between one-level model applications.

On the other hand, the use of XML W3C Schemas for knowledge modeling based on the dual-model ISO 13606 standard was tested and validated by Rinner *et al.* [31], allowing semantic validation of knowledge components, conditional to the definition of a "fully generic validation" provided by the Reference Model. On the other hand, the authors describe the technical difficulties regarding the specific transformations (or constraints) on the ISO 13606 Reference Model classes that are needed to express a given healthcare concept. That suggests the need for a knowledge modeling editor, which is a common concern for multilevel modeling projects [23,33].

In the MLHIM specifications, a Constraint Definition Designer has been developed, using Mind Maps, which are proven efficient as concept definition tools [7]. XMind (XMind Ltd.), an open source Mind Map editor, was used to build the CCD template for the MLHIM Constraint Definition Designer. In this template, instead of using the Mind Map nodes to directly constraint the domain concepts, they were defined as classes of the MLHIM Reference Model, being the constraint definitions applied by the specific arrangement of some classes required for a specific concept, and by defining constraints on the attributes of those classes, expressed as sub-items of a given Mind Map node. The resulting XMind file should be transformed into a XSD file, which is the CCD for a particular healthcare concept and it can be validated against the MLHIM Reference Model. Furthermore, it can be combined to other CCDs to inform the development of Graphic User Interfaces for MLHIM-based applications. The MLHIM Constraint Definition Designer is available at https://launchpad.net/cdd.

3 Application Development

This section proposes a set of best practices for the multilevel modeling-based application development. The underlying reasoning expressed here is that interoperability and semantic coherence are ensured by open specifications based on multilevel modeling, such as the MLHIM specifications, which include a generic, stable, standard-compliant Reference Model and the rules for the Concept Constraint Definitions on the Reference Model. In order to increase the probability of building CCDs that are valid against the MLHIM Reference Model, the Constraint Definition Designer was devised as a CCD editing tool.

Any other particular feature such as the combination of CCDs in templates, the definition of GUI, the choice of the object-oriented programming language and the correspondent application framework, the choice of the No-SQL or object-oriented database to persist the data and the query algorithms are considered as implementation choices and do not interfere on the technical aspects regarding interoperability and semantic coherence, already addressed in a comprehensive manner by the MLHIM specifications.

However, in order to guarantee that the interoperability and the semantic coherence ensured by the specifications will be attained by multilevel modeling-based systems, it is necessary to develop high quality applications. Given this reasoning, a non exhaustive set of best practices for healthcare application development based on multilevel modeling is presented below.

3.1 Application Framework and Data Persistence

The MLHIM specifications provide the Open Source Health Information Platform (OSHIP) (https://launchpad.net/oship), a open source implementation of the MLHIM specifications. OSHIP is implemented in Eclipse Modeling Framework (EMF) (The Eclipse Foundation), which allows code export to the main object-oriented languages, such as Java, Python and Ruby. This raw code can be wrapped into many application development frameworks based on those programming languages, allowing a wider adoption of the specifications, since the application framework is not an additional learning curve for the developers.

Usually, the chosen application framework will guide the choice of the type of database to be adopted for the persistence of data. It is important to notice that, since the MLHIM specifications are object-oriented, the more obvious choice would be an object-oriented database for data persistence. However, object-oriented databases may present performance or query issues. Taking that technical difficulty into account, No-SQL or hierarchical databases can be chosen instead, in order to circumvent those technical complexities in real life applications, although the persistence of data origined from MLHIM-based applications is trivial if an object-oriented database is chosen.

3.2 Communication Layer

The communication layer is proposed to be built on any software component based on distributed technology, available in object-oriented programming languages, running on distributed or wireless communication networks. The association between the layers of modeling and data communication can be done through a model-driven approach. In this approach, the overall architecture of the system will be specified at a high level, using an Architecture Description Language (ADL), implicitly or explicitly annotated with the CCDs used in the data layer, allowing the development of the Graphical User Interfaces, which are likely to be persisted in a database, and which are communicated through the network from each system component. This specification will therefore be the basis for the generation of code for different parts of the application.

The proposed ADL library to be adopted is the ADL Acme (http://www.cs.cmu.edu/~acme/). The library AcmeLib can be used as a basis for building code generators for different parts of the application.

3.3 System Integration

The adoption of multilevel modeling for the development of healthcare applications brings a great deal of flexibility for application developers. There is no need for the development of a monolithic Electronic Medical Record (EMR) for the entire healthcare setting, irregardless of its size and complexity (primary care, outpatient clinic, hospital). Multilevel modeling allows the development of purpose-specific applications, no matter how restricted the data is (which includes applications for specialized scientific research); the data can be exchanged from any application to any other, and the data extracts of both are still valid. However, some institutions might require a higher level of system integration due to the specificities of its particular workflow.

In order to allow system integration, the use of a Service Oriented Architecture (SOA) is proposed. SOA allows better integration between different languages and platforms, which is necessary since MLHIM-based applications can be developed on any object-oriented language. Additionally, SOA makes easier the management of scalability, reusability, distribution and storage of applications. The development of this application integration is proposed by the adoption of the Representational State Transfer (REST) architectural style. In order to achieve system integration via REST, it is proposed the use open source libraries for the development of clients and servers based on REST with portability for the main object-oriented languages (e.g., Restfulie, RIP). Those libraries allow code breaking so that it creates a very flexible service that is able to evolve with minimal change on the client side. Thus, it is possible to integrate different applications developed in different languages and platforms in a cost-effective way, reducing risks and costs associated with traditional system integration tasks.

3.4 Decision Support

The C Language Inference Processing System (CLIPS) is proposed as the inference mechanism for the development of decision support algorithms in MLHIM-based applications. CLIPS provides important benefits for the development of decision support engines in healthcare systems for the following reasons: (a) it supports a forward-chaining (or data-driven) mode of processing inference rules, which means that whenever new data become available, all available states of the system are checked again, and (b) it provides a powerful applications programming interface in C/C++, which allows, for the processing of a rule, the addition of routines that are interactive with the user and the management of dependencies between the actions of dynamic control. Both features are essential for modeling rules for dynamic scenarios such as healthcare. In addition, CLIPS supports multiple persistence layers, allowing a fully object-oriented system design, essential for information systems based on multilevel modeling.

3.5 Data Aggregation and Business Intelligence

It is proposed that the preparation of reports based on aggregated data from the local level to regional and national levels be based on the Statistical Data and Metadata Exchange - Health Domain (SDMX-HD) specifications, which are being developed under the auspices of the World Health Organization (WHO) for the standardization of aggregate data formats, in order to facilitate the exchange of statistical measurements and health indicator definitions. These specifications are inspiring some of the requirements of ISO 14639 - eHealth Architecture (now at the stage of Draft Technical Report).

For the execution of Business Inteligence (BI) tasks, open source solutions such as the Pentaho BI tool (http://www.pentaho.com) (Pentaho Corporation) are suggested for the preparation of pre-defined management reports, monitoring reports, custom data analysis, data aggregation and formatting data for export to legate management systems required by national and regional healthcare authorities that are not compliant to the SDMX-HD specifications.

4 Conclusions

The issues regarding interoperability and semantic coherence are more relevant for health information systems than for any other economic sector of the society. That is so due to the need for the maintenance of the citizen's longitudinal health record all through his life. However, the extreme conceptual, spatial and temporal dynamics of the healthcare activities require a high level of diversity between information systems for different healthcare settings and purposes.

Traditional one-level data model applications, which fit the needs of almost any other economic activity of human society are being used for 45 years in the healthcare sector and have not been able to provide a citizen-centered, interoperable and semantically coherent health record.

Over the last 20 years, multilevel modeling specifications have been developed in order to address those important issues. Over that development process, it became evident that true interoperability will only be achieved if the multilevel modeling specifications were openly available. The implementation of the multilevel specifications in open source software has the potential to increase the competition for the development of good quality software, which is critical in the healthcare sector.

This paper presented the state of the art of the open source multilevel modeling that are currently available, which demonstrates the practical possibility of development of open source healthcare applications based on multilevel modeling. Thus, by contributing to those projects, the open source community can help improving the quality of healthcare on a global basis.

References

1. Almeida, F., Oliveira, J., Cruz, J.: Open standards and open source: enabling interoperability. Int. J. Soft. Eng. App. 2(1) (2011), doi:10.5121/ijsea.2011.2101
2. Ammenwerth, E., Shaw, N.: Bad informatics can kill: is evaluation the answer? Methods Inf. Med. 44(1), 1–3 (2005)
3. Barretto, S.A., Warren, J., Goodchild, A., Bird, L., Heard, S., Stumptner, M.: Linking guidelines to Electronic Health Record design for improved chronic disease management. In: AMIA Annu. Symp. Proc., pp. 66–70 (2003)
4. Beale, T., Heard, S.: openEHR Architecture overview. openEHR Foundation, London (2008)
5. Blobel, B.: Comparing concepts for electronic health record architectures. Stud. Health Technol. Inform. 90, 209–214 (2002)
6. Cantiello, J., Cortelyou-Ward, K.H.: The American Recovery and Reinvestment Act: lessons learned from physicians who have gone electronic. Health Care Manag (Frederick) 29(4), 332–338 (2010)
7. Chen, R., Klein, G.: The openEHR Java reference implementation project. Stud. Health Technol. Inform. 129(Pt 1), 58–62 (2007)
8. D'Antoni, A.V., Zipp, G.P., Olson, V.G., Cahill, T.F.: Does the mind map learning strategy facilitate information retrieval and critical thinking in medical students? BMC Med. Educ. 10, 61 (2010)
9. Dutton, W.: Key enablers for eTransformation? eID, interoperability and open source. Eur. J. ePractice 6, 2 (2009)
10. Eccles, M., Mason, J.: How to develop cost-conscious guidelines. Health Technol. Assess. 5(16), 1–69 (2001)
11. Gao, S., Mioc, D., Yi, X., Anton, F., Oldfield, E., Coleman, D.J.: Towards web-based representation and processing of health information. Int. J. Health Geographics 8, 3 (2009)
12. Garde, S., Chen, R., Leslie, H., Beale, T., McNicoll, I., Heard, S.: Archetype-based knowledge management for semantic interoperability of electronic health records. Stud. Health Technol. Inform. 150, 1007–1011 (2009)

13. Garde, S., Knaup, P., Hovenga, E., Heard, S.: Towards semantic interoperability for electronic health records. Methods Inf. Med. 46(3), 332–343 (2007)
14. Hägglund, M., Scandurra, I., Moström, D., Koch, S.: Bridging the gap: a virtual health record for integrated home care. Int. J. Integr. Care 7, 26 (2007)
15. Hammond, W.E., Bailey, C., Boucher, P., Spohr, M., Whitaker, P.: Connecting information to improve health. Health Aff. (Millwood) 29(2), 284–288 (2010)
16. Haughton, J.: Look up: the right EHR may be in the cloud. Major advantages include interoperability and flexibility. Health Manag. Technol. 32(2), 52 (2011)
17. Haux, R.: Medical informatics: past, present, future. Int. J. Med. Inform. 79(9), 599–610 (2010)
18. Hudson, D.L.: Cohen ME. Uncertainty and complexity in personal health records. In: Conf. Proc. IEEE Eng. Med. Biol. Soc., pp. 6773–6776 (2010)
19. Hulse, N.C., Rocha, R.A., Del Fiol, G., Bradshaw, R.D., Hanna, T.P., Roemer, L.K.: KAT: A flexible XML-based knowledge authoring environment. J. Am. Med. Inform. Assoc. 12, 418–430 (2005)
20. Kalra, D., Beale, T., Heard, S.: The openEHR Foundation. Stud. Health Technol. Inform. 115, 153–173 (2005)
21. Kashfi, H.: An openEHR-based clinical decision support system: a case study. Stud. Health Technol. Inform. 150, 348 (2009)
22. Kelley, J.: The interoperability hang-up. When it comes to information exchange, how should precede what. Health Manag. Technol. 32(2), 32–34 (2011)
23. Kohl, C.D., Garde, S., Knaup, P.: Facilitating secondary use of medical data by using openEHR archetypes. Stud. Health Technol. Inform. 160(Pt 2), 1117–1121 (2010)
24. Maldonado, J.A., Moner, D., Bosca, D., Fernandez-Breis, J.T., Angulo, C., Robles, M.: LinkEHR-Ed: a multi-reference model archetype editor based on formal semantics. Int. J. Med. Inform. 78(8), 559–570 (2008)
25. Martinez-Costa, C., Menarguez-Tortosa, M., Fernandez-Breis, J.T.: An approach for the semantic interoperability of ISO EN 13606 and OpenEHR archetypes. J. Biomed. Inform. 43(5), 736–746 (2010)
26. Michelsen, L., Pedersen, S.S., Tilma, H.B., Andersen, S.K.: Comparing different approaches to two-level modelling of electronic health records. Stud. Health Technol. Inform. 116, 113–118 (2005)
27. Neeser, A.E.: XML: The open source solution to interoperability. Open Lib. Class J. 1(2) (2009),
    http://www.infosherpas.com/ojs/index.php/openandlibraries/article/view/31/40 (accessed on March 6, 2011)
28. Norlin, C., Kerr, L.M., Rocha, R.A.: Using clinical questions to structure the content of a web-based information resource for primary care physicians. In: AMIA Symp. Proc., pp. 482–486 (2009)
29. Paterson, T., Law, A.: An XML transfer schema for exchange of genomic and genetic mapping data: implementation as a web service in a Taverna workflow. BMC Bioinformatics 10, 252 (2009)
30. Qian, Y., Tchuvatkina, O., Spidlen, J., Wilkinson, P., Gasparetto, M., Jones, A.R., Manion, F.J., Scheuermann, R.H., Sekaly, R.P., Brinkman, R.R.: FuGEFlow: data model and markup language for flow cytometry. BMC Bioinformatics 10, 184 (2009)
31. Reynolds, C.J., Wyatt, J.C.: Open Source, open standards, and health care information systems. J. Med. Internet Res. 13(1), e24 (2011)

32. Rinner, C., Janzek-Hawlat, S., Sibinovic, S., Duftschmid, G.: Semantic validation of standard-based electronic health record documents with W3C XML schema. Methods Inf. Med. 49(3), 271–280 (2010)
33. Seibel, P.N., Krüger, J., Hartmeier, S., Schwarzer, K., Löwenthal, K., Mersch, H., Dandekar, T., Giegerich, R.: XML Schemas for common bioinformatic data types and their application in workflow systems. BMC Bioinformatics 7, 490 (2006)
34. Sundvall, E., Qamar, R., Nystrom, M., Forss, M., Petersson, H., Karlsson, D., Ahlfeldt, H., Rector, A.: Integration of tools for binding archetypes to SNOMED CT. BMC Med. Inform. Decis. Mak. 8(suppl. 1), S7 (2008)
35. Yoder, J.W., Balaguer, F., Johnson, R.: Architecture and design of adaptive object-models, http://citeseerx.ist.psu.edu/viewdoc/download?doi=10.1.1.65.9077&rep=rep1&type=pdf
36. Zhao, L., Lee, K.P., Hu, J.: Generating XML Schemas for DICOM structured reporting templates. J. Am. Med. Inform. Assoc. 12, 72–83 (2005)

Open Source Software for Model Driven Development: A Case Study

Jonas Gamalielsson[1], Björn Lundell[1], and Anders Mattsson[2]

[1] University of Skövde, Skövde, Sweden
{jonas.gamalielsson,bjorn.lundell}@his.se
[2] Combitech AB, Jönköping, Sweden
anders.mattsson@combitech.se

Abstract. Model Driven Development (MDD) is widely used in the embedded systems domain, and many proprietary and Open Source tools exist that support MDD. The potential for sustainability of such tools needs to assessed prior to any organisational adoption. In this paper we report from a case study conducted in a consultancy company context aiming to investigate Open Source tools for MDD. For the company it was interesting to explore the two Open Source modelling tools Topcased and Papyrus for potential adoption. The focus for our case study is on assessing the health of the ecosystems for the two investigated Open Source projects by means of quantitative analysis of publically available data sources about Open Source projects. The health of ecosystems is an important prerequisite for a long term sustainable OSS (Open Source Software) tool-chain in the MDD area, which can aid strategic decision making for potential adoption within a company context. We have established details on the extent to which developers and users are active in two specific OSS ecosystems, and identified organisational influence for both ecosystems. We find that the investigated tools are promising regarding the health of their ecosystems, and a natural next step for the company would be to proceed with a pilot study in order to analyse the effectiveness of the investigated tools in company contexts.

1 Introduction

Despite the existence of many tools supporting Model Driven Development (MDD), there is currently a lack of reported case studies from MDD practice in industry (Mattsson et al. 2009). Some of these tools are licensed as Open Source Software (OSS). There are a number of different motivations for utilising OSS within companies, and companies have adopted several different business models for engaging in Open Source (Bonaccorsi and Rossi 2006; Lundell et al. 2010; van der Linden et al. 2009). In fact, the majority of today's innovative products and solutions are developed on the basis of Open Source software (Ebert 2008). In recent years, there has been an increased interest in OSS usage amongst practitioners in Swedish companies, with an increased emphasis on active involvement in OSS projects beyond passive use of OSS products (Lundell et al. 2010). In this paper we report from an analysis of two MDD tools (Topcased (Topcased.org 2011a) and Papyrus

S.A. Hissam et al. (Eds.): OSS 2011, IFIP AICT 365, pp. 348–367, 2011.

(Eclipse.org 2011a)) provided as Open Source software, with the goal to aid strategic decision making for potential adoption within a company context.

Some of the contributors in the OSS projects are affiliated with companies that provide support and consultant services related to the OSS projects at hand. On the other hand, other contributors are affiliated with companies in the secondary software sector (Fitzgerald 2006), i.e. companies with extensive software development in domains such as avionics, automotive, telco, etc., but whose main goal is not to provide services to OSS projects. For companies in such domains, developed solutions often require long term maintenance, something which developers in this area see "as fundamentally important" (Lundell et al. 2011). For example, support for products in the avionics domain will often last for more than 70 years (Robert 2007). Further, previous results from the embedded systems area show that support from large OSS communities is "considered superior compared to proprietary alternatives in some cases." (Lundell et al. 2011)

In recent years we have seen a commodification in software development (van der Linden et al. 2009), and many companies in the secondary software sector develop and are dependent upon a large amount of software which do not give a competitive advantage. For such commodity software, many large and small companies seek to leverage from the Open Source licensing model in their collaboration on commodity software. An example of such commodity software is the Eclipse platform, which constitutes the basis for many Open Source licensed development tools (West 2003).

Before an organisation adopts an Open Source project it is important to evaluate its ecosystem in order to make sure that it is healthy and that the project is likely to be sustainable and maintained for a long time (van der Linden et al. 2009). One important means in such an evaluation is to quantitatively assess the health of an Open Source community (Crowston and Howison 2006). A number of studies have investigated large, well known Open Source projects through quantitative analysis, including the Linux kernel (Moon and Sproull 2000), Apache (Mockus et al. 2002), Mozilla (Mockus et al. 2002), Gnome (German 2004) and KDE (Lopez-Fernandez et al. 2006). Such quantitative assessment includes, but is not limited to, analysis of data from: Software Configuration Management (SCM) systems such as CVS and SVN (Lopez-Fernandez et al. 2006; Gamalielsson and Lundell 2010), mailing lists (Kamei et al. 2008; Gamalielsson et al. 2010), and bug tracking systems (Crowston and Howison 2005; Mockus et al. 2002).

Our study is undertaken as a case study conducted in a medium sized consultancy company active in the embedded systems domain. The goal has been to investigate the two Eclipse-based Open Source projects Topcased and Papyrus for potential company adoption. Specifically, the aim is to reveal insights concerning the health of the ecosystems of the two OSS projects as this constitutes an important basis for strategic decision making within the company.

2 Research Approach

In this paper we report from a case study conducted within the consultancy company Combitech AB (hereafter referred to as Combitech), which is a company working with advanced systems design and software development, electronic engineering, process optimisation, and staff training. It has approximately 800 employees and covers a broad spectrum of business areas such as defence, aviation, automotive, medical and telecoms.

The company has a long experience of systematic method work and model-based system development. In several development projects, UML is used (e.g., Mattsson et al. (2009)), but other modelling techniques are used as well. The company has experience from use of three major case tools supporting both UML and time-discrete modelling: Rose Realtime® (from IBM), Rhapsody® (from IBM and previously i-Logix), and TAU® (from Telelogic). Combitech has an interest in exploring the potential of the Eclipse platform and Open Source tools to complement (and possibly replace) its current tool suite.

For Combitech it was interesting to explore the two Open Source modelling tools Topcased (Topcased.org 2011a) and Papyrus (Eclipse.org 2011a) for potential adoption. A motivation for conducting a study involving Papyrus and Topcased is that both tools are based on the Eclipse Modelling Framework (EMF). With an organisational adoption of EMF, it will be possible to exchange models between the many tools supporting EMF, thus minimising the risk for lock-in and enabling integration of tools to support company development practices with effective tool chains. Further, initial use of Topcased within the company context was a positive experience, which increased the interest in further exploration.

Topcased is a software environment primarily intended for development of critical embedded systems encompassing both software and hardware. It promotes the use of model-driven engineering and formal methods. Topcased stems from an industrial R&D project (Lundell et al. 2008), and is released as Open Source by a partner group where members originate from various organisations. According to Eclipse.org (2010), Topcased "is backed by some major industrial players". At time of writing (January 2011), the project contains approximately 8000 source code files. 71% of these are Java files, in total containing 1.2 MLOC (excluding comments and blank rows) (Ohloh.net 2011a).

Papyrus is a graphical editing tool for models in languages such as UML, SysML and MARTE. Papyrus was included in the MDT (Model Development Tools) subproject of the Eclipse platform in January 2009. The project currently contains approximately 9400 source code files, where 73% of these are Java files which in total contain 2.3 MLOC (Ohloh.net 2011b).

To guide the decision on whether to engage in any of the ecosystems for the two projects, Combitech wanted to establish that the projects are likely to be sustainable and maintained for a long time. This was considered important since any organisational adoption of an OSS MDD tool-chain implies a long-term commitment which affects working practice within the company. Specifically, three aspects of ecosystem health were considered important: activity in ecosystems; company influence in ecosystems; and interaction between ecosystems for the two OSS projects (Topcased and Papyrus). The company interest in these three aspects of ecosystem health is in line with previous research (Crowston and Howison 2006; German 2004; Gamalielsson et al. 2010).

To investigate the activity in ecosystems, we analysed the contributions in terms of committed SCM artefacts of the Open Source projects over time (from start of projects to August 2010). To investigate the company influence in ecosystems, we analysed over time the extent to which different contributors are affiliated with different companies. The interaction between ecosystems was assessed by studying contributors that are active in both projects. The data for Topcased was collected from the Gforge website for the project (http://gforge.enseeiht.fr/projects/topcased, accessed 10 September 2010), and for Papyrus the data was collected from the Eclipse website for the project (http://www.eclipse.org/projects/project_summary .php?projectid=modeling.mdt.papyrus, accessed 10 September 2010).

3   Results

This section presents results on the activity in the two OSS projects Topcased and Papyrus, an analysis of how different contributors are affiliated with different companies, and the interaction between the two ecosystems.

3.1   Activity in Ecosystems

The activity in ecosystems is an important factor that reflects the health and long term sustainability of OSS projects, and we therefore studied each project separately in terms of the extent of contributions to SCM repositories, mailing lists, and forums over time.

The number of commits for Topcased as a function of time is shown in Figure 1. The first commits were made in November 2004 and the last month studied is August 2010 (70 months in total). It can be observed that a major part of the commits are code commits (red trace), constituting on average 60% of the total number of commits

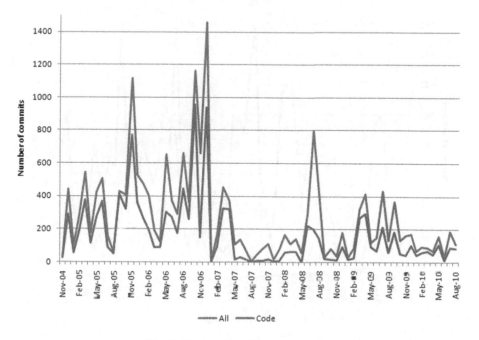

Fig. 1. Topcased: number of commits

(blue trace). Another fact is that some peaks (e.g. November 2005, November 2006 and July 2008) co-occur with events in the project version history shown in Table 1. It can be noted that there has been a large number of releases since the start of the project.

The number of messages contributed to the developer- and user mailing lists of Topcased is illustrated in Figure 2. Like in Figure 1, the peaks can be related to the version history in Table 1. It can be noted that there is elevated activity in the developer list around May 2005 (before the initial release) and also at the time of the initial release (October 2005) and version 2.0.0 (July 2008). Similarly, peaks in the user mailing list often co-occur with events in the version history (e.g. September 2006, May 2007 and May 2009). Some peaks in Figure 2 also co-occur with peaks in Figure 1 (e.g. October 2006, July 2008 and April/May 2009). The average number of postings each month is 21 for the developer mailing list and 67 for the user mailing list, which shows that there is more activity on average in the user mailing list.

The number of commits for Papyrus as a function of time is shown in Figure 3. SCM data for this project spans from January 2009 to August 2010 (20 months in total). It can be noted that code artefacts are dominating (79% of total number of commits). The commit activity is high during the entire period of study, and the three largest peaks occur in February 2009, December 2009 and July 2010. The first two of these peaks do not co-occur with events in the version history of Papyrus (see table 2), but the last peak co-occurs with the release of version 0.7. So far, there has only been a few releases of the tool.

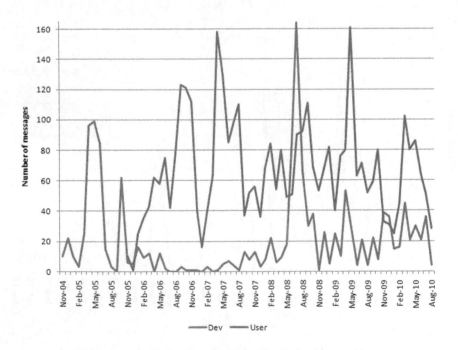

Fig. 2. Topcased: number of messages in developer- and user mailing lists

Table 1. Topcased: version history

Date (Y-M-D)	Release
2005-10-11	Initial
2005-11-17	0.6.0
2006-01-24	0.7.0
2006-03-17	0.8.0
2006-06-07	0.9.0
2006-09-20	0.10.0
2006-11-08	0.11.0
2007-05-29	1.0.0M4.1
2007-08-11	1.0.0
2007-11-26	2.0.0M3
2007-12-07	1.2.0
2008-01-20	2.0.0M4
2008-02-20	1.3.0
2008-03-03	2.0.0M5
2008-04-19	2.0.0M6
2008-07-18	2.0.0
2008-09-01	2.1.0
2008-11-06	2.2.0
2009-01-07	2.3.0
2009-03-16	2.4.0
2009-05-07	2.5.0
2009-07-27	3.0.0 & 2.6.0
2009-09-28	3.1.0
2009-12-03	3.2.0
2010-03-22	3.3.0
2010-06-02	3.4.0
2010-06-21	3.4.1
2010-07-23	4.0.0

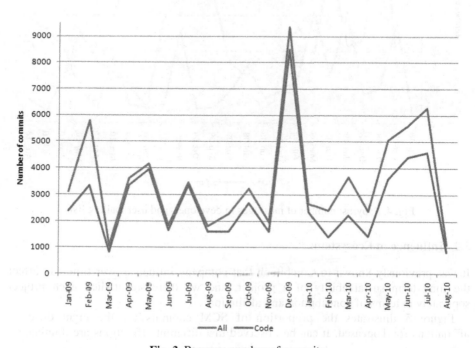

Fig. 3. Papyrus: number of commits

The number of messages contributed to the developer mailing list and forum of Papyrus is illustrated in Figure 4. It can be observed that the two largest peaks for the developer mailing list co-occur with the initial- (November 2008) and second milestone release (October 2009) of Papyrus. The peaks for the forum are less distinct. The average number of postings each month is 47 for the developer mailing list and 27 for the forum, which shows that there is more activity on average in the developer mailing list.

Table 2. Papyrus: version history

Date (Y-M-D)	Release
2008-11-25	Initial
2009-09-18	0.7M1
2009-10-16	0.7M2
2010-07-14	0.7

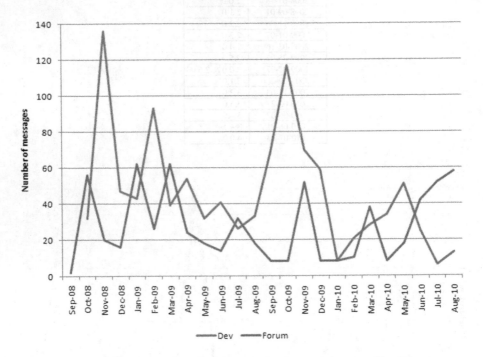

Fig. 4. Papyrus: number of messages in developer- and user mailing lists

3.2 Influence in Ecosystems

It was previously known to Combitech that company influence over time can affect the long term sustainability of a project, and we therefore studied each project separately in terms of contribution and affiliation.

Figure 5 illustrates the proportion of SCM commits for the eight different affiliations for Topcased. It can be observed that different affiliations are dominating in different time periods. Affiliations are also ranked from most influential over all time (A1) to least influential (A8). Affiliation A1 (blue) is dominating from November 2004 until August 2008, whereas A2 (red) and A3 (green) together largely dominate from September 2008 until August 2010. The remaining affiliations (A4 through A8) are considerably less influential. The orange colour represents unknown affiliations.

Figure 6 shows the corresponding proportion of SCM commits for the six different affiliations for Papyrus. Like for Topcased, affiliations are ranked from most influential over all time (A9) to least influential (A12). It can be noted that A9 is the most influential affiliation over all time, and is dominating during most months. A2 (which was also active in Topcased) is active from April 2009 until August 2010, and is dominating during some months. A4 (also active in Topcased) is the third most influential affiliation, and is dominating in April and May 2009.

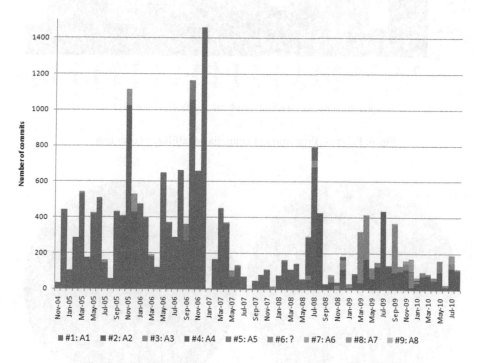

Fig. 5. Topcased: Proportion of commits per affiliation over time

Figure 7 illustrates the total proportion of SCM contributions for different affiliations in Topcased (left pie chart) and Papyrus (right pie chart). It is evident that for both projects, one single affiliation is clearly dominating (A1 for Topcased with 76% of the commits, and A9 for Papyrus with 68% of the commits). For both Topcased and Papyrus, the three most influential affiliations account for approximately 95% of the commits.

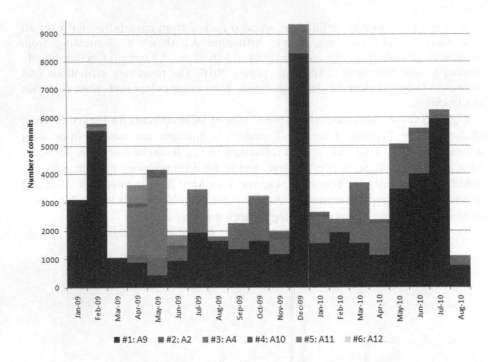

Fig. 6. Papyrus: Proportion of commits per affiliation over time

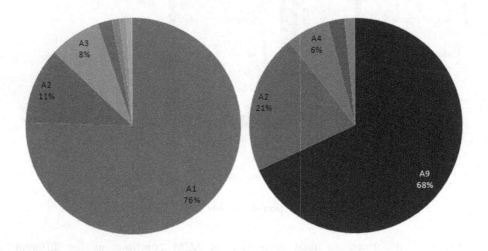

Fig. 7. Total affiliation commit influence (left pie: Topcased, right pie: Papyrus)

The affiliations can be classified into different affiliation types: Small and Medium Enterprise (SME), Large Company (LC) and Research Institute (RI), see Table 3.

Table 3. Affiliation type (rows) for affiliations (columns)

	A1	A2	A3	A4	A5	A6	A7	A8	A9	A10	A11	A12
SME	X			X			X	X			X	X
LC		X	X		X	X						
RI									X	X		

The proportion of SCM commits for different affiliation types is illustrated in Figure 8 for Topcased. It can be observed that SME (blue) is dominating from November 2004 until August 2008, whereas "large company" (red) largely dominates from September 2008 until August 2010.

Figure 9 shows the corresponding proportions of SCM commits for Papyrus. It is evident that the most influential affiliation is "research institute" (brown), which is active during all months and dominating during the majority of the months. SME (blue) is most active, and dominating, in April and May 2009. "Large company" (red) is active from April 2009 until August 2010, and is occasionally dominating.

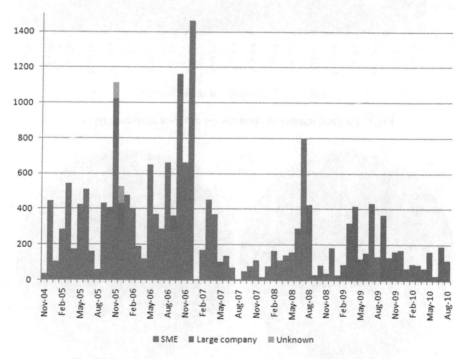

Fig. 8. Topcased: number of commits for different affiliation types

Figure 10 illustrates the total proportion of SCM contributions for different affiliation types in Topcased (left pie chart) and Papyrus (right pie chart). It is evident that for both projects, one single affiliation type is dominating (SME for Topcased with 78% of the commits, and "research institute" for Papyrus with 71% of the commits). It is interesting to note that "large company" has the same proportion of commits (21%) for both projects.

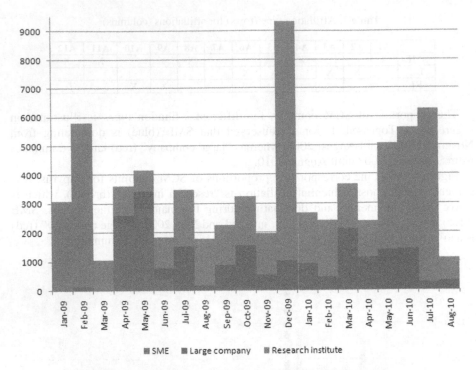

Fig. 9. Papyrus: number of commits for different affiliation types

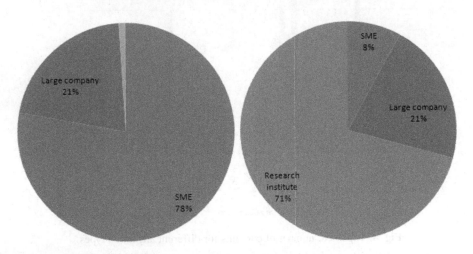

Fig. 10. Total affiliation type commit influence (left pie: Topcased, right pie: Papyrus)

The influence can also be studied at committer level. The number of SCM commits from the top committers (those who have made largest proportion of commits over all time) in Topcased as a function of time, is shown in Figure 11. It can be observed that the single most influential committer (blue trace) is dominating in several periods and

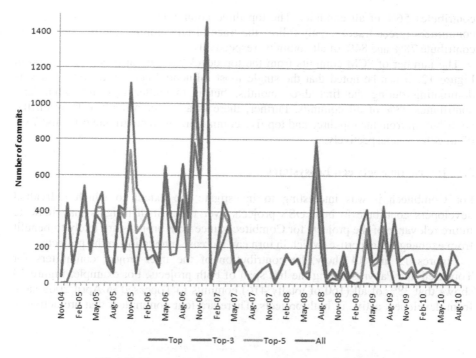

Fig. 11. Topcased: number of commits for top committers

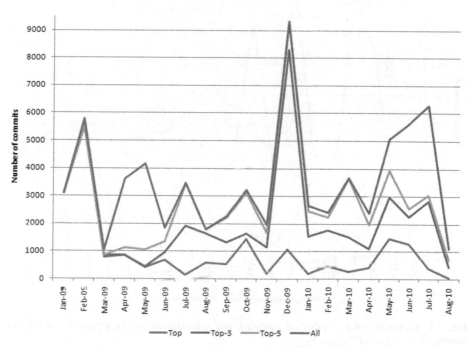

Fig. 12. Papyrus: number of commits for top committers

contributes 56% of all commits. The top three committers (red trace) and top five committers (green trace) largely follow the trace of all commits (purple colour), and contribute 78% and 84% of all commits, respectively.

The number of SCM commits from the top developers in Papyrus is illustrated in Figure 12. It can be noted that the single most influential committer (blue trace) is dominating during the first three months, but is thereafter less influential, and contributes 28% of all commits. Further, there is a bigger difference between the contribution from the top three and top five committers, who contribute 60% and 78% of all commits, respectively.

3.3 Interaction between Ecosystems

For Combitech it was interesting to investigate the extent to which individual developers contribute to both OSS projects over time. Such insights may indicate future relevance of the projects for Combitech since both projects are likely to benefit from exchange of expertise, which in turn can increase innovation in the projects.

Figures 13 and 14 show the contribution of the inter-project committers for Topcased and Papyrus during the life span of both projects. For example, Figure 13 shows that 38% of all (160 of totally 416) commits to Topcased in April 2009 stem from developers active in both projects. There are four committers that are active in

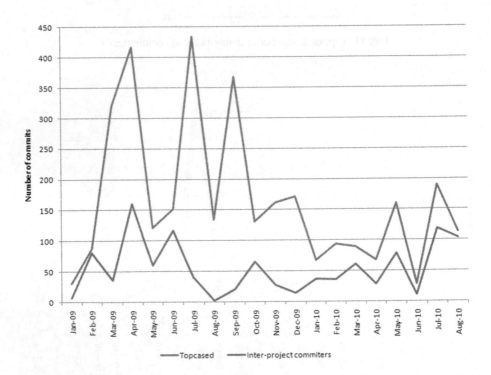

Fig. 13. Topcased: total number of commits and contribution by inter-project committers (during the life span of Papyrus)

both projects during the life span of the projects. These committers are all affiliated with A2 (a large company), and contribute 33% of all commits in Topcased (from January 2009, i.e. during the life span of both projects) and 21% of all commits in Papyrus. At affiliation level, A4 (an SME) is also active in both projects (as indicated in e.g. Figure 7), but with different committers in the two projects. The results indicate that there is considerable interaction between ecosystems, as a large proportion of the commits each month is provided by developers active in both projects. However, from our limited analysis of the time period for both projects it is difficult to establish any current and future trends in interactions between projects.

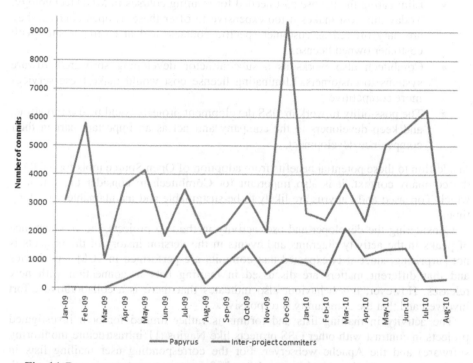

Fig. 14. Papyrus: total number of commits and contribution by inter-project committers

4 Analysis

There are a number of reasons for considering adoption of Open Source tools in a commercial context. For Combitech, development process optimisation based on MDD technology is an important service. This service is today mostly based on proprietary tools. Complementing this with Open Source tools would potentially have several benefits:

- Possibility to influence the development of the tool by contributing to the development. This could enable better support for the processes that Combitech is offering its customers to deploy.

- Possibility to sell new services around the tool, for example custom made adaptations and bug fixing support.
- Today when offering tool experts, Combitech has a natural disadvantage towards the vendor of the tool which would not exist with an Open Source tool.
- The license cost for proprietary tools targeted towards the embedded market with full code generation capabilities is very high and can be a barrier hard to overcome for companies considering employing MDD. Being able to offer tooling with no license cost could be a door-opener to new customers.
- Eliminating the license cost needed for training courses in MDD technology. Today this cost makes it too expensive to offer these as open courses, they are only offered as customer specific courses held at customer sites with customer owned licenses.
- Combitech also works as a subcontractor developing software/hardware systems to customers. Eliminating license cost would make these services more competitive.
- The possibility to work in OSS development projects could be used to attract and keep developers to the company and act as an important part in their competence development.

In addition to these potential benefits from adoption of Open Source tools for MDD in the company context, it is also important for Combitech to consider the extent to which Topcased and Papyrus are likely to be sustainable and maintainable for a long time.

Considering the developer and user activity in the two projects, the co-occurence of peaks in the activity diagrams and events in the version history of the projects is not surprising, since it is expected that commits are performed in SCM repositories and that different matters are discussed in mailing lists in connection with new releases. However, this behaviour also indicates that there is a collaborative effort involved and that the community is responsive.

The activity in mailing lists and forums is rather limited for both investigated projects in contrast with other OSS projects like Nagios (IT infrastructure monitoring software) and the Apache webserver. For the corresponding user mailing lists in Nagios and Apache there are between 10 and 33 times as many messages each month according to Gamalielsson et al. (2010) and data from Gmane (http://gmane.org, accessed 7 January 2011). For the corresponding developer mailing lists there are between two and 19 times as many messages each month. However, Nagios and Apache, two projects aimed at the monitoring and operation of IT infrastructure, are more established projects with a broader user base. Another reason for the limited mailing list activity for the investigated tools may be extensive internal usage within organisations and also that other communication channels are used apart from public mailing lists and forums. Further, the maturity level of the projects may affect the activity in mailing lists.

Our analysis of the two OSS projects suggests that there is significant development involvement by companies and research institutes. This is in stark contrast with several other Open Source projects, where this kind of involvement is more limited, for example the projects Nagios, Mono and Evolution (Gamalielsson et al. 2010;

Martinez-Romo et al. 2008). For the Topcased project there was a clear shift in affiliation amongst participants at a given point in time, where an SME that had been dominating since the start of the project was replaced by two dominating large companies. This shift of domination has also been observed in the Open Source project "Open XML/ODF translator" (Gamalielsson and Lundell 2010), where an SME was initially dominating but later replaced by another (somewhat larger) SME and a large company. A large company is backing up this project, and has also provided the specification for the tool. In the case of the investigated Papyrus project, committers affiliated with SMEs were also active early in the project, and large companies became more influential later on. The findings regarding contribution by affiliation may suggest that large companies engage in OSS projects with increased maturity and after initial releases of the tools.

From the results it is also evident that few committers contribute a large proportion of the commits. This is in agreement with the previous finding that "the vast majority of mature OSS programs are developed by a small number of individuals" (Krishnamurthy 2002), where it was found that the median number of developers was 4. As a comparison, the top five developers for Topcased and Papyrus contributed 84% and 78% of the commits, respectively. Further, Papyrus has more committers than the average incubator project on the Eclipse platform, as evidenced when comparing our results to a previous analysis of 23 incubator projects (Duenas et al. 2007).

There has been regular and frequent releases of the tools, and from the results we can not observe any decline in activity. On the contrary, it is likely that the interest and user base for the tools will increase in the future. One reason for this is that Papyrus is still in the "incubation phase" (Eclipse.org 2011b), but based on the current status of the project it is likely to be promoted to the mature phase soon, which in turn most likely will lead to an increased user base. It has also been reported that there is an on-going integration and consolidation effort of a number of Eclipse-based tools and technologies, which includes but is not limited to, Topcased and Papyrus (Eclipsecon.org 2010). An indication of this is our finding that several committers from the same large company contribute significantly to both projects. It is also a fact that an SME is active in both projects, but with different committers. The integration and consolidation effort could potentially lead to increased uptake within companies in the embedded systems domain.

Organisations that want to use/adopt OSS-based MDD tools need to engage in the ecosystems of the projects (Eclipse.org 2010). For our analysis of the two tools we found that both projects have a strong commercial involvement. Earlier experience (Lundell et al. 2010) from the Swedish context has identified a difference in the "collaboration climate between projects that have a strong commercial involvement and those that are more 'traditional', community-driven OS projects". Professional developers in Swedish companies developing with Open Source have also expressed concerns regarding a potential risk that a project is too strongly goal driven, leading to that "the fun and enthusiasm for those participating on a voluntary basis can disappear" (Lundell et al. 2010). In order to investigate the collaborative climate in each ecosystem, it would necessitate genuine interaction with each project in a pilot study. Without such it would be difficult to assess whether the collaborative climate in the Topcased and Papyrus ecosystems is congruent with existing work practices in the company.

In summary, our analysis suggests that both projects have healthy ecosystems, with an active base of developers and users. Despite a seemingly commercial drive, both projects seem to have developers representing an appropriate mix of different kinds of organisations involving the secondary software sector. This, in turn, can significantly contribute to long term sustainability for the OSS projects. For Combitech it is interesting to note that a number of practitioners representing large companies in the embedded systems domain have shown interest in Topcased, as evidenced by practitioner presentations at an industrial conference (Topcased.org 2011b). Further, in a consultancy report on the analysis of OSS-based MDD tools prepared on behalf of a large company in the embedded systems domain, a number of tools (including Topcased and Papyrus) were found promising and a more in-depth investigation of these was recommended (Eclipse.org 2010). In the same report it is also stated that of "particular interest and relevance is the TOPCASED project ...". In terms of functionality, the report also points out that "MDT Papyrus is still quite far from being ready for prime time, and would perhaps require additional attention and investment ..." (Eclipse.org 2010). This view is something which needs to be investigated in a pilot project within the organisation before adoption.

Our specific assessment of the health for two Open Source communities may have broader implications for evaluation and assessment of software systems. Evaluation of software systems has many dimensions and it is widely acknowledged to be a complex activity (Lundell and Lings 2004). Our specific strategy used for assessing health of Open Source ecosystems has certain similarities with, and may contribute to, previously proposed approaches (e.g. OpenBRR, QSOS, OMM) for evaluation and assessment of Open Source projects (Petrinja et al. 2010). Common to all these are that they are based on an a-priori evaluation framework (Lundell and Lings 2004). For example, QSOS features an intrinsic durability category which includes metrics such as activity on releases and number of developers. Similarly, OMM has number of commits and bug reports as metrics. Our activity analysis goes beyond previously presented approaches in that we elaborate on more intricate aspects such as affiliation and organisation types in our activity analysis. Further, a number of research projects (e.g. Flossmetrics and QualOSS) have systematically analysed OSS projects and as part of this proposed certain metrics (Daffara 2009).

5   Conclusion and Future Work

Prior to the initiation of this case study, the company explored the functionality offered in a number of Open Source MDD tools for the Eclipse platform for potential company adoption. The focus for this case study has been to explore the sustainability of the ecosystems for the two Open Source projects Topcased and Papyrus.

Overall, we found that both projects are active as evidenced by contributions in SCM repositories, mailing lists and forums. There seems to be an appropriate mix of professional organisations (SMEs, large companies and research institutes) involved in the development of the projects, and developers from one large company are active in both projects. From this we note that both Open Source projects have promising health in their respective ecosystems. Our analysis shows that it is likely that the external user base of the projects will increase over time as the projects mature and as a consequence of the on-going integration and consolidation effort of Eclipse-based tools and technologies. However, having studied two Open Source projects that have existed for less than a decade, only time can tell whether they are still active after several decades.

A limitation in our study concerning assessment of the user base is that we only consider publicly available project data in our analysis. From previous studies it is known that many large companies adopt and use Open Source software internally within their own organisational context. This is of particular importance in this domain with its very long life cycles. Further, we note that there is a lack of documented evidence that report on usage of OSS-based MDD tools in company contexts. One notable exception is the Topcased project for which it has been claimed that the Topcased tool "is in productive use in Airbus" (Eclipse.org 2009), and more specifically "Airbus Industry is committed to using TOPCASED in its A350 program" (Eclipse.org 2010). Further, the functionality and reliability of the Topcased tool has been found sufficient in other usage contexts where critical systems are being developed. For example, it has been used in the context of validation of satellite flight software (Pouly et al. 2011) and railway safety systems (Hase 2011). Motivated by such experiences from other companies in the secondary software sector and the results from our investigation of the Topcased community, the tool is currently being explored by developers at Combitech.

Based on our results, a natural next step for the company will be to proceed with a pilot study in order to analyse the extent to which the functionality offered in the investigated tools support the preferred working practice used in development projects in the company. Since the company perceives the health of the Topcased ecosystem to be promising and based on earlier positive experience of the tool, such a pilot study involving Topcased usage within the company context is seen as a natural next step. From this, the company will be able to further investigate the collaborative climate through genuine interaction between the company context and the broader Topcased ecosystem.

Acknowledgement. This research has been financially supported by the ITEA2 project OPEES (www.opees.org) through Vinnova (www.vinnova.se).

References

Bonaccorsi, A., Rossi, C.: Comparing motivations of individual programmers and firms to take part in the open source movement: from community to business. Knowledge Technology and Policy 18, 40–64 (2006)

Crowston, K., Howison, J.: The social structure of Free and Open Source software development. First Monday 10(2) (2005)

Crowston, K , Howison, J.: Assessing the Health of Open Source Communities. IEEE Computer 39(5), 89–91 (2006)

Daffara, C.: The SME guide to Open Source Software, 4th edn. FLOSSMETRICS report, European Commission project FP6-033982 (July 4, 2009),
http://www.flossmetrics.org/sections/deliverables/docs/WP8/
D8.1.1-SMEs_Guide.pdf (accessed June 4, 2011)

Duenas, J.C., Parada, G., H., A., Cuadrado, F., Santillan, M., Ruiz, J.L.: Apache and Eclipse: Comparing Open Source Project Incubators. IEEE Software 24(6), 90–98 (2007)

Ebert, C.: Open source software in industry. IEEE Software 25, 52–53 (2008)

Eclipse.org. Eclipse Automotive Interest Group- meeting minutes (2009), http://dev.eclipse.org/mhonarc/lists/auto-iwg/pdfTcIV3Ghb68.pdf (accessed June 4, 2011)

Eclipse.org. Papyrus (2011a), http://www.eclipse.org/modeling/mdt/papyrus (accessed June 4, 2011)

Eclipse.org. Papyrus (2011b), http://www.eclipse.org/modeling/mdt (accessed June 4, 2011)

Eclipse.org. An Extended Survey of Open Source Model-Based Engineering Tools (2010), http://wiki.eclipse.org/images/d/dc/Report.external.bvs.pdf (accessed June 4, 2011)

Eclipsecon.org. Papyrus: Advent of an Open Source IME at Eclipse (2010), http://www.eclipsecon.org/2010/sessions/sessions?id=1385 (accessed June 4, 2011)

Fitzgerald, B.: The Transformation of Open Source Software. MIS Quarterly 30(3), 587–598 (2006)

Gamalielsson, J., Lundell, B., Lings, B.: The Nagios community: An extended quantitative analysis. In: Ågerfalk, P., Boldyreff, C., González-Barahona, J.M., Madey, G.R., Noll, J., et al. (eds.) OSS 2010. IFIP AICT, vol. 319, pp. 85–96. Springer, Heidelberg (2010)

Gamalielsson, J., Lundell, B.: Open Source Software for Data Curation of Digital Assets: a case study. In: Lugmayr, A., et al. (eds.) Proceedings of Mindtrek 2010 of 14th International Digital Media & Business Conference (MindTrek): Envisioning Future Media Environments, pp. 53–56. ACM, New York (2010)

German, D.: The GNOME project: a case study of open source global software development. Journal of Software Process: Improvement and Practice 8(4), 201–215 (2004)

Hase, K.R.: "openETCS": An Open Source Approach for Railway Safety Systems Adopting TOPCASED for CENELEC EN 50126/50128 Safety Case. In: First Topcased Days Toulouse 2011, Toulouse, France, February 2-4 (2011), http://gforge.enseeiht.fr/docman/view.php/52/4289/A2-DeutscheBahn.pdf (accessed June 4, 2011)

Kamei, Y., Matsumoto, S., Maeshima, H., Onishi, Y., Ohira, M., Matsumoto, K.: Analysis of Coordination Between Developers and Users in the Apache Community. In: Russo, B., et al. (eds.) Open Source Development, Communities and Quality, pp. 81–92. Springer, Boston (2008)

Krishnamurthy, S.: Cave or Community? An Empirical Examination of 100 Mature Open Source Projects. First Monday 7(6) (2002)

van der Linden, F., Lundell, B., Marttiin, P.: Commodification of Industrial Software: A Case for Open Source. IEEE Software 26(4), 77–83 (2009)

Lopez-Fernandez, L., Robles, G., Gonzalez-Barahona, J.M., Herraiz, I.: Applying Social Network Analysis Techniques to Community-driven Libre Software Projects. International Journal of Information Technology and Web Engineering 1, 27–48 (2006)

Lundell, B., Bermejo, J., Labezin, C., Sempert, F., Valentin, M.-L., Laprevote, A., van der Linden, F., Pablos, J.J.: Open Source Software Workshop, ITEA 2 Symposium, Rotterdam, October 21 (2008)

Lundell, B., Lings, B.: On understanding evaluation of tool support for IS development. Australasian Journal of Information Systems (AJIS) 12(1), 39–53 (2004)

Lundell, B., Lings, B., Lindqvist, E.: Open Source in Swedish companies: where are we? Information Systems Journal 20(6), 519–535 (2010)

Lundell, B., Lings, B., Syberfeldt, A.: Practitioner perceptions of Open Source software in the embedded systems area. Journal of Systems and Software (in press, 2011)

Martinez-Romo, J., Robles, G., Ortuño-Perez, M., Gonzalez-Barahona, J.M.: Using Social Network Analysis Techniques to Study Collaboration between a FLOSS Community and a Company. In: Russo, B., et al. (eds.) Open Source Development, Communities and Quality, pp. 171–186. Springer, Boston (2008)

Mattsson, A., Lundell, B., Lings, B., Fitzgerald, B.: Linking Model-Driven Development and Software Architecture: A Case Study. IEEE Transactions on Software Engineering 35(1), 83–93 (2009)

Mockus, A., Fielding, R.T., Herbsleb, J.D.: Two case studies of open source software development: Apache and Mozilla. ACM Transactions on Software Engineering and Methodology 11(3), 309–346 (2002)

Moon, Y.J., Sproull, L.: Essence of distributed work: The case of the Linux kernel. First Monday 5(11) (2000)

Ohloh.net. Topcased (2011a), http://www.ohloh.net/p/topcased (accessed June 4, 2011)

Ohloh.net. MDT Papyrus (2011b), http://www.ohloh.net/p/mdt-papyrus (accessed June 4, 2011)

Petrinja, E., Sillitti, A., Succi, G.: Comparing OpenBRR, QSOS and OMM Assessment Models. In: Ågerfalk, P., Boldyreff, C., González-Barahona, J.M., Madey, G.R., Noll, J., et al. (eds.) OSS 2010. IFIP AICT, vol. 319, pp. 224–238. Springer, Heidelberg (2010)

Pouly, J., Rolland, J.F., Faure, T., Hyounet, P., Zanon, O.: Automatic generation of tests from UML models to validate satellite flight software. In: First Topcased Days Toulouse 2011, Toulouse, France, February 2-4 (2011), http://gforge.enseeiht.fr/docman/view.php/52/ (accessed June 4, 2011)

Robert, S.: New trends and needs for Avionics Systems. In: ARTEMIS Conference, Berlin (May 2007), https://www.artemisia-association.org/downloads/SYLVIE_ROBERT_AC_2007.pdf (accessed June 4, 2011)

Topcased.org. Topcased – The Open Source Toolkit for Critical Systems (2011a), http://www.topcased.org (accessed June 4, 2011)

Topcased.org. First Topcased Days Toulouse 2011, Toulouse, France, February 2-4 (2011b), http://www.topcased.org/index.php/content/view/53 (accessed June 4, 2011)

West, J.: How Open is Open Enough? Melding Proprietary and Open Source Platform Strategies. Research Policy 32(7), 1259–1285 (2003)

The Third Generation of OSS: A Three-Stage Evolution from Gift to Commerce-Economy

Toshihiko Yamakami

ACCESS, CTO Office,
1-10-2 Nakase, Mihama-ku, Chiba-shi, Chiba-ken, JAPAN 261-0023
http://www.access-company.com

Abstract. Linux is penetrating into mobile software as the basis for a mobile middleware platform. It is accelerating the increasing visibility of open source software (OSS) components in mobile middleware platforms. Considering the 10-million lines of code of OSS-based industrial platforms such as a mobile middleware platform, engagement in foundations is inevitable for large-scale packages of OSS for industrial solutions. The author discusses the driving factors toward a foundation-based OSS and the transition of the underlying economy types to analyze the transitions to the third-generation OSS.

Keywords: Open source software (OSS), evolution of OSS, industrial OSS, foundation-based OSS.

1 Introduction

Linux has penetrated into a wide range of digital appliances, e.g. mobile handsets, digital TVs, game consoles, and HD recorders. It facilitates the reuse of PC-based rich user experience data service software with the high speed network capabilities of an embedded software environment. As Linux-based software is widely adopted for digital appliances, the original weak points of Linux in an embedded environment have been addressed, namely its real time processing and battery life capabilities. The author reviews the patterns of evolution in the past literature, and proposes the concept of third-generation OSS, a foundation-based OSS. The author discusses the driving factors toward a foundation-based OSS and the transition of the underlying economy types to analyze the transitions to the third-generation OSS.

2 Purpose and Related Work

2.1 Purpose of Research

The purpose of this research is to identify and understand the transitions in OSS with regard to base economy types.

S.A. Hissam et al. (Eds.): OSS 2011, IFIP AICT 365, pp. 368–378, 2011.
© IFIP International Federation for Information Processing 2011

2.2   Related Work

OSS was separated from the concept of free software in the late 1990's in order to revisit the commercial issues of using OSS. It is a paradoxical to publish source code, the core competence of the software industry, so openly. Fitzgerald et al discussed the contradictions, paradoxes and tensions of OSS in [4]. Fitzgerald named OSS 2.0, the open source phenomenon has undergone a significant transformation from its free software origins to a more mainstream, commercially viable form [3]. This paper deals with a follow-up to OSS 2.0.

Raymond discussed open source from the business model perspective in this famous open source work series [8].

OSS has continued to evolve. Watson presented the second generation of OSS, or professional OSS [13] in contrast to the three types of first generation OSS: community OSS, sponsored OSS, and corporate distribution.

Letellier discussed the third generation of OSS [5] from the perspective of its organizational structure.

The long-term factors of OSS have also attracted the attention of researchers. Subramaniam discussed success factors using longitudinal data on OSS projects [9] and presented the impacts of different license types. Yu discussed time series analysis techniques to study the time dependence of open-source software activities using mailing lists, bug reports, and revision history [15] and presented diversity in cyclic-ness and in seasonal dependency.

As the size of OSS software has grown, organizational governance has emerged. Examples include the Eclipse Foundation [12] and Apache Software Foundation [11]. There are new industrial organizations emerging for industry-specific software: for example, mobile handset software-related foundations including the LiMo Foundation [6], the OHA [7], and the Symbian Foundation [10].

Capra et al analyzed analyses the impact of firms' participation on popularity and internal software design quality for 643 SourceForge.net projects [2].

Yamakami presented multiple views of the generations of OSS in order to understand the diverse evolutions of OSS [14].

The originality of this paper lies in its examinations of the transitions of the base economy types in order to identify and understand the evolution of OSS.

3   Landscape of OSS

3.1   Generational Views of OSS

OSS has been successful at penetrating into the world, including the software industry. It is no longer a question any more whether we should use OSS, because the current industrial best practices adopt the OSS-based development in many industrial domains. Examples include mobile multimedia software for mobile handsets.

OSS has become a reasonably complete stack, therefore, many user applications can be built using only OSS. This user application covers many industrial applications and enterprise applications.

Augustin discussed generations of OSS dating back to 1974 from the viewpoint of stacks in OSS. Augustin presented a 5-generation view as depicted in Fig. 1 [1].

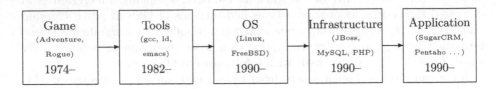

Fig. 1. Augustin's 5 generation-view

The first generation of OSS consists of games distributed by mailing lists. The second generation consists of tools for development environments. The third generation is the OS (operating system). The fourth generation consists of infrastructure elements such as databases and web server scripting languages. The fifth generation is applications. It is a generational analysis by domains or completion of computing stacks.

Watson presented a view in terms of a business model. Watson discussed the emergence of professional OSS, as the second generation of OSS. The two generations of OSS are depicted in Fig. 2.

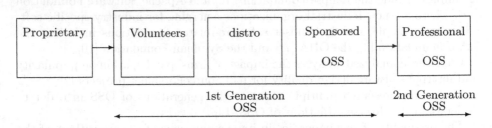

Fig. 2. Watson's 2 generation-view

The second generation that emerges is professional OSS, in which companies contribute their assets to open source and explore a wide range of business models based around that. Full-time employees are engaged in OSS to leverage their business models.

The author has a sense that this increased interaction between OSS and business was similar to that between the Internet and business in the first half of the 1990s. In the case of the Internet, it was gradually recognized that a fusion with business was the way that would lead the Internet to its full potential. The author believes that a similar conclusion will be drawn in the case of OSS.

3.2   Emerging New Aspects of OSS: Foundation Dimension

The completeness of OSS components for industrial solutions allows for a complete platform consisting of OSS modules. Large-scale OSS platforms have emerged these days, such as Symbian, and Android. Such industry-scale OSS solutions require the following, depicted in Table 1.

Table 1. Requirements for industry-scale OSS solutions

Item	Description
Governance	Governance to manage quality and performance of the entire platform.
Neutrality	Neutrality to serve as an industrial platform that can be supported by a wide range of stakeholders.
OSS community-friendliness	Friendliness toward upstream OSS communities to enable coordination and collaboration with upstream communities.
Ecosystem considerations	Ecosystem to enable different industrial stakeholders to participate, enabling both the maintenance of common parts and customized differentiation for each stake holder.

Considering the 10-million lines of code of OSS-based industrial platforms such as a mobile middleware platform, engagement in foundations is inevitable for large-scale packages of OSS for industrial solutions.

3.3   Perspective for Three-Stage Evolution of Large-Scale OSS

The author believes that the shift towards foundation-based OSS is a natural consequence of the shifts in base economy types of OSS.

The shift in base economy types of OSS is depicted in Fig. 3.

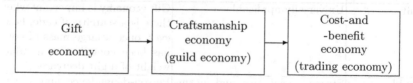

Fig. 3. 3-generation view of the base economy types of OSS

In the early days of OSS, the basic economy type is the gift economy. They give, so we gave. Gifts in turns are the basic constructs of the economy. This is similar to the primitive economy of early civilization. There are neither economic rules nor quantitative measures to be used for trades. It is the starting point of an economy.

If a project and the community surrounding it persist, then social norms, values, and social ties are developed. Internal rules and guidelines are developed for a community. In this stage, an economy of Guilds is used. A guild is an association of craftsmen in a particular trade. Confraternities of workers were organized in a manner that was something between a trade union, a cartel and a secret society. The community is the core part of this economy.

Then, OSS collides with the real world economy, the basic economy type is shifted to a cost-benefit economy. This is the common trading economy, used in the modern world. There are two types of economy. One is the OSS-centered economy. When Red Hat started a business based on giving copy-left software away and providing expert services for a fee, many people thought that copy-left would not be a sustainable business. However, Red Hat still persists and has proven that the radical copy-left is still viable when a related business is successfully built.

The other is the professional OSS economy. The dual license is one example. It is allowed to provide multiple licenses including OSS license and commercial license. The dual license is to provide codes with an OSS license and a comercial license. Full-time employees with fully crafted business models enable this type of economy, which is a common business model with business model engineering. Another example is proprietary add-ons.

The misfits and fits with the current industrial landscape are illustrated in Table 2.

3.4   Implications from the Transitions of Base Economy Types

These transitions in base economy types can explain the evolution of licenses.

Table 2. Misfits and fits with the current industrial landscape

Economy type	Fits	Misfits
Gift	Universally applicable	Gift economy is important when there is a scarcity of code, however, once a large mass of code has been contributed, a value-weight of a gift decreases.
Guild	Strong social ties and norms help in the management of a large mass of code	Person-dependence may create obstacles to consistent handling of large-scale packages of OSS code for industrial solutions.
Commerce	Large-scale OSS projects require alignment to roadmaps, structured governance, skilled project management, coordination among different stakeholders, and industrial support.	Not all OSS communities accept these commerce-driven activities due for the historical reason and the volunteer-oriented nature of maintenance.

Licensing is an important aspect of OSS. Even so, there are a large number of licenses in OSS, the author identifies a kind of evolution in licenses, depicted in Fig. 4.

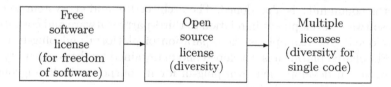

Fig. 4. 3-generation view of licenses

Early examples of OSS licenses include the GPL (GNU Public License) and BSD licenses. GPL is an important license that is based on free software, and pursues the freedom of software. It is unique in that it represents a philosophy rather than software development practices.

Then next generation consists of open source software licenses. The term OSS was coined when the community discussed licensing with the publication of Netscape software in 1998. There was some misunderstanding of licenses and many projects started to create new licenses for their source code, which lead to a significant number of OSS licenses. The OSS licenses represented the diversity of OSS projects.

As people continued to learn and explore OSS licensing issues, it was recognized that there is no reason that one piece of software should have only one license. Software code can have as many licenses as needed. The dual license in OSS is a departure from a rigid and fixed licensing system. It allows a certain flexibility of business development to a company, as long as that company has a copyright for the entire code. This is the basis for professional OSS.

As features are extended and communities grow with enhancements in the IT infrastructure, the volume of code simply continues to grow. This exposes OSS projects to the challenges of large-scale software development. In order to understand these challenges, the following generations are observed during the evolution of large-scale OSS, depicted in Fig. 5.

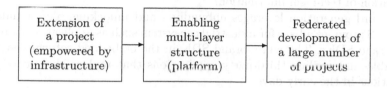

Fig. 5. 3-generation view of large-scale software development

In the first generation, a community grows and extends a number of small projects in the community. The IT infrastructure enables the management of larger-scale software projects. Accumulation of community experience, project management experience, and the increased capabilities of development environments enable the development of large-scale software.

In the second generation, the idea of developing an entire software package within the community is abandoned. The architecture and ecosystem to enable further software development in relation to third-party software and corporations are developed. One example is the separation of platform and plug-in components. The shared platform is developed and maintained by the community. Each corporation can develop their own plug-in for its purposes, including business purposes. Examples include Eclipse.

In the third generation, a foundation is established to govern a large number of projects that are loosely connected. Each project is isolated in terms of functions and project management. The foundation has a higher level of orchestration. Examples include the Apache software foundation and the GNOME foundation.

The generations of diffusion using OSS are depicted in Fig. 6.

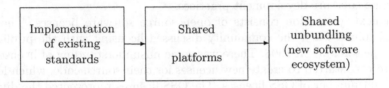

Fig. 6. 3-generation view of diffusion using OSS

The first generation accommodates the diffusion of open standards. The implementation of open standards in OSS leverages the acceptance of a standard. This also fits OSS because open standards provide clear requirements, which eliminates overhead in OSS development with less ambiguity compared to other types of software.

The second generation leverages the diffusion of shared platforms. Splitting software into a shared platform and plug-ins provides efficient development of software as long as the architecture is properly designed. Eclipse is one example. When a platform is based on OSS, it provides transparency of governance, neutrality of delivery control, and public participation. It also facilitates open distribution of technical information.

The third generation leverages unbundling and unlocking. The foundation-based OSS provides a base for industrial ecosystem such as application stores and white-brand SDKs. The white-brand SDK in the embedded software engineering enables unbundling third-party applications that replace the second-party applications in the early days.

The transitions of corporate engagement are depicted in Fig. 7.

In the first stage, enterprises are engaged in users of OSS.

In the second stage, enterprises place resources for OSS projects, where stable development and delivered quality of code influences their businesses. One example is Eclipse. IBM hosted Eclipse in its early stage of and worked in the Eclipse

Fig. 7. 3-generation view of corporate engagement in OSS

consortium. Then, a non-profit foundation that owned all the code would work much better. IBM donated the entirety of the Eclipse code to the Eclipse Foundation in order to form an industrial framework to support Eclipse development. Another example is Linux. IBM, HP, SGI, Intel, and other industrial players provided human resources toward the improvement of Linux to commercial-grade quality.

In the third stage, many leading industrial players play an important role in industry-backed OSS-based foundations. They pay extra fees for an Advisory Board to manage the industrial governance of large-scale OSS projects. The extra fees can be used to reduce the fees of other regular members to host a wide range of support for diverse stakeholders.

4 Discussion

4.1 Advantages of the Proposed Model

The proposed transition model addresses the shifts in underlying schemes during the evolution of OSS.

Over the decades, OSS has gained experience, and has achieved a high level of quality, performance, and completeness. It has also increased its code size, diversity and heterogeneity. This has brought about changes in the landscape of OSS.

In the early stage of OSS, there was some anti-proprietary feeling in many OSS communities. After professional OSS emerged as well as a dual licensing scheme, an increasing number of OSS projects accepted their co-existence with enterprise involvement. The scale of current OSS projects requires some management and governance skills from large organizations such as global industrial players.

The proposed model highlights this underlying scheme change over decades of OSS evolution.

4.2 Implications

OSS is a multi-faceted phenomenon, as shown in Fig. 8.

One aspect of OSS is philosophy and social movement. From a dimension such as this, there is still hesitation to accept the commerce-based economy that underlies OSS.

Although there is some anti-proprietary feeling in many OSS communities, with respect for the volunteer attitudes in the communities, large organizations,

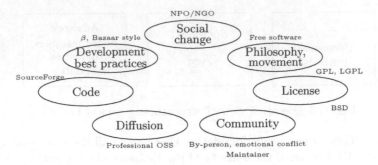

Fig. 8. Seven views of OSS

Table 3. Driving forces of commerce-based economy in OSS communities

Item	Description
Code scale	Large-scale code requires highly structured management.
Industry as user	Coverage of the completed software stacks lead the entire industry to become a customer of OSS component packages, and must have representatives from industry in OSS communities.
Organizational governance	Large-scale OSS solutions require highly organized governance to represent neutrality and compose reasonable roadmaps for stakeholders.
Global awareness	OSS is increasing in width and diversity, therefore, global awareness, usually provided by a global industry leading company, is important to OSS management.

mainly consisting of global industrial players are increasing their presence in OSS communities. The driving forces of this trend are depicted in Table 3.

4.3   Limitations

This research is descriptive and qualitative without any quantitative measures. In particular, this paper lacks any quantifiable measures for the transitions between economy types. Those quantitative approaches are beyond the scope of this paper.

This paper focuses on large-scale and industrial OSS solutions. Although, they are increasingly visible, however, they are a part of an OSS world of diversity and heterogeneity.

OSS projects are diverse and this paper lacks the in-depth analysis of each OSS project.

The impacts on social norms and trusts in OSS communities from these transitions are not covered in this paper.

The detailed business ecosystem engineering driving the proposed transitions is not addressed in this paper.

5   Conclusion

Over the decades, OSS has accomplished many great achievements and penetrated into the entire software industry. OSS has come to occupy the mainstream of the software industry. The success of OSS includes quality, performance, coverage and completeness as well as many established OSS-based foundations with global governance.

This success also highlights the necessity for revisiting the underlying economy types.

The author proposes a transition model from free economy to commerce-based economy. The transition is driven by the completeness of OSS for industrial solutions and the engagement of industry in OSS projects.

This trend does not fit with the free-economy-oriented-ness inherited from the early stage of OSS. However, multiple driving forces are visible to support the transition toward a commerce-based economy as the underlying principle in OSS projects.

The awareness of stages raised by the proposed model will help build mutual understanding between OSS communities; and industrial engagement, which has lead to productive evolution in many OSS projects.

The author presents multiple generation views that fit the proposed transition model in order to provide supplemental support for the transition discussed.

The transition has become visible over a long span of time, even decades. And it happens due to the maturity of OSS projects, and due to a mutual understanding between OSS communities and enterprises. The resulting institutional characteristics such as global awareness and harmonization with industrial solutions can contribute to further productive collaboration between communities and industries in OSS projects.

References

1. Augustin, L.: Why now is the time for open source applications. In: Plenary Speech of OSS 2010, Notre Dame, IN, USA (May 2010)
2. Capra, E., Francalanci, C., Merlo, F., Rossi-Lamastra, C.: Firms' involvement in open source projects: A trade-off between software structural quality and popularity. J. Syst. Softw. 84, 144–161 (2011),
   http://dx.doi.org/10.1016/j.jss.2010.09.004
3. Fitzgerald, B.: The transformation of opensource software. MIS Quarterly 30, 587–598 (2006)
4. Fitzgerald, B., Agerfalk, P.J.: The mysteries of open source software: Black and white and red all over? In: HICSS 2005: Proceedings of the Proceedings of the 38th Annual Hawaii International Conference on System Sciences, p. 196.1. IEEE Computer Society, Washington, DC (2005)
5. Letellier, F.: Open source software: the role of nonprofits in federating business and innovation ecosystems (January 2008) (a submission for AFME 2008),
   http://flet.netcipia.net/xwiki/bin/download/Main/publications-fr/
   GEM2008-FLetellier-SubmittedPaper.pdf

6. LiMo Foundation: LiMo Foundation Home page (January 2007),
   http://www.limofoundation.org/
7. Open Handset Alliance: Open Handset Alliance web page (2007),
   http://www.openhandsetalliance.com/
8. Raymond, E.S.: The magic cauldron (August 2000),
   http://www.catb.org/~esr/writings/cathedral-bazaar/magic-cauldron/
9. Subramaniam, C., Sen, R., Nelson, M.L.: Determinants of open source software
   project success: A longitudinal study. Decis. Support Syst. 46(2), 576–585 (2009)
10. Symbian Foundation: Symbian Foundation web page (2008),
    http://www.symbianfoundation.org/
11. The Apache Software Foundation: The Apache Software Foundation web page
    (1999), http://www.apache.org/
12. The Eclipse Foundation: The Eclipse Foundation web page (2004),
    http://www.eclipse.org/
13. Watson, R.T., Boudreau, M.C., York, P.T., Greiner, E., Donald Wynn, J.: The
    business of open source. CACM 51(4), 41–46 (2008)
14. Yamakami, T.: Generations of oss in evolutionary paths: Toward an understanding
    of where oss is heading. In: IEEE ICACT 2011, pp. 1599–1603. IEEE, Los Alamitos
    (2011)
15. Yu, L., Ramaswamy, S., Lenin, R.B., Narasimhan, V.L.: Time series analysis of
    open-source software projects. In: ACM-SE 47: Proceedings of the 47th Annual
    Southeast Regional Conference, pp. 1–6. ACM, New York (2009)

Standing Situations and Issues of Open Source Policy in East Asian Nations: Outcomes of Open Source Research Workshop of East Asia

Tetsuo Noda[1], Terutaka Tansho[1], and Shane Coughlan[2]

[1] Shimane University
nodat@soc.shimane-u.ac.jp
tansho@riko.shimane-u.ac.jp
[2] Regional Director Asia of Open Invention Network
shane@opendawn.com

Abstract. East Asia nations have made some progress with this technology, and started to introduce OSS for e-government systems during the early part of this century. Many countries granted it a central role in their policies. The reasons for this include adoption of software based on standard specification, liberation from vender lock-in, or opposition to the market control of proprietary software. However, the primary reason is to reduce adoption costs for e-government systems. While this policy work is useful, there is a great deal more that needs to be done. The OSS adoption policy in each nation of East Asia must be accompanied by technological progress in domestic IT service industries or US multinationals will expand at the cost of local businesses. If this continues unchecked it will create a new form of lock-in for East Asian nations. Some Asian nations are trying to promote their domestic IT service industries, putting their OSS adoption policy to practical use, and this workshop will provide case studies of that work. It will also provide a forum for discussing current challenges and opportunities around both policy and practical implementation issues across Asia.

1 Introduction

As part of Shimane University's research into Open Source matters, the Research Project Promotion Institute[1] has run a project called 'Stabilization and Business Models for Open Source Software' since 2008. One deliverable has been the hosting of a seminar entitled 'Open Source Research Workshop in East Asia' on November 26th and 27th 2010. This seminar was attended by Japanese and academic thought-leaders as well as a diverse range of international participants.

East Asian nations have made some progress with Open Source technology in the last decade, building on the early introduction of OSS for e-Government systems at

---

[1] Shimane University is executing forward empirical and theoretical research on the productivity of the business model's construction is to be done by the cooperation of the industrial-government-academic-community complex.
http://albatross.soc.shimane-u.ac.jp/oss/index.html

S.A. Hissam et al. (Eds.): OSS 2011, IFIP AICT 365, pp. 379–384, 2011.

the turn of the last century. Many countries granted OSS a central role in their policies for reasons like adherence to standard specification, freedom from vender lock-in, or opposition to the market control of proprietary software. However, the primary reason is to reduce adoption costs for e-Government systems.

While the initial governmental policy work has been useful, a great deal more needs to be done before the value proposition offered by OSS is realized. One key example is that the OSS adoption policy in each nation of East Asia must be accompanied by progress in domestic IT service industries to prevent multinationals expanding at the cost of local businesses. This is a consideration given that Open Source originates from the West coast of the USA and is still primarily developed and enhanced by US corporations. It could even be said that the current technical evolution of OSS is driven mainly by companies originating from the United States. While the inherent benefits of OSS extend beyond the boundaries of enterprises, organizations and nations, and OSS has the potential to foster new business markets in regions other than North America, the current status quo has the potential for a new form of lock-in for East Asian nations.

Some East Asian nations are trying to promote domestic IT service industries by putting their OSS adoption policy to practical use, and this workshop provided case studies of that work. It also provided a forum for discussing current challenges and opportunities around both policy and practical implementation issues across East Asian nations.

The intention was to extract the aspects of Open Source adoption policy that are not accompanied by the technological progress in domestic IT service industries, and from this derive an indication of the role government should play in East Asia and in other developing countries generally.

At the workshop Open Source thought-leaders from Japan, China, South Korea, Vietnam, The Gambia and Ireland discussed the nuances and known outcomes of Open Source adoption policies, and contributed to the publication of a proceeding as a special issue of our bulletin[2] in conjunction with the event. As the contentions of the researchers are contained in the proceeding, the focus of this paper is to explain the presentations of the researchers and the main points of discussion at the workshop.

2   Open Source Policy in East Asian Nations

2.1   Open Source Policy in Japan

Mr. Shunichi Tashiro, the chief officer of Open Software Center at the Japanese Industrial-Technology Promotion Agency[3], introduced Open Source Policy in Japan and the activity of his department.

---

[2] Journal of Economics Memoirs Of The Faculty Of Law And Literature, Shimane University No.37 Special Issue "Open Source Policy and Promotion of IT Industries in East Asia" Nov.2010.

[3] The Open Software Center is an organization within the Information-Technology Promotion Agency (IPA), Japan, one of Independent Administrative Agencies 1 in Japan, and is operated under the budget of the Japanese government. http://www.ipa.go.jp/index-e.html

He explained that software is built on platforms, and that it ultimately cannot work without interoperability with other software. Openness of the platforms and interfaces are crucial to provide freedom to software developers and users, and therefore to foster a healthy, competitive environment in the national and international software industry. The Open Software Center was founded in 2006 to promote Open Source software and open standards as an important aspect of information services, development methods, assessment criteria, standardization systems, and research studies. The core proposition is that OSS enhances knowledge sharing and sustains a collaborative development environment.

The key issue is probably 'Open Standards' rather than Open Source. Open Standards are technical standards for which specifications are publicly available, that any part can use and that any party can participate in regarding further development. These are critical to ensure the interoperability of software and freedom of action for software developers and users.

One example of practical engagement with this issue in Japan is the standardization of the programming language Ruby, which is now in the screening process of JIS (Japanese Industrial Standards) and will soon obtain JIS certification before submission to ISO (International Organization for Standardization) in 2011. This will encourage the Japanese-created Ruby language to be adopted in e-Government systems and enterprise environments across the world.

In this way a Japanese government agency is supporting the promotion of the domestic information service industry while also contributing positively to the global IT market.

2.2 Open Source and the Software Industry in China

Dr. Dongbin Wang, the researcher of Center of China Study, Tsinghua University[4], introduced the history of the Internet and Open Source in China.

Before China joined in WTO in 2001, software piracy was very popular and lead to a large loss of potential revenue every year for the industry. One high profile example is that Microsoft's software products were frequently copied and illegally installed on computers across the private and public sectors, a situation that somewhat ironically also helped Microsoft become the de-facto norm in computing for the local market. However, after 2001 the Chinese government began to increase the legislation and enforcement around software piracy, and companies like Microsoft accompanied this shift in the local market norms by taking a much stricter line regarding unauthorized coping of their products. One side effect of this was for many cyber cafes to react by switching to Linux and other Open Source software to reduce licensing costs.

It is possible to conclude that intellectual property rights protection is a double-edged sword for Open Source development and adoption. Recently many large companies in China have formally preferred Linux as a pre-installation operation system, but due to the rampant piracy of Windows in the past there is an on-going

---

[4] Center for China Study (CCS), Tsinghua University is a leading academic think-tank for policy making in China. Its researches cover most fields of China Studies, which include Development and China Study, Chinese Economy and Development Strategy. http://www.facebook.com/group.php?gid=48859734768

familiarity and desire for Microsoft and other proprietary products. In effect, the over commercialization and effective commoditization of proprietary software presents a key obstacle to Open Source expansion in China.

2.3 Open Source Software - Education, Practice and Applications at the University of Engineering and Technology

Dr. Nam Hai Nguyen & Dr. Quang Hieu Le from the University of Engineering and Technology school of the Vietnam National University, Hanoi(UET)[5] provided information on policies and challenges experienced by the Vietnamese government regarding OSS development in their country.

The national government has long emphasized a focus on and priority for OSS development and deployment. For example, in 2006 the government mandated that procurement of IT products and services should give preference to investment, adoption and application of OSS and OSS-based software products, especially those with quality and functionality equal to proprietary software supplied by domestic enterprises

UET has engaged with the trend towards OSS by exploring best practices in the field and building educational knowledge to share with students and with society-at-large. Through these activities the faculty members of UET play an important role in advancing the widespread use of OSS in teaching and research activities, and in practical deployment outside of the confines of academia.

To further both the activities of UET, and the Vietnamese engagement with OSS generally, efficient international cooperation is of crucial importance. UET is a leading proponent of such cooperation, is seeking to build knowledge-sharing bridges with the broader global community.

3 Free and Open Source Software Governance: Turning Potential into Deliverables

The presentation examined some of the key governance approaches and resources that help turn the potential of FOSS into a deliverable, whether that deliverable is a product, revenue stream or altruistic solution to a shared problem.

The concept of governance has become increasingly important because - while Free and Open Source Software (FOSS) offers tremendous potential to create technology platforms and develop business opportunities - the best methods to obtain results from FOSS adoption have not always been clear. Simply being 'open' or using third party code appears to have limited value on its own, and does not address the management requirements or legal imperatives that stakeholders face. Deriving ongoing value from FOSS requires understanding the ideas and norms that underpin the field. FOSS has a premise that third-party sharing and contribution delivers enhanced value over proprietary approaches to managing software, and that FOSS can deliver this value through copyright licenses that allow everyone to use, study, share and improve software code. As this idea has reached the mainstream, the licenses are increasingly the

---

[5] http://www.vnu.edu.vn/en/contents/index.php?ID=932

subject of legal and management scrutiny, and best practices have inevitably emerged for adopting, developing and deploying code subject to their terms.

It was explained how organizations ranging from the Linux Foundation to Free Software Foundation Europe have built networks, encouraged publications and developed tools to assist their own audiences and the broader community of all potential FOSS stakeholders, and some of these governance projects or organizations were explored in brief case studies.

4 Discussion and Conclusion: The Potential Crowding-Out Effect of Government Policy

The majority of the presentations therefore engaged with the role of each Asian government or educational institutional in furthering Open Source policy and its practical application. One immediately observable point from their exploration is that each nation discussed places importance on Open Source adoption and supports it politically.

However, the point was raised that excessive policy engagement by central government might spoil the development of Open Source communities by creating a crowding-out effect. There are many stakeholders involved and the interests of these stakeholders are not necessarily entirely in sync. The development methods of Open Source software often include collaboration on the Internet through the participation of a number of developers, corporations and other stakeholders. This environment is competitive not only towards proprietary software but also within itself, and as such requires the flexibility that markets and their associated competitiveness provide.

As the motivation factors of Open Source software developers and investors vary, they tend to participate in creating code and therefore platforms with broadly applicable value. Government engagement needs to take this into account, and avoid causing the risk of constrained, prescriptive environments for Open Source development. This applies in the educational field too. While it is clearly important to explore and integrate OSS in educational environments, especially in the context of computer engineering, such coverage needs to be broad. The field needs to be presented in a context that does not ignore the way that OSS often appears, grows and delivers value outside of the traditional constraints and assumptions of formal engineering education. The discussion suggested that governmental policy around OSS is not clear cut. While positive policy is to be encouraged, the value offered by OSS needs to be understood clearly to ensure it is not inadvertently smothered by well-meaning but ineffective mandates. This indicates that perhaps there are 'new' issues of Open Source adoption policy in East Asian nations to consider, and all participants at the workshop agreed to proceed with examination of how these issues can be resolved in a future workshop.

References

1. Noda, T., Tansho, T.: Regional Industrial Promotion through Open Source Software by Local Government in Japan. In: Proceeding of the First International Workshop on Building Sustainable Open Source Communities (2009)

2. Noda, T., Tansho, T.: Open Source Introduction Policy and Promotion of Regional Industries in Japan: Open Source Software. In: Ågerfalk, P., Boldyreff, C., González-Barahona, J.M., Madey, G.R., Noll, J. (eds.) OSS 2010. IFIP AICT, vol. 319, pp. 425–426. Springer, Heidelberg (2010)
3. Wang, D.: Open Source and Software Industry in China. Journal of Economics Memoirs of the Faculty of Law and Literature (November 2010); Shimane University No.37 Special Issue, Open Source Policy and Promotion of IT Industries in East Asia
4. Yi, S., Noda, T., Oh, J.: Comparative study on the implementation and performance of open source activating policy between Korea and Japan. National Information Agency Society, Korea (2009)
5. Yi, S., Noda, T.: Enhancing the Circulation: Some Implications on the OSS Policy and IT Industry Promotion. Journal of Economics Memoirs of the Faculty of Law And Literature (November 2010); Shimane University No.37 Special Issue Open Source Policy and Promotion of IT Industries in East Asia

Towards Sustainable Open Source

Imed Hammouda[1] and Björn Lundell[2]

[1] Tampere University of Technology, Finland
[2] University of Skövde, Sweden

Open source software is gaining momentum in several forms. In addition to the huge increase in the number of open source projects started and the remarkable rise of FLOSS adoption by companies and governments, new models of participation in the movement are emerging rapidly. For instance, companies are increasingly releasing some of their proprietary software systems as open source on one hand and acquiring open source software on the other hand. For all these forms of involvement, a central question is how to build and maintain a sustainable ecosystem around the open source projects. Sustainability issues of open source extends beyond the technical challenges of building project infrastructure covering other important aspects related to business, economic, legal, social, and cultural dimensions. Long term sustainability will be the theme of OSS 2012 to be held in Tunisia. We think that the OSS community could start discussing the theme by exchanging related experiences, sharing relevant concerns, and proposing topics of interest.

S.A. Hissam et al. (Eds.): OSS 2011, IFIP AICT 365, p. 385, 2011.
© IFIP International Federation for Information Processing 2011

Improving US Department of Defense Technology Acquisition and Development with Open Source

Guy Martin[1] and Aaron Lippold[2]

[1] CollabNET, USA
[2] Forge.mil, USA

In late 2008, DISA (Defense Information Systems Agency), the global IT arm of the US Department of Defense, embarked upon a project to create an internal collaboration and software application lifecycle management system. Beyond simply fielding yet another tool, the Forge.mil effort was designed to fundamentally change the way the DoD developed and acquired software technology and systems. The method of this change was the application of Open Source principles inside of the larger DoD community, including ideas such as: meritocracy, code sharing, as well as Agile and collaborative software development. This workshop will explore where the program has succeeded, as well as areas that need to be improved. It is hoped that participants will be able to bring perspectives from their work in the external Open Source world to this discussion.

S.A. Hissam et al. (Eds.): OSS 2011, IFIP AICT 365, p. 386, 2011.
© IFIP International Federation for Information Processing 2011

Author Index